W9-CSP-229

ACS SYMPOSIUM SERIES **506**

Phenolic Compounds in Food and Their Effects on Health I
Analysis, Occurrence, and Chemistry

Chi-Tang Ho, EDITOR
Rutgers, The State University of New Jersey

Chang Y. Lee, EDITOR
Cornell University

Mou-Tuan Huang, EDITOR
Rutgers, The State University of New Jersey

Developed from a symposium sponsored
by the Division of Agricultural and Food Chemistry
of the American Chemical Society
at the Fourth Chemical Congress of North America
(202nd National Meeting of the American Chemical Society),
New York, New York,
August 25–30, 1991

American Chemical Society, Washington, DC 1992

Library of Congress Cataloging-in-Publication Data

Phenolic compounds in food and their effects on health I: analysis, occurrence, and chemistry / Chi-Tang Ho, editor, Chang Y. Lee, editor, Mou-Tuan Huang, editor.

  p.   cm.—(ACS symposium series, ISSN 0097–6156; 506)

"Developed from a symposium sponsored by the Division of Agricultural and Food Chemistry of the American Chemical Society at the Fourth Chemical Congress of North America (202nd National Meeting of the American Chemical Society), New York, N.Y., August 25–30, 1991."

Includes bibliographical references and indexes.

ISBN 0–8412–2475–7

1. Phenols in the body—Congresses. 2. Phenols—Health aspects—Congresses.

I. Ho, Chi-Tang, 1944-   . II. Lee, Chang Y., 1935-   . III. Huang, Mou-Tuan, 1935-   . IV. American Chemical Society. Division of Agricultural and Food Chemistry. V. Chemical Congress of North America (4th: 1991: New York, N.Y.) VI. American Chemical Society. Meeting (202nd: 1991: New York, N.Y.) VII. Series.

QP801.P4P45   1992
612'.0157632—dc20   92–23283
                                               CIP

The paper used in this publication meets the minimum requirements of American National Standard for Information Sciences—Permanence of Paper for Printed Library Materials, ANSI Z39.48–1984. ∞

Copyright © 1992

American Chemical Society

All Rights Reserved. The appearance of the code at the bottom of the first page of each chapter in this volume indicates the copyright owner's consent that reprographic copies of the chapter may be made for personal or internal use or for the personal or internal use of specific clients. This consent is given on the condition, however, that the copier pay the stated per-copy fee through the Copyright Clearance Center, Inc., 27 Congress Street, Salem, MA 01970, for copying beyond that permitted by Sections 107 or 108 of the U.S. Copyright Law. This consent does not extend to copying or transmission by any means—graphic or electronic—for any other purpose, such as for general distribution, for advertising or promotional purposes, for creating a new collective work, for resale, or for information storage and retrieval systems. The copying fee for each chapter is indicated in the code at the bottom of the first page of the chapter.

The citation of trade names and/or names of manufacturers in this publication is not to be construed as an endorsement or as approval by ACS of the commercial products or services referenced herein; nor should the mere reference herein to any drawing, specification, chemical process, or other data be regarded as a license or as a conveyance of any right or permission to the holder, reader, or any other person or corporation, to manufacture, reproduce, use, or sell any patented invention or copyrighted work that may in any way be related thereto. Registered names, trademarks, etc., used in this publication, even without specific indication thereof, are not to be considered unprotected by law.

PRINTED IN THE UNITED STATES OF AMERICA

1992 Advisory Board

ACS Symposium Series

M. Joan Comstock, *Series Editor*

V. Dean Adams
Tennessee Technological
University

Mark Arnold
University of Iowa

David Baker
University of Tennessee

Alexis T. Bell
University of California—Berkeley

Arindam Bose
Pfizer Central Research

Robert F. Brady, Jr.
Naval Research Laboratory

Margaret A. Cavanaugh
National Science Foundation

Dennis W. Hess
Lehigh University

Hiroshi Ito
IBM Almaden Research Center

Madeleine M. Joullie
University of Pennsylvania

Mary A. Kaiser
E. I. du Pont de Nemours and
Company

Gretchen S. Kohl
Dow-Corning Corporation

Bonnie Lawlor
Institute for Scientific Information

John L. Massingill
Dow Chemical Company

Robert McGorrin
Kraft General Foods

Julius J. Menn
Plant Sciences Institute,
U.S. Department of Agriculture

Vincent Pecoraro
University of Michigan

Marshall Phillips
Delmont Laboratories

A. Truman Schwartz
Macalaster College

John R. Shapley
University of Illinois
at Urbana—Champaign

Stephen A. Szabo
Conoco Inc.

Robert A. Weiss
University of Connecticut

Peter Willett
University of Sheffield (England)

QP801
P4 P45
1992
v. 1
CHEM

Foreword

THE ACS SYMPOSIUM SERIES was first published in 1974 to provide a mechanism for publishing symposia quickly in book form. The purpose of this series is to publish comprehensive books developed from symposia, which are usually "snapshots in time" of the current research being done on a topic, plus some review material on the topic. For this reason, it is necessary that the papers be published as quickly as possible.

Before a symposium-based book is put under contract, the proposed table of contents is reviewed for appropriateness to the topic and for comprehensiveness of the collection. Some papers are excluded at this point, and others are added to round out the scope of the volume. In addition, a draft of each paper is peer-reviewed prior to final acceptance or rejection. This anonymous review process is supervised by the organizer(s) of the symposium, who become the editor(s) of the book. The authors then revise their papers according the the recommendations of both the reviewers and the editors, prepare camera-ready copy, and submit the final papers to the editors, who check that all necessary revisions have been made.

As a rule, only original research papers and original review papers are included in the volumes. Verbatim reproductions of previously published papers are not accepted.

M. Joan Comstock
Series Editor

Contents

Contents

Phenolic Compounds in Food
and Their Effects on Health II

Antioxidants and Cancer Prevention

CHEMICAL AND BIOLOGICAL ACTIVITIES
OF PHENOLIC ANTIOXIDANTS

Preface

SCIENTIFIC AND COMMERCIAL INTEREST in phenolic compounds in food has been extremely active in recent years. Apart from the purely academic study of their natural occurrence, distribution, and quality attributes in food, phenolic compounds are becoming increasingly important in applied science. Phenolic compounds aid in the maintenance of food, fresh flavor, taste, color, and prevention of oxidation deterioration. In particular, many phenolic compounds are attracting the attention of food and medical scientists because of their antioxidative, antiinflammatory, antimutagenic, and anticarcinogenic properties and their capacity to modulate some key cellular enzyme functions.

Dietary factors play an important role in human health and in the development of certain diseases, especially cancer. The frequent consumption of fresh fruits and vegetables is associated with a low cancer incidence. It is not known with certainty which components in fruits and vegetables contribute to inhibiting tumor development in humans. Part of the activity may be due to the presence of vitamins A, C, and β-carotene in fruits and vegetables. However, almost all fresh fruits and vegetables contain rich amounts of naturally occurring phenolic compounds such as flavonoids. The amount of ingestion of plant phenolics is proportional to the consumption of fruits and vegetables. In addition, the synthetic phenolic antioxidants, such as butylated hydroxyanisole and butylated hydroxytoluene, are widely used as food antioxidants. Some beverages also contain high amounts of naturally occurring phenolic compounds, for example catechin polyphenols in teas and red wine and chlorogenic acid in coffee. Other phenolic compounds in food include curcumin, a food coloring agent present in turmeric and curry, and carnosol and carnosic acid in rosemary leaves used as a spice.

Humans daily ingest a large amount of phenolic compounds. Do these phenolic compounds contribute many benefits to human health or do they have any adverse effects? In order to probe these questions, the two volumes of *Phenolic Compounds in Food and Their Effects on Health I* and *II* present the recent research data and review lectures by numerous prestigious experts from the international symposium on phenolic compounds in food and health. Contributors from academic institutions, government, and industry were carefully chosen to provide different insights and areas of expertise in these fields.

This subject is presented in two volumes. Volume I consists of occurrence, analytical methodology, and polyphenol complexation in food and the effect of phenolic compounds on the flavor, taste, color, texture, and nutritional quality of food, as well as the utilization of phenolic antioxidants in foods. Volume II includes the source of phenolic compounds, the chemical and biological properties of phenolic compounds in food, and their health effects. Special emphasis is placed on the biological influence of phenolic compounds on the modulation of tumor development in experimental animal models, and possibly in humans.

We are indebted to the contributing authors for their creativity, promptness, and cooperation in the development of this book. We also sincerely appreciate the patience and understanding given to us by our wives, Mary Ho, Ocksoo Kim Lee, and Chiu Hwa Huang. Without their support, this piece of work would not have materialized. We thank Harold Newmark, James Shaw, and Thomas Ferraro for their help in review, suggestions, and editing.

We acknowledge the financial support of the following sponsors: Campbell Soup Company; CPC International, Inc.; Givaudan Corporation; Glaxo, Inc.; Hoffmann–La Roche, Inc.; Johnson & Johnson Consumer Products, Inc.; Kalsec, Inc.; Lipton Foundation; Merck Sharp & Dohme Research Laboratories; Takasago USA; Tea Council of the USA; The Procter & Gamble Company; The Quaker Oats Company; Warner–Lambert Foundation; and the Division of Agricultural and Food Chemistry of the American Chemical Society.

CHI-TANG HO
Rutgers, The State University of New Jersey
New Brunswick, NJ 08903

CHANG Y. LEE
Cornell University
Geneva, NY 14456

MOU-TUAN HUANG
Rutgers, The State University of New Jersey
Piscataway, NJ 08854

June 30, 1992

PERSPECTIVES

Chapter 1

Phenolic Compounds in Food
An Overview

Chi-Tang Ho

Department of Food Science, Cook College, Rutgers, The State
University of New Jersey, New Brunswick, NJ 08903

Phenolic compounds including simple phenols and phenolic acids,
hydroxycinnamic acid derivatives and flavonoids are bioactive
substances occurring widely in food plants. Phenolic compounds are
closely associated with the sensory and nutritional quality of fresh and
processed plant foods. The enzymatic browning reaction of phenolic
compounds, catalyzed by polyphenoloxidase, could cause the
formation of undesirable color and flavor and the loss of nutrient in
fruits and vegetables. Many phenolic compounds in plants are good
sources of natural antioxidants. It is a great interest in recent years that
many phenolic compounds in foods have inhibitory effects on
mutagenesis and carcinogenesis.

The term 'phenolic' or 'polyphenol' can be defined chemically as a substance which
possesses an aromatic ring bearing one or more hydroxy substituents, including
functional derivatives (esters, methyl ethers, glycosides etc.) (*1*). Most phenolics
have two or more hydroxyl groups and are bioactive substances occurring widely in
food plants that are eaten regularly by substantial numbers of people.

Occurrence of Phenolic Compounds

The phenolic compounds which occur commonly in food material may be classified
into three groups, namely, simple phenols and phenolic acids, hydroxycinnamic acid
derivatives and flavonoids.

The Simple Phenols and Phenolic Acids. The simple phenols include
monophenols such as *p*-cresol isolated from several fruits (e.g. raspberry,
blackberry) (*2*), 3-ethylphenol and 3,4-dimethylphenol found to be responsible for
the smoky taste of certain cocoa beans (*3*) and diphenols such as hydroquinone
which is probably the most widespread simple phenol (*4*).
    A typical hydroquinone derivative, sesamol, is found in sesame oil (*4*).
Several derivatives of sesamol, such as sesaminol, found in sesame oil have been
evaluated to have strong antioxidant activity (Osawa, Chapter 10, Vol. II).

0097–6156/92/0506–0002$06.00/0
© 1992 American Chemical Society

Vanillin (4-hydroxy-3-methyoxybenzaldehyde) is the most popular flavor. The determination of vanillin in vanilla beans is discussed by Hartman et al. (Chapter 4, Vol. I).

Gallic acid, a triphenol, is present in an esterified form in tea catechins (Balentine, Chapter 8, Vol. I). Gallic acid may occur in plants in soluble form either as quinic acid esters (5) or hydrolyzable tannins (Okuda et al., Chapter 12, Vol. II).

The Hydroxycinnamic Acid Derivatives. Hydroxycinnamic acids and their derivatives are almost exclusively derived from *p*-coumaric, caffeic, and ferulic acid, whereas sinapic acid is comparatively rare. Their occurrence in food has recently been reviewed by Herrmann (6). Hydroxycinnamic acids usually occur in various conjugated forms, more frequently as esters than glycosides.

The most important member of this group in food material is chlorogenic acid, which is the key substrate for enzymatic browning, particularly in apples and pears (7).

A number of chapters in this book discuss the occurrence of hydroxycinnamic acids in foods which include:

Umbelliferous vegetables	Roshdy et al. (Chapter 6, Vol. I)
Citrus fruits	Naim et al. (Chapter 14, Vol. I)
Brassica oilseed	Shahidi (Chapter 10, Vol. I)
Corn flour	Gibson and Strauss (Chapter 20, Vol. I)
Raspberry	Rommel et al. (Chapter 21, Vol. I)
Plums	Fu et al. (Chapter 22, Vol. I)

The Flavonoids. The most important single group of phenolics in food are flavonoids which consist mainly of catechins, proanthocyanins, anthocyanidins and flavons, flavonols and their glycosides.

Although catechins seem to be widely distributed in plants, they are only rich in tea leaves where catechins may constitute up to 30% of dry leaf weight. A number of chapters in Volume II of this book discuss current research on antioxidative and cancer chemopreventive properties of tea and its catechin components. Lunder (Chapter 8, Vol. II) has shown that the antioxidative activity of green tea extract could be related to the content of epigallocatechin. Osawa (Chapter 10, Vol. II) was able to demonstrate that epicatechin gallate and epigallocatechin gallate not only inhibit the free radical chain reaction of cell membrane lipids, but also inhibit mutagenicity and DNA damaging activity. Laboratory studies conducted by Ito et al. (Chapter 19, Vol. II), Conney et al. (Chapter 20, Vol. II), Wang et al. (Chapter 21, Vol. II), Chung et al. (Chapter 22, Vol. II), Laskin et al. (Chapter 23, Vol. II) and Yoshizawa et al. (Chapter 24, Vol. II) are presented. They have shown that tea and tea catechin components can inhibit tumorigenesis and tumor growth in animals.

Proanthocyanidins, or condensed tannins, are polyflavonoid in nature, consisting of chains of flavan-3-ol units. They are widely distributed in food such as apple, grape, strawberry, plum, sorghum and barley (8). Proanthocyanidins have relatively high molecular weights and have the ability to complex strongly with carbohydrates and proteins. Comprehensive treatment of polyphenol complexation is given by Haslam et al. in this book (Chapter 2, Vol. I). Hagerman (Chapter 19, Vol. I) and Butler and Rogler (Chapter 23, Vol. I) also present overviews on these topics.

Anthocyanins are almost universal plant colorants and are largely responsible for the brilliant orange, pink, scarlet, red, mauve, violet and blue colors of flower petals and fruits of higher plants (9). Anthocyanins as food colorants have recently been reviewed (10).

Flavones, flavonols and their glycosides also occur widely in the plant kingdom. Their structural variations and distribution have been the subjects of several comprehensive reviews in recent years (11-13). It has been estimated that humans consuming high fruit and vegetable diets ingest up to 1 g of these compounds daily (Leighton et al., Chapter 15, Vol. II). The most common and biologically active dietary flavonol is quercetin. A number of presentations in Volume II of this book discuss current research on the effect of quercetin on mutagenesis and carcinogenesis. Quercetin was found by Verma (Chapter 17, Vol. II) to inhibit both initiation with 7,12-dimethylbenz[a]anthracene (DMBA) and tumor promotion with 12-O-tetradecanoylphorbol-13-acetate (TPA) of mouse skin tumor formation. Starvic et al. (Chapter 16, Vol. II) suggest that quercetin and other polyphenols such as ellagic acid and chlorogenic acid may play a dual protective role in carcinogenesis by reducing bioavailability of carcinogens, and by interfering with their bio-transformation in the liver. By using an experimental model of colon cancer, Deschner (Chapter 18, Vol. II) was able to demonstrate that under conditions of low dietary fat intake, quercetin and rutin have displayed considerable activity in suppressing the hyperproliferation of colonic epithelial cells, thereby reducing focal areas of dysplasia and ultimately colon tumor incidence.

Effect of Phenolic Compounds on Food Quality

Phenolic compounds are closely associated with the sensory and nutritional quality of fresh and processed plant foods (14). The enzymatic browning reaction of phenolic compounds, catalyzed by polyphenoloxidase, is of vital importance to fruit and vegetable processing due to the formation of undesirable color and flavor and the loss of nutrients. For examples, polyphenoloxidase was found to be responsible for the browning of grapes (15), and catechin, polyphenoloxidase and oxygen were reported to be required for the browning of yams (16). The enzymatic browning reaction in fruits often has been considered to be a linear function of the phenolic content and polyphenoloxidase activity. In this book, however, Lee (Chapter 24, Vol. I) show that the rate of browning in fruit products is not a linear function of the total phenolic content and that browning of fruits depends on the concentration and nature of polyphenol compounds that are co-present.

A common approach for the prevention of the browning of food and beverages has been the use of antibrowning agents. The most widespread antibrowning agents used in the food and beverage industries are sulfites. Due to the health risks of sulfiting agents (17), the Food and Drug Administration has banned or limited the use of sulfites in certain foods (18). McEvily et al. (Chapter 25, Vol. I) discuss the isolation and characterization of several 4-substituted resorcinols from figs. These novel 4-substituted resorcinols were shown to be potent polyphenol-oxidase inhibitors.

Oxidative changes of polyphenols during processing are important for the development of color and flavor in certain foods. Browning of polyphenols is a natural process of cocoa fermentation (19). For the manufacture of black tea, the tea leaves are crushed, causing polyphenoloxidase-dependent oxidative polymerization and leading to the formation of theaflavins and thearubigins, the orange and red pigments of black tea. This subject is reviewed in the chapter by Balentine.

Phenolic compounds may contribute directly to desirable and undesirable aromas and tastes of food. Recently, Ha and Lindsay (*20*) reported that highly characterized sheep-mutton aromas in ovine fats were contributed by *p*-cresol, 2-isopropylphenol, 3,4-dimethylphenol, thymol, carvacrol, 3-isopropylphenol and 4-isopropylphenol. They also observed that cresols, especially *m*-cresol, appeared to contribute to beef flavors. Several discussions of flavor characteristics of phenolic compounds are included in this book. Maga (Chapter 13, Vol. I) reviews the roles of hemicellulose, cellulose and lignin thermal degradation on the formation of phenolic compounds which are the major contributors to wood smoke aroma. Fisher (Chapter 9, Vol. I) and Omar (Chapter 12, Vol. I) present overviews on the contribution of phenolic compounds to the aroma and taste of certain spices and plant extracts. Naim et al. (Chapter 14, Vol. I) report that 4-vinylguaiacol is one of the major detrimental off-flavors that form under typical processing and storage of citrus products. Their studies reveal that 4-vinylguaiacol is formed from ferulic acid following the release of ferulic acid from bound forms.

Phenolic Compounds as Natural Antioxidants

Antioxidants are added to fats and oils or foods containing fats to prevent the formation of various off-flavors and other objectionable compounds that result from the oxidation of lipids. BHA and BHT, the most widely used synthetic antioxidants, have unsurpassed efficacy in various food systems besides their high stability, low cost, and other practical advantages. However, their use in food has been falling off due to their suspected action as promoters of carcinogenesis as well as being due to a general rejection of synthetic food additives (*21*).

The most important natural antioxidants which are commercially exploited are tocopherols. Tocopherols have a potent ability to inhibit lipid peroxidation *in vivo* by trapping peroxyl radicals. Their antioxidative mechanism and structure-activity relationship are discussed by Hughes et al. (Chapter 13, Vol. II). Unfortunately, tocopherols are much less effective as food antioxidants. The search and development of other antioxidants of natural origins is highly desirable. Such new antioxidants would also be welcome in combatting carcinogenesis as well as the aging process.

Most natural antioxidants are phenolic in nature. Some of the food materials containing phenolic antioxidants studied and reported herein include:

Chili pepper	Nakatani (Chapter 5, Vol. II)
Ginger	Nakatani (Chapter 5, Vol. II)
Green tea	Lunder (Chapter 8, Vol. II), Osawa (Chapter 10, Vol. II)
Pepper	Nakatani (Chapter 5, Vol. II)
Oregano	Nakatani (Chapter 5, Vol. II)
Osbeckia chinensis	Osawa (Chapter 10, Vol. II)
Rice hull	Osawa et al. (Chapter 9, Vol. II)
Rosemary	Nakatani (Chapter 5, Vol. II)
Sesame seeds	Osawa (Chapter 10, Vol. II)
Soybean	Fleury et al. (Chapter 7, Vol. II)
Thyme	Nakatani (Chapter 5, Vol. II)

Phenolic antioxidants not only inhibit the autoxidation of lipids, but sometimes, they also have the ability to retard lipid oxidation by inhibiting lipoxygenase activity. It is believed that the metabolism of arachidonic acid to lipid peroxides and various other oxidative products is significant in carcinogenesis (22). It appears to play an important role in tumor promotion because inhibitors of arachidonic acid metabolism have been observed to inhibit this promotion (23). Four green tea catechin components having strong antioxidant activity also exhibited various degrees of lipoxygenase-inhibitory activities (Ho, C.-T.; Shi, H., unpublished data). (-)-Epigallocatechin gallate, (-)-epicatechin gallate and epigallocatechin displayed IC_{50} values toward soybean 15-lipoxygenase enzyme ranging from 10-21 μM (Table I). (-)-Epicatechin is, on the other hand, relatively inactive. It is also interesting to note that two of the oxidative dimers of tea catechins, the theaflavin monogallate B and theaflavin digallate, which are important polyphenols of black tea, have even stronger lipoxygenase-inhibitory activities than the catechin monomers. The other two structurally closely-related theaflavins, theaflavin and theaflavin monogallate A, have no activity at all (Table I). The detailed chemical structures of these tea polyphenols can be found in the chapter by Balentine (Chapter 8, Vol. I).

Table I. Inhibition of Soybean Lipoxygenase by Tea Polyphenols

Compound	IC_{50} (μM)
(-)-Epicatechin (EC)	140
(-)-Epicatechin gallate (ECG)	18
(-)-Epigallocatechin (EGC)	21
(-)-Epigallocatechin gallate (EGCG)	10
Theaflavin	3604
Theaflavin monogallate A	366
Theaflavin monogallate B	0.62
Theaflavin digallate	0.25

Conclusion

Phenolic compounds are ubiquitous in plant foods, and therefore, a significant quantity is consumed in our daily diet. They are closely associated with the sensory and nutritional quality of fresh and processed plant foods. The antioxidant activities of phenolic compounds have been recognized for decades, and research and development on the use of natural substances or food ingredients containing phenolic antioxidants will continue to be of great interest to the food industry.

Biological activities of phenolic compounds have become well known in recent years. The most important biological activity of phenolic compounds is probably their many observed inhibitory effects on mutagenesis and carcinogenesis. This topic is covered in great depth in the overview chapter by Huang and Ferraro (Chapter 1, Vol. II) and many chapters in Volume II of this book.

Acknowledgements

This publication, New Jersey Agricultural Experiment Station Publication No. D-10205-1-92, has been supported by State Funds. We thank Mrs. Joan Shumsky for her secretarial aid.

Literature Cited

1. Harborne, J. B. In *Methods in Plant Biochemistry, Vol. 1: Plant Phenolics*; Harborne, J. B., Ed.; Academic Press: London, UK, 1989, pp. 1-28.
2. Van Straten, S. *Volatile Compounds in Food*; Central Institute for Nutrition and Food Research: Zeist, The Netherlands, 1977.
3. Guoyt, B.; Gueule, D.; Morcrette, I.; Vincent, J. C. *Coffee, Cocoa, Tea* **1986**, *30*, 113-120.
4. Van Sumere, C. F. In *Methods in Plant Biochemistry, Vol. 1: Plant Phenolics*; Harborne, J. B., Ed.; Academic Press: London, UK, 1989, pp. 29-73.
5. Nishimura, H.; Nonaka, G. I.; Nishioka, I. *Phytochem.* **1984**, *23*, 2621-2623.
6. Herrmann, K. *CRC Crit. Rev. Food Sci. Nutri.* **1989**, *28*, 315-347.
7. Eskin, N. A. M. *Biochemistry of Foods*; Academic Press: San Diego, CA, 1990, pp. 401-432.
8. Haslam, E. *Plant Polyphenols*; Cambridge University Press: Cambridge, UK, 1989.
9. Harborne, J. B. *Comparative Biochemistry of the Flavonoids*; Academic Press: London, 1967.
10. Francis, F. J. *CRC Crit. Rev. Food Sci. Nutri.* **1989**, *28*, 273-314.
11. Harborne, J. B.; Mabry, T. J. *The Flavonoid—Advances in Research*; Chapman and Hall: London, UK, 1982.
12. Harborne, J. B.; Mabry, T. J. *The Flavonoid—Advances in Research*; Chapman and Hall: London, UK, 1988, Vol. 2.
13. Markham, K. R. In *Methods in Plant Biochemistry, Vol. 1: Plant Phenolics*; Harborne, J. B., Ed.; Academic Press: London, UK, 1989, pp. 197-235.
14. Macheix, J. J.; Fleuriet, A.; Billot, J. *Fruit Phenolics*; CRC Press: Boca Raton, FL, 1990.
15. Sapis, J. C.; Macheix, J. J.; Cordonnier, R. E. *J. Agric. Food Chem.* **1983**, *31*, 342-345.
16. Ozo, O. N.; Caygill, J. C. *J. Sci. Food Agric.* **1986**, *37*, 283.
17. Taylor, S. L.; Higley, N. A.; Bush, R. K. *Adv. Food Res.* **1986**, *30*, 1-76.
18. Anonymous. *Food Institute Report* **1990**, *63*, 9.
19. Quesnel, V. C.; Jugmohunsingh, K. *J. Sci. Food Agric.* **1970**, *21*, 537-541.
20. Ha, J. K.; Lindsay, R. C. *J. Food Sci.* **1991**, *56*, 1197-1202.
21. Namiki, M. *CRC Crit. Rev. Food Sci. Nutri.* **1990**, *29*, 273-300.
22. Powles, T. J; Bockman, R. S; Honn, K. V; Ramwell, P. *First International Conference on Prostaglandins and Cancer, Prostaglandins and Related Lipids*; New York: Liss, 1982, Vol. 2.
23. Belman, S; Solomon, J; Segal, A; Block, E; Barany, G, *J. Biochem. Toxicol.* **1989**, *4*, 151-160.

RECEIVED June 30, 1992

Chapter 2

Polyphenol Complexation

A Study in Molecular Recognition

Edwin Haslam, Terence H. Lilley, Edward Warminski, Hua Liao, Ya Cai,
Russell Martin, Simon H. Gaffney, Paul N. Goulding, and Genevieve Luck

Department of Chemistry, University of Sheffield, Sheffield S3 7HF,
United Kingdom

Natural polyphenols (*syn* vegetable tannins) are complex higher plant secondary metabolites. According to earlier definitions they are water soluble, possess relative molecular masses in the range 500 to 3,000 and besides giving the usual phenolic reactions, they have the ability to precipitate some alkaloids, gelatin and other proteins from solution. These complexation reactions have importance and wide ranging practical applications - in the manufacture of leather, in foodstuffs and beverages, in herbal medicines and in chemical defence and pigmentation in plants. The principal factors which influence the reversible association of polyphenols with other substrates are :- (i) solubility, solvation and desolvation; (ii) molecular size and conformational flexibility; (iii) general salt and specific metal ion (e.g. Ca^{2+}, Al^{3+}) effects; (iv) 'hydrophobic effects'; (v) hydrogen bonding *via* phenolic groups; (vi) the presence of tertiary amide groups in the co-substrate. The evidence for some of these observations is outlined as are methods to inhibit or enhance polyphenol-protein precipitation.

" There are no Applied Sciences...........there are only applications of Science and this is a very different matter.....................The study of the application of Science is very easy to anyone who is master of the theory of it ."

Louis Pasteur

Plant Polyphenols, Polyphenol Complexation

Plants metabolise an extraordinary array of phenolic compounds and E.C.Bate-Smith made some early seminal observations on the phenolic constituents of the leaves and other vegetative tissues of vascular plants based on material from nearly 200 families (1). He found that three classes of phenolic metabolite overwhelmingly predominate in vascular plants and amongst these were the proanthocyanidins (known earlier as leucoanthocyanins or condensed tannins). Less widely occurring in plants are the derivatives of gallic and hexahydroxydiphenic acid (known earlier as hydrolysable

0097–6156/92/0506–0008$11.75/0
© 1992 American Chemical Society

tannins) and together these two groups constitute those substances referred to in this text as **polyphenols** (known earlier as vegetable tannins).

Most phenolic metabolites are sequestered safely in the vacuole of the plant cell. Many, and in particular polyphenols, nevertheless have the potential to react with proteins and other cytoplasmic components. Such interactions may be of great significance as the cell structure is degraded, e.g. as the tissue senesces, is harvested or is damaged by predation. In some instances the reactions of the phenolic constituents which result when a plant is harvested or deliberately damaged are crucial in the preparation of certain foodstuffs - teas, coffee and cocoa with their characteristic taste, flavour and appearance.

This ability to associate with other metabolites, although it manifests itself most strikingly in the case of **polyphenols**, is nevertheless a property *per se* of the phenolic nucleus itself, e.g. the caffeine - potassium chlorogenate complex (2). There seems little doubt that the efficacy of polyphenols as complexing agents derives principally from their relative molecular size and from the possession within the same molecular species of a multiplicity of phenolic groups and associated aromatic nuclei. These give rise not only to a number of important physical and chemical properties but also to the ability to act as multidentate ligands in complexation reactions.

The processes of polyphenol complexation may be reversible or irreversible and may involve co-substrates - proteins, polysaccharides, alkaloids, anthocyanins etc. from within the same organism or extrinsic to that organism. Thus leather is formed as a result of the interaction of plant polyphenols with the protein collagen of hides and animal skins; the astringency of many fruits and beverages derives from the interaction of the salivary proteins in the mouth with polyphenolic substances in the fruit or beverage; the hard exoskeleton (cuticle) of many insects is formed by the oxidative polymerisation of phenolics and proteins in the freshly secreted cuticle. *Reversible* complexation of polyphenols may be considered as a two stage process in the first of which the polyphenol and the co-substrate, by the deployment of various non-covalent forces, are in equilibrium with soluble complexes, Figure 1.

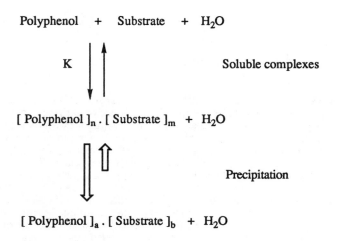

Polyphenol   +   Substrate   + H_2O

K   Soluble complexes

[Polyphenol $]_n$. [Substrate $]_m$ + H_2O

Precipitation

[Polyphenol $]_a$. [Substrate $]_b$ + H_2O

Figure 1. Reversible Polyphenol Complexation.

As the position of this equilibrium changes then, as a second stage, these soluble complexes may well aggregate and precipitate from solution. The whole process is however *usually* reversible and under suitable conditions the precipitated complexes may be redissolved. Thus addition of finite amounts of a protein to a polyphenol solution may produce a precipitate, as however further protein is added to the solution the protein- polyphenol complex frequently redissolves - an observation first made by Sir Humphry Davy (3) using isinglass in 1803 ! Similar effects can be observed with polyphenol-caffeine complexes.

The close juxtaposition of polyphenol and co-substrate which complexation brings about and the influence of external agencies (e.g. oxygen, metal ions, acid) may well promote secondary irreversible processes in which covalent bond formation takes place between the polyphenol* and co-substrate*, or in which entirely new types of product are formed, Figure 2.

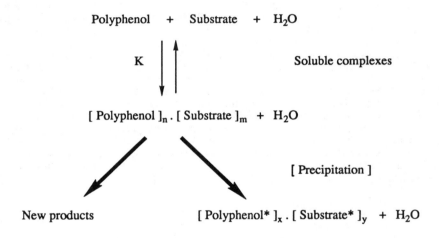

Figure 2. Irreversible Polyphenol Complexation.

The process of complexation then becomes an *irreversible* one. Polyphenols are thus particularly prone to oxidation - under the influence of enzymes, metal ions or autocatalytically in mildly basic media - to give *ortho*-quinones which are extremely reactive intermediates. Likewise polyphenolic proanthocyanidins are susceptible to acid catalysed rupture of the interflavan bond in media of pH 4.0 and less. This leads to the formation of highly electrophilic carbocation species which, like the *ortho*-quinones, are able to react rapidly with nucleophilic groups ($-NH_2$, $-SH$) in proteins with the formation of covalently bound polyphenol-protein complexes. Thus although it is frequently possible to examine reversible polyphenol complexation reactions in solution in the laboratory, in practice in the very many situations in which polyphenol complexation reactions are encountered *in vivo* the association processes may well have features which are thermodynamically reversible and others which are irreversible. As a broad rule of thumb in such situations the proportion of irreversibly bound polyphenolic molecules may be expected to increase with time. Irreversible polyphenol complexation reactions are widely encountered. Some typical examples are :

(a) enzymic and non-enzymic browning of fruits and fruit juices, and in the fermentation processes leading to the preparation of teas and cocoa.
(b) permanent haze formation in beers and lagers.
(c) ageing of red wines which leads to changes in pigmentation and astringency.
(d) Humic acid formation as organic matter is degraded in the generation of humus in soils.
(e) Necrosis - the protection of plant tissues against infective agents, predators and parasites.
(f) Sclerotisation - the development of the hard exoskeleton in insects.
(g) 'fixed tannage' in leather manufacture.

The major route whereby quinones are generated from phenols in plant extracts is probably by enzymic oxidation following the action of polyphenol oxidases which have a virtually universal distribution in the plant kingdom (A.Mayer, 4). Whatever their physiological function is deemed to be these enzymes mediate the oxidation of phenolic substrates to give highly reactive and often highly coloured quinone intermediates. These materials and their derivatives appear in many food products either as a deliberate result of processing methods (tea, cocoa - where the action of a catechol oxidase is an important initial step in the derivation of the desired product) or by undesirable and frequently accidental 'browning' reactions. The oxidation of several complex polyphenolic substrates has been shown to occur by enzyme (polyphenol oxidase) catalysed coupled oxidations in which a simple phenolic molecule acts as the initial enzyme substrate and the *o*-quinone generated then participates in a coupled redox system with the complex polyphenol. Caffeoyl tartaric acid is thus believed to act in this manner in the coupled oxidation of red wine polyphenolics ,(5), Figure 3.

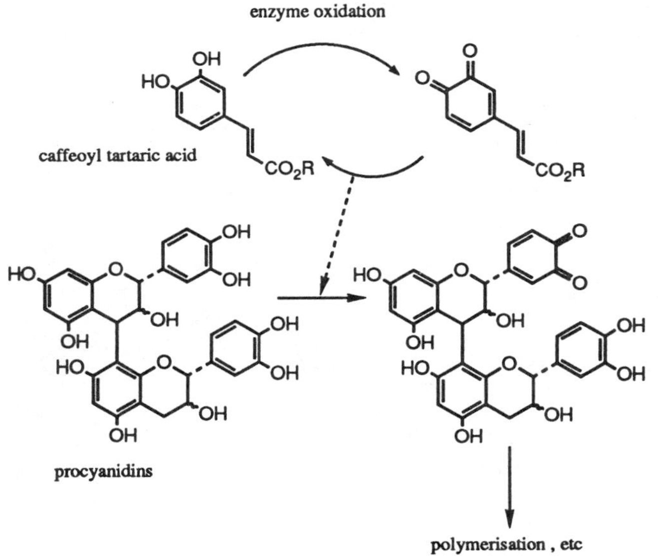

Figure 3. Coupled enzymic oxidation of polyphenols

In the case of tannin-protein complexes the covalent bonds most likely to form are those between the *o*-quinone intermediates and nucleophilic -NH₂ and -SH groups on the protein; because of the reactivity of the quinones these new covalent linkages are likely to be generated in a random and indiscriminate manner, Figure 4. Whatever their precise structure and dispositon their importance is that they change irreversibly the physical, chemical and biological characteristics of the protein. These types of reaction must also occur between proteins and very simple phenols which by themselves *per se* would have little affinity for the naked protein. Parenthetically the best authenticated example of this type of condensation is in the formation of the exoskeleton of insects and this provides some important guiding principles to the general chemical mechanisms involved.

Insects are said to owe their numerical success, in part, to the development of a hard exoskeleton - the cuticle - which provides protection against predators, parasites and dessication. When the cuticle is freshly secreted it is soft and pale in colour but it soon becomes hard and dark by a process of 'quinone tanning', (6, 7). Phenolic substrates in the cuticle are oxidised to give *o*-quinones which react with nucleophilic groups on the structural proteins to bring about cross-linking and ultimately a highly condensed polymer. The tanned saturnid silks which form the cocoon of saturnid moths have likewise been characterised as sclerotins. The fibroin and sericin which are originally secreted contain varying amounts phenolic glucosides - the 5-O-glucoside of gentisic acid and the 3-O-glucoside of hydroxyanthranilic acid. The presence of moisture allows these to interact under the influence of various enzymes. The glucosides are first cleaved by a glucosidase and the liberated phenols are subsequently oxidised by an oxidase to give the derived *o*-quinones. These react as described to introduce exogenous cross-links between the polypeptide chains.

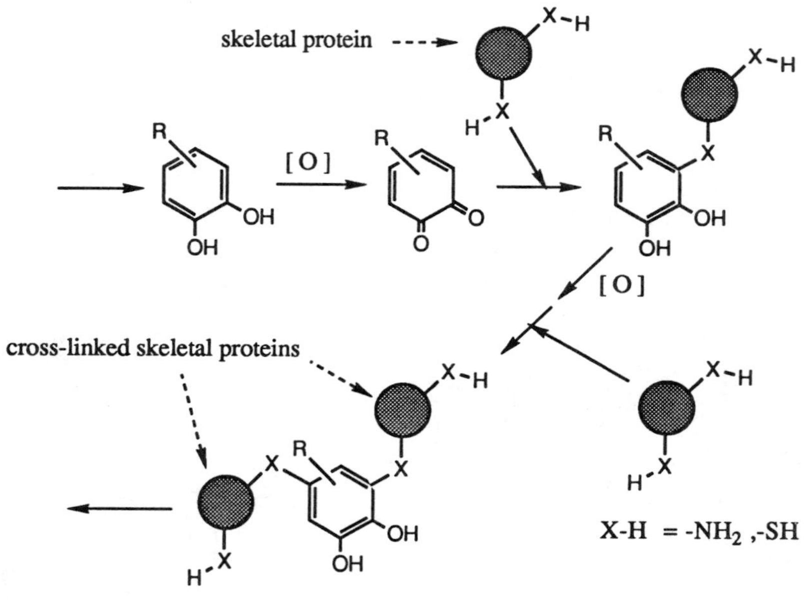

Figure 4. 'Quinone tanning'

   The oxidatively mediated condensation of proteins and polyphenolic substrates are undoubtedly of immense importance in the plant kingdom and in the ways in which man treats plant materials as raw materials paricularly in the food and beverages industry and in leather technology. The very nature of the chemical reactions involved and their speed means however that in most cases all that is possible is to point out the general rather than the specific nature of the chemical processes which are taking place.

Polyphenol Complexation - Some Practical Case Studies

Polyphenols are phytochemical chameleons. The interaction of polyphenols with other molecular species underlies man's use of these materials in a number of practical spheres. In other situations these same interactions may well be undesirable and uncceptable. Although the most commonly recognised and often the most significant examples involve proteins and to a lesser extent polysaccharides the association of polyphenols with small molecules - caffeine and anthocyanins - is also very important and some examples are given below.

Leather.   One of man's earliest uses of plant materials, such as plant galls, rich in polyphenolics, was in the conversion of raw animal hides and skins into leather. The protein collagen has a fibrous appearance at all levels of optical resolution and forms a diversity of patterns in the tissues of the animal body. It is distinguished by its unusually high content of the amino acids glycine (gly), proline (pro) and hydroxyproline (hypro) which together account for over 50% of the amino acid content of the protein. Sequences of the type - [gly. pro. hypro. gly] - form regular repeating units in the protein primary structure. Rich and Crick (8) proposed a triple helical structure (collagen II) for the protein in which 3 polypeptide chains (~ 1,000 amino acids each chain) are wound around each other, much as in an old-fashioned electrical flex.
   In collagen hydrolysates considerable quantities (~ 20%) of various amino acids such as glutamic acid (glu), arginine (arg) and asparagine (asn) with acidic, basic and amide functional side chains are also present. If these are incorporated into the collagen II structure by replacement of pro and hypro residues then their longer less compact side chains might be expected to influence the subsequent close-packing and stability of intermolecular regions in the fibre. Electron micrographs of stained collagen preparations indeed show a characteristic 'band - interband' repeat. Bear (9) was the first to suggest that the readily stained 'bands' contained high proportions of the bulky polar amino acids (e.g. glu, arg, asn) which prevent proper close packing of the collagen molecules in these regions of the fibrils and result in an amorphous and readily stained area of the fibre. Conversely the 'interbands' are thought to contain the non-polar residues (pro, hypro, gly). The close and regular packing of the collagen molecules in these regions gives rise to the crystalline portions of the fibril, Figure 5.
   It is generally agreed that the essence of 'vegetable tannage' is to protect the molecular form of the collagen fibres. Present views suggest that this is acheived by packing the amorphous regions of the fibrils with plant polyphenols, since the close-packed crystalline parts do not, it is argued, require protection. A great deal of evidence suggests that this is complete when the collagen has absorbed about half its own weight of polyphenols. It is interesting to note that after tannage a certain proportion of polyphenols may be removed from the finished leather by solvent extraction (presumably that portion held in association with the collagen by the formation of non-covalent bonds). Other polyphenolic material however remains bound irreversibly to the collagen and this is brought about over the longer term by covalent bond formation between the polyphenol and the protein under the influence of external agencies such as acid, air and light.

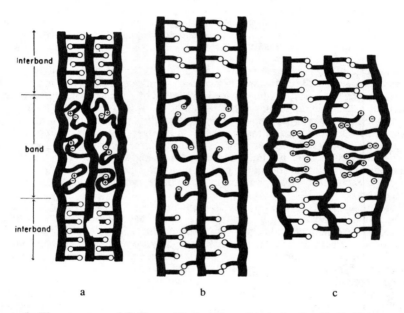

Figure 5. The structure of Collagen fibrils (Bear, 9); (a) - dry fibril, (b) - swollen in water, (c) swollen in acid.

Presumably because of this link to vegetable tannage the precipitation of gelatin (soluble collagen) from aqueous media has been a traditional test for the presence of polyphenols in plant extracts. Relatively recently Hagerman and Butler (10) were the first to demonstrate that most proline rich proteins (PRP's) had a particularly strong affinity for polyphenols. The choice of gelatin may therefore be seen as an unusually serendipitous one. It does nevertheless raise the question of the role of proline in the promotion of the association of proteins with polyphenols. Is it derived from the loose, open and flexible conformation of PRP's or does it derive from the structure of prolyl peptides themselves ? Thus in the case of vegetable tannage association, it is believed, takes place preferentially with those regions of the collagen fibrils which are generally *devoid* of pro and hypro residues. Is this the state in solution complexation with gelatin also or do the imino acids here play a more positive role in binding ?

Herbal Medicines.   The discovery that certain plants and herbs had curative or palliative effects in the treatment of illness and disease was of paramount importance in the promotion of the scientific study of plants. In addition to their extensive early use in tanning and dyeing and in the manufacture of inks infusions of plant galls have traditonally also provided powerful astringents useful in the treatment of cases of dysentery, diarrhoea, cholera and as an injection in gonorrhea. A number of oriental medicinal plants and crude drugs are rich in polyphenolics based on gallic acid and some of these traditional herbal remedies survive and find their place in modern pharmacopoeias. Typical examples are :- Geranii Herba (*Geranium maculatum, G. thunbergii*) - dried rhizome and leaves; Meadowsweet (*Filipendula ulmaria*) - leaves and flowers; Raspberry (*Rubus idaeus*) - leaves and fruit; Witch Hazel (*Hammamelis virginiana*) - dried bark or fresh dried leaves : Paeonia radix (*Paeonia albiflora* var. *trichocarpa*) - root and rhizome; Bearberry (*Arctostaphylos uva ursi*) - dried leaves; Agrimony (*Agrimonia* sp.) - roots and dried aerial parts of the plant. Although the uses of polyphenols as medicinal agents may be summarised under several headings many of their actions appear to devolve, at least in the first instance, on their ability to complex with proteinaceous or carbohydrate containing species.

Astringency, Fruit and Beverages.   Polyphenolic substances contribute significantly, if not uniquely to the astringency of wines, fruits and fruit juices, teas and other beverages. Polyphenols thus have a harsh astringent taste and produce in the palate a feeling of constriction, dryness and roughness. As a sensation of taste, that of astringency - generally recognised as a feeling of extreme dryness and 'puckeriness' - is not confined to a particular region of the mouth or tongue, but is experienced invariably as a diffuse stimulus. Moreover it may take a significant time to develop. A mucous membrane covers all the exposed surfaces of the mouth which are moistened by the secretions of the salivary glands. According to Bate-Smith (11 ,12) the primary reaction whereby astringency develops is *via* precipitation of proteins and mucopolysaccharides in the mucous secretions.
    " Loss of astringency is one of the major changes which takes place during the ripening of many edible fruits. It is generally agreed that this property (*astringency*) is due to the presence of tannins, but although some astringent fruit show a reduction in tannins on ripening, others do not. Even in those fruits where both tannins and astringency are reduced on ripening , the biochemistry underlying these changes has not been fully studied, and it is not at all obvious how the two are interrelated ". This was a view cogently expressed by Goldstein and Swain in 1963 (13). The question remains a fundamental and intriguing one. Variety, maturity and climate are known to influence the astringency of fruit, and present evidence suggests that it is at present preferable to regard each fruit as an individual case, rather than attempt an all-embracing explanation for the phenomenon. In the case of the Japanese persimmon fruit (*Diospyros kaki*) the astringency is believed to be due to the presence of polyphenols of the proanthocyanidin type (kaki tannin). Matsuo and Ito (14) have

shown that the loss of astringency upon maturation is due to the immobilisation of the kaki tannin within the fruit caused by reaction with acetaldehyde, generated during ripening. In this case the evidence points unequivocably to the fact that the acetaldehyde reacts promiscuously with the 'A' ring of the flavan units, cross-linking the proanthocyanidin oligomers , leading to a loss of solubility and to a decrease in astringency, Figure 6.

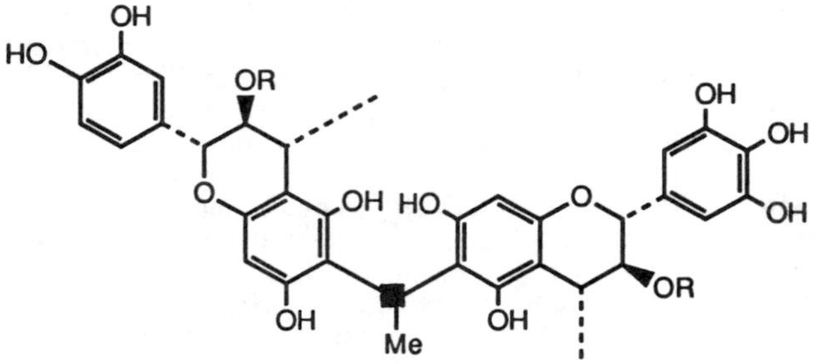

Figure 6. Cross-linking of proanthocyanidin oligomers by acetaldehyde - loss of astringency in persimmon fruit on ripening (14).

In the case of other fruit, such as raspberry and blackberry (*Rubus* sp.) in which the principal polyphenolics are based upon gallic and hexahydroxydiphenic acid, such explanations of the changes in fruit taste and flavour upon ripening are not tenable. During the ripening process in these fruit there appears to be no substantial quantitative or qualitative changes in the phenolic content . It has been suggested that in these cases during the ripening process the cellular structure of the fruit changes and soluble pectin fragments are released. These, it has been proposed, modify and disrupt the ability of polyphenols to bind to glycoproteins in the mouth when the fruit is tasted (15).

Ageing of Red Wines. The technical literature of the ageing of red wines has become remarkably extensive since the investigations of Louis Pasteur. It remains however essentially empirical and speculative in nature because of the great overall complexity of red wine composition. New wine is an elaborate chemical system which has the capability of sustaining a range of chemical reactions and a range of physico-chemical equilibria which are all interrelated and are controlled by a series of factors - such as pH, ethanol content, temperature, oxygen access and the SO_2 regime. The particular state of the evolution of a red wine is an integration of all such effects. However there is little doubt that during the ageing process the polyphenols (flavan-3-ols, proanthocyanidins) and the pigments (anthocyanins) are of crucial importance in the development of taste, flavour and appearance.
While a certain amount of tannin (polyphenols) is both desirable and essential in the making of a red wine to give body, longevity and backbone, it is possible to have too much. Where vintages have an apparent excess of polyphenols the reason is invariably the weather, particularly when there are extreme variations. Thus, apart from ripening the grapes, a spell of intense summer heat thickens their skins - a major source of both pigments and polyphenols. If these factors are combined with a relatively small concentrated yield then this produces a claret that embarks with a deep colour and a high polyphenol content. As the wine matures over the next decade the question is whether the claret will retain sufficient " fruit and flavour " to balance the polyphenol content or whether it will retain a hard unyielding backbone of polyphenol

throughout. Likewise the colour changes from a purplish-red in the young wine to the more fiery amber of the mature wine and ultimately to the tawny hues of a long-aged wine. These two features are perhaps the most readily appreciated aspects of wine ageing and yet they are not at all well understood from a chemical point of view.

Somers has studied these problems in great detail (16, 17). His conclusions are that during the ageing of red wines the anthocyanins responsible for the initial colour of the wine are displaced progressively and irreversibly by more stable polymeric pigments which may account for up to 50% of the observed colour density within the first year. The new pigment forms are much less sensitive to changes in pH than the original anthocyanins and Somers has suggested, on the basis of earlier model reactions, that they are formed by the chemical interaction of polyphenols (various oligomeric procyanidins) and anthocyanin pigments to give products in which the anthocyanin is incorporated into the oligomeric polyphenol, Figure 7.

In another related study (18) it was concluded that the loss of polyphenols and the concommitant loss of astringency in red wines may also take place *via* the familiar acid catalysed (pH ~ 3.0 - 4.0) bond breaking and bond making processes which typify procyanidin chemistry. Fission of the interflavan bonds in procyanidin oligomers is essentially random to produce carbocations and procyanidin oligomers of varying but smaller size. The recombination of the carbocations with procyanidin oligomers (the reverse of the fission reaction, 19) produces over a period of time, and on a statistical basis, procyanidins of increased molecular size which ultimately precipitate from solution.

Haslam and Cai have also shown in model systems that substrates such as the phenolic flavan-3-ols, (+)-catechin and (-)-epicatechin and their derivatives, react in aqueous ethanol solution with anthocyanins such as malvin to give a series of related yellow, (λ_{max} 430 - 440 nm), water soluble pigments. The reactions take place by a combination of both substrates and the structure of the products is as yet undefined.

A complete understanding of the manner in which each particular substrate in a new red wine - ethanol, anthocyanins, flavan-3-ols, proanthocyanins, etc. - participates and changes during maturation is not yet possible. They remain intriguing chemical problems. The evidence suggests that the changes in colour and the loss in astringency are processes which are probably linked and involve, at least in part, the chemical interaction of polyphenols and anthocyanins in pre-formed complexes.

Anthocyanin co-pigmentation.   Fruit and floral colours derive from a small group of pigments - principally, carotenoids, betacyanins (in the Centrospermae) and anthocyanins and other flavonoids. The anthocyanins are pre-eminent and they may vary the colour of flower petals from salmon and pink, through scarlet, magenta and violet to purple and blue. Under the physico-chemical conditions appertaining to those in typical cell vacuoles (e.g. weakly acidic) and in the absence of other substrates, most anthocyanins exist substantially in stable colourless forms - the carbinol bases, Figure 8. The question of how anthocyanins give rise to such a striking range of floral colours is therefore an important one.

The observation that the colour of isolated anthocyanins could be varied by the presence of other substances (co-pigmentation) was first made by Willstatter and Zollinger (20) and Robinson and Robinson (21, 22). Both groups noted that 'tannin' induced a bathochromic shift in the visible absorbtion spectra of anthocyanins. Sir Robert and Lady Robinson observed that co-pigmentation was almost universal in flower colours and they concluded that " the phenomenon............is entirely the result of the formation of weak additive complexes between pigment and co-pigment ". By themselves co-pigments have non or very little visible colour but when they are added to an anthocyanin solution they not only greatly enhance the colour of that solution but also produce a shift towards higher wavelength of the visible absorbtion maximum. Co-pigmentation has been shown to be dependent on pH, temperature and metal salts

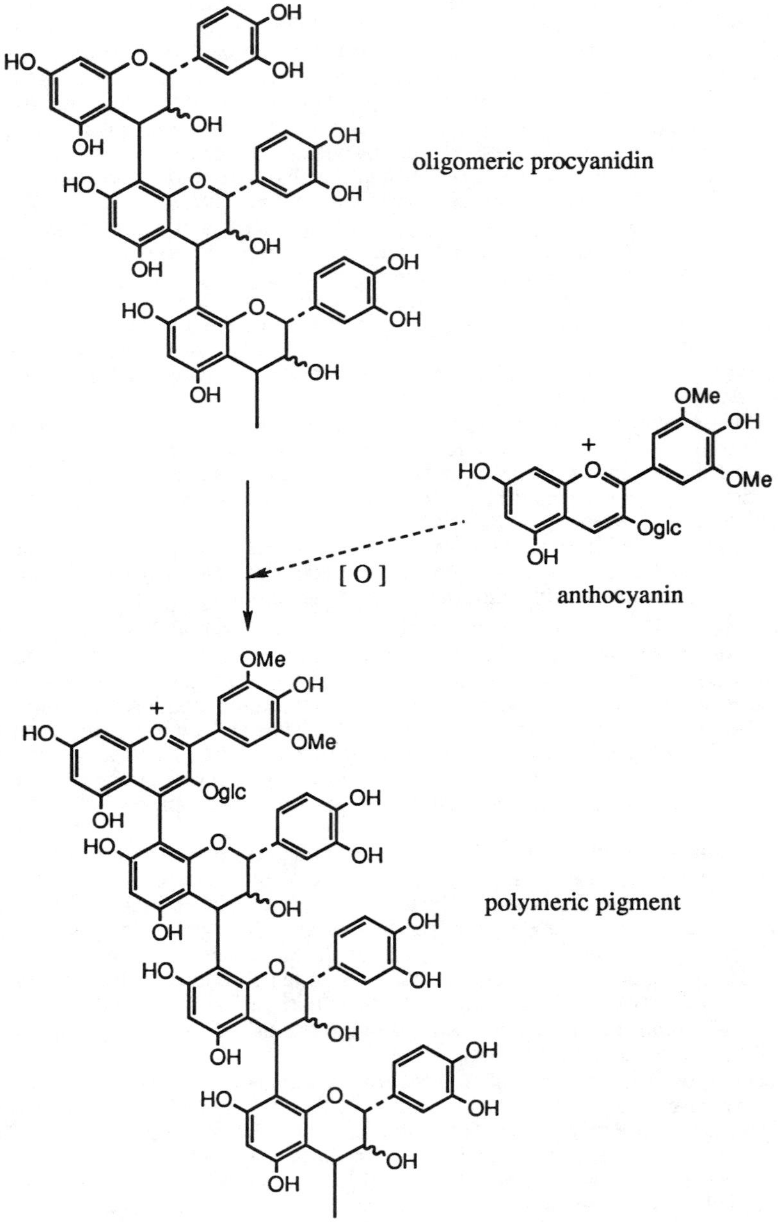

Figure 7. Polymeric pigment formation in red wines (Somers, 16, 17)

and may take place inter- or intra-molecularly. Various groups of compounds
including flavonoids, hydroxycinnamoyl esters (e.g. chlorogenic acid), polyphenols
based upon gallic acid, caffeine and theophylline may act as co-pigments, (23, 24,
25). Brouillard and his colleagues (26, 27) suggested that the phenomenon arises
intermolecularly by hydrophobically reinforced 'π-π' stacking of anthocyanin and co-
pigment molecules in the aqueous environment, that the flavylium ion - co-pigment
complex does not hydrate, and therefore at a given pH more flavylium ions are present.

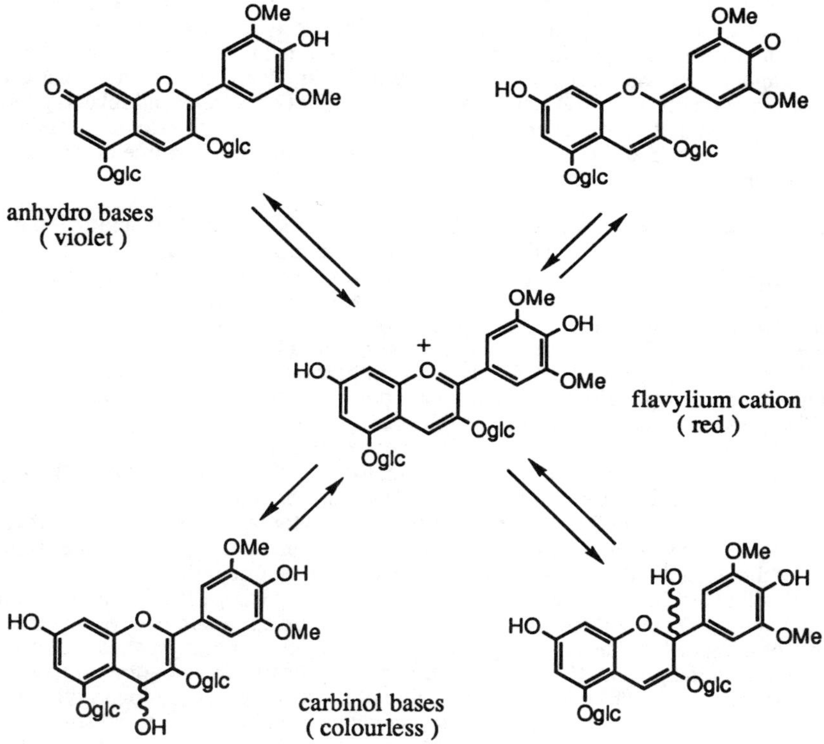

Figure 8. Principal anthocyanin equilibria in weakly acidic media.

Sorghum and Nutrition.   Considerable attention has focussed on various aspects of
Sorghum bicolor (L) Moench - the traditional cereal crop for much of Africa - but
particularly on the agronomic benefits, and the dietary and anti-nutritional effects of
sorghum polyphenols. More than other common cereals sorghum relies on intrinsic
chemical defence - in the form of polyphenolic metabolites - to repel herbivores, to
resist pathogens and inhibit the growth of potential competitors. Indeed the presence
of relatively high levels of condensed proanthocyanidins (28) in sorghum cultivars is
so frequently associated with unpalatability and resistance to bird damage that the
terms " bird resistant " and " high tannin " have come to be synonymous. Although the
mechanisms involved are still something of a matter for debate feeding trials have also

consistently pointed to the anti-nutritional effects of " high tannin " sorghums in animal diets. In this context perhaps the most striking and original observations made with " high tannin " sorghums are the way in which certain tannin (polyphenol) consuming animals defend themselves against these effects and act to diminish their severity. Butler and his collaborators (29) have shown that mammalian herbivores (but not carnivores) produce unique proline rich (up to ~ 45%) salivary proteins which have a very high affinity for tannins (polyphenols). In rats and mice these tannin complexing proteins are virtually absent until they are induced by dietary sorghum tannin. The parotid glands enlarge and within 3 days there is a dramatic increase in salivary proline rich salivary proteins (PRP's). The rats and mice showed an initial loss in weight and only when the PRP's were synthesised did the animals gain weight in the normal way. In all other animals examined by Butler and his colleagues, including humans and ruminants, these same PRP's appear to be constitutive and are present in amounts which reflect the approximate level of tannins and related phenolics in their normal diet. It has been suggested that these proline rich salivary proteins (PRP's), with their ability to bind polyphenols, constitute the first line of defence against tannins in the digestive tract.

Black Teas. The formation of the caffeine - polyphenol complex ('cream') in a cup of tea is thought to be a reversible physico-chemical process in neutral or acidic media. It is believed to occur *via* the initial formation of soluble caffeine - polyphenol complexes which subsequently aggregate and then precipitate, c.f. Figure 1. It is the green, tender and immature, rapidly growing shoots of the tea plant (*Camellia sinensis*) - the tea flush - that provide the raw material for tea manufacture. Millin and Rustidge (30) and others (31 ,32) have given general analyses of fresh tea leaf composition. The principal components are polysaccharides (~20%), proteins (~15%) and variable amounts of phenolic flavan-3-ols (15 - 30%). Since most of the distinctive sensory characteristics of tea - such as colour, taste and aroma - are associated directly or indirectly with the oxidative transformations of phenolic flavan-3-ols of green tea leaf during fermentation these sustances have attracted much attention over the past 40 years. They occur in the cytoplasmic vacuoles of leaf cells and analysis shows that (-)-epigallocatechin-3-O-gallate always predominates (up to 50% of the phenolic flavan-3-ols) accompanied by lesser quantities of (-)-epigallocatechin, (-)-epicatechin and (-)-epicatechin-3-O-gallate. Parenthetically it is interesting to note that the tea plant, *Camellia sinensis,* is amongst a very small group of plants which display a significant variation on the normal patterns of phenolic flavan-3-ol metabolism. These plants metabolise alongside the monomeric flavan-3-ols their gallate esters to the virtual exclusion of the more generally found oligomeric and polymeric proanthocyanidins. After many comprehensive and exhaustive studies of green tea leaf Nonaka, Kawahara and Nishioka (33) reported for the first time in 1983 the isolation of small quantities of four proanthocyanidin gallate esters (~ 0.07% fresh weight) from green tea. All these compounds generally lack the properties of the so-called vegetable tannins (they thus very weakly associate with proteins) but they form complexes with caffeine and may be partially precipitated from solution by the alkaloid. The tea phenolic flavan-3-ols nevertheless are oxidatively transformed into products [theaflavins and thearubigins, Roberts (34), Figure 9] which possess some of the characteristics of tannins and lend to manufactured tea many of its distinctive features as a beverage.

Theaflavin (M_R = 564) and its associated monogallate esters (M_R = 716) and digallate ester (M_R = 870) are all benzotropolone derivatives formed by mixed oxidation of the principal phenolic flavan-3-ol substrates, (34, 35). The thearubigins are responsible for much of the colour of a black tea infusion and they contribute significantly to the qualities of 'strength and mouthfeel'. Much speculation has surrounded the mechanism of their formation and their structure and they appear to be characteristic of many of the complex ill-defined phenolic polymers which invariably result when fresh plant tissues are damaged, from the random *in vivo* enzyme catalysed

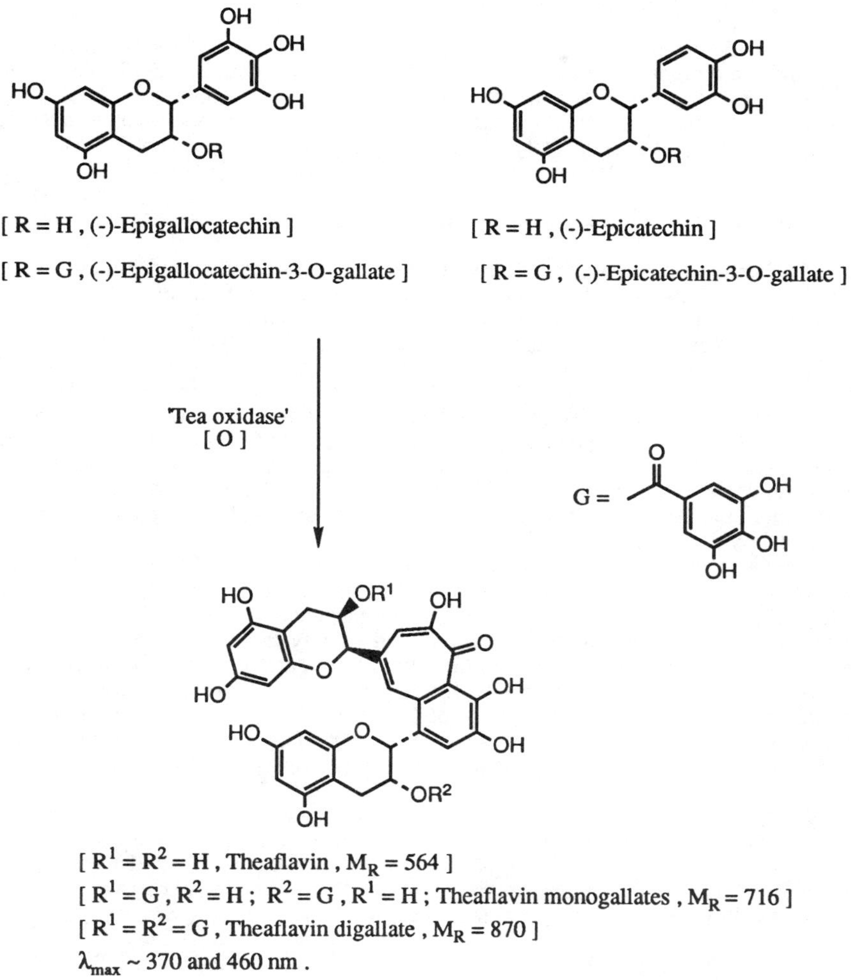

[R = H , (-)-Epigallocatechin]

[R = G , (-)-Epigallocatechin-3-O-gallate]

[R = H , (-)-Epicatechin]

[R = G , (-)-Epicatechin-3-O-gallate]

'Tea oxidase'
[O]

$G =$

[$R^1 = R^2 = H$, Theaflavin , $M_R = 564$]

[$R^1 = G$, $R^2 = H$; $R^2 = G$, $R^1 = H$; Theaflavin monogallates , $M_R = 716$]

[$R^1 = R^2 = G$, Theaflavin digallate , $M_R = 870$]

$\lambda_{max} \sim 370$ and 460 nm .

Figure 9. Black Tea Polyphenols - Oxidative generation of Theaflavins

oxidation of phenolic substrates *via ortho*-quinone intermediates, for which no adequate structural formulations are yet possible (*vide supra*). These polymers often contain acidic groups probably derived by aryl ring fission but a certain fraction of aromatic nuclei survive intact in the polymers.

The physicochemical properties of the theaflavins and the thearubigins have surprisingly not yet been studied in any comprehensive or systematic way but the assumption is often made by implication that these substances are tannins and possess many of the characteristics of the natural polyphenols themselves. Likewise 'tea creaming' is an important attribute of black tea infusions. It has all the physical and chemical characteristics of a reversible complex (Figure 1) formed between the alkaloid and the polyphenols in the extract. The amount of tea cream which forms in a cup of tea is reported to be directly proportional to the thearubigin, theaflavin and caffeine content of the infusion but real quantitative data is still not available.

Polyphenol Complexation - Experimental Observations

Emil Fischer, at the turn of the century, made some characteristically brilliant and definitive contributions to the study of the constitution of the gallotannins from Chinese and Aleppo galls. His work and that of Karrer and Freudenberg stimulated a great deal of interest amongst chemists. Several important polyphenols were isolated in crystalline form. These initial enthusiasms waned however as the great complexity of many plant extracts (often natural, but frequently induced by the many and varied post-mortal processes required to derive them) was realised. Crystallinity was the exception rather than the rule and by the 1950's the area had become one of the dark, impenetrable and neglected areas of Organic chemistry. Its renaissance coincided with the advent of the vast new armamentarium of physical methods of analysis and separation in the 1950's and 1960's. Today the composition of many plant extracts can be adequately defined in terms of their polyphenolic (tannin) and simple phenolic constituents. The structure and biosynthesis of these metabolites is well described and there is thus, for the first time, a firm base from which to embark upon studies of the biological properties of plant polyphenols and in particular their complexation reactions (36). Themes related to structure and activity can now begin to be sketched and suggestions as to their molecular basis and interpretation can also be made. If there are still exceptions to this generalisation then they relate to polyphenolic metabolites obtained from the wood and bark of trees. Here post-mortal processes multiply the complexities of normal metabolism. The nature and total composition of many of these commercially important polyphenolic extracts remains uncertain, to say the very least.

Natural Plant Polyphenols. With this caveat in mind it is now possible to describe in broad terms the nature of natural plant polyphenols. They are secondary metabolites widely distributed in various sectors of the higher plant kingdom. They are distinguished by the following general features :-

>(a) -*Water solubility*. Although when pure some natural polyphenols may be difficultly soluble in water, in the natural state polyphenol-polyphenol interactions usually ensure some minimal solubility in aqueous media.

>(b) - *Molecular weights*. Natural polyphenols cover a substantial molecular weight range from 500 to 3 - 4,000. Suggestions that metabolites occur which retain the ability to act as tannins and possess molecular weights up to 20,000 must be doubtful in view of the solubility proviso.

(c) *Structure and Polyphenolic Character* - Polyphenols per 1,000 relative molecular mass possess some 5 - 7 aromatic nuclei and 12 - 16 phenolic hydroxyl groups .They are based upon two broad structural themes :-

(i) - *Condensed Proanthocyanidins* - in which the fundamental structural unit is the phenolic flavan-3-ol nucleus. They exist as oligomers (soluble) and polymers (insoluble) of this unit linked principally through the 4 and 8 positions. In most plant tissues the polymers are of greatest quantitative significance but usually there is also found a range of soluble molecular species, from the monomers, dimers, trimers to higher oligomers.

(ii) - *Galloyl and Hexahydroxydiphenoyl Esters and their derivatives* .These metabolites are almost invariably found as multiple esters of D-glucose and many can be envisaged as derived from β-1,2,3,4,6-pentagalloyl-D-glucose as the key biosynthetic intermediate. Derivatives of hexahydroxydiphenic acid are thus assumed to be formed by oxidative coupling of vicinal galloyl ester groups in a preformed galloyl glucose ester.

galloyl ester   hexahydroxydiphenoyl ester

Studies of the reversible association of polyphenols with proteins have a long history; one of the first scientific papers on this subject is that of Sir Humphry Davy in 1803, (3). Davy's work was undertaken at the instigation of the Directors of the Royal Institution and was directed towards an understanding of the age-old process of vegetable tannage whereby 'astringent vegetable matter' (Davy's description) converts animal hides and skins to leather.

The early work in this field demonstrated some of the macroscopic features of reversible polyphenol complexation and it permitted some wholly empirical definitions of the term 'vegetable tannin' to be advanced. However until such times as structurally defined plant polyphenols became available the molecular mechanisms underlying polyphenol complexation remained poorly understood. Various studies of polyphenol complexation have been carried out in recent years and from these structure-activity relationships have been delineated. Although as a sub-group the proanthocyanidins are most commonly responsible for the range of reactions normally attributed to 'tannins' in plants many of these studies have been most conveniently pursued with a series of biosynthetically inter-related esters of gallic and hexahydroxydiphenic acid. These are accessible in homogeneous forms and differ systematically in solubility, phenolic content, molecular size and flexibility, and conformation. Some typical examples are shown in Appendix 1.

In the ensuing discussion particular reference is made to recent work from the authors' laboratory, where the *reversible association* of polyphenols with proteins, polysaccharides, caffeine and related heterocycles, anthocyanins, anthracyclinones, methylene blue and α, β and γ cyclodextrins has been investigated. Various physico-chemical techniques have been employed including :- ^1H and ^{13}C nmr spectroscopy, microcalorimetry, equilibrium dialysis, studies of enzyme inhibition and protein precipitation. Particular attention has focussed upon the relationships between structure and function and facets of this work are discussed below.

Solubility. Various observations point to the crucial importance of substrate solubility in polyphenol complexation reactions and hence to the significance of solvation and desolvation processes in the aqueous environment. This, of course, is one of the major areas of uncertainty in the study of association phenomena in aqueous media, largely because of our ignorance of the quantitative nature of the role of water itself as a solvent. In the case, for example, of proteins the surface of the folded molecule minimally perturbs the surrounding aqueous environment as the macromolecule with its hydration monolayer " meshes " with the bulk water. Apparently simple questions as to how the water surrounding the protein enters into or modulates the processes of complexation are not amenable to a ready answer.

Chemistry, like life, is however about differences. Intermolecular recognition thus depends on the differences between interactions of the isolated host and guest molecules and their environments (for polyphenols this is invariably an aqueous one), and the interactions developed in the host - guest complex. Inextricably linked with solubility and solvation in aqueous media are 'hydrophobic effects'. Increasingly indeed solvation and related 'hydrophobic effects' are seen as the major driving forces in the intermolecular association reactions of polyphenols, with direct hydrogen bonding contributing little binding energy but enhancing specificity (37, 38). Thus 'hydrophobic effects' result from the differences in the energetics of interaction between hydrophobic groups on the host and guest and the water medium and their interaction with each other, which thereby reduces the total surface area of the hydrophobic groups exposed to water, Figure 10.

The importance of 'hydrophobic effects' and water solubility can be most readily seen by reference to β-1,2,3,4,6-pentagalloyl-D-glucose and three polyphenols which are believed to be derived biosynthetically from it. Geraniin is the principal polyphenol of plants of *Geranium* and some *Acer* species, (39, 40). It is yellow, highly crystalline and, when pure, is virtually insoluble in water. It is quantitatively extracted from aqueous media by ethyl acetate and has K [octan-1-ol / water] > 100, (a measure of its 'hydrophobicity'). Testing the efficacy of geraniin as a 'tannin' because of its low solubility in water when pure is difficult. However bearing in mind the mutual solubilisation which takes place in water in crude plant extracts the evidence indicates that geraniin is, *in vivo,* a very effective 'tannin'. This fact is indeed borne out by model precipitation studies with methylene blue (Okuda, 41). Geraniin is formally six hydrogen atoms less than its presumed biosynthetic precursor β-1,2,3,4,6-pentagalloyl-D-glucose.

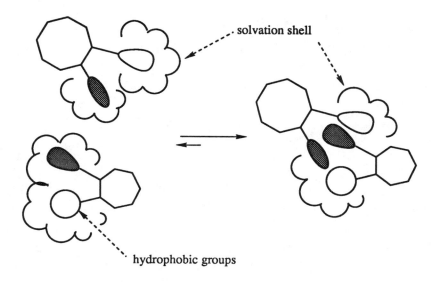

Figure 10. 'Hydrophobic effects'

The latter is amorphous, has a limited solubility in water (~ 1.0 mM at 20°C), is readily extracted from water by ethyl acetate, and has K [octan-1-ol / water] = 32. Both *in vivo* and *in vitro* it is a highly effective 'tannin'.

Vescalagin and castalagin are unique open-chain diastereoisomeric polyphenols from *Quercus* and *Castanea* species. Like geraniin they are formally six hydrogen atoms less than their presumed biosynthetic pecursor β-1,2,3,4,6-pentagalloyl-D-glucose. Both are nicely crystalline, highly soluble in water, **not** extracted therefrom by ethyl acetate, and show K [octan-1-ol / water] = 0.1. Both, in the conventional *in vitro* quantitative tests for 'tannins', react very poorly indeed. It is interesting to note that vescalagin and castalagin possess fifteen phenolic hydroxyl groups identical to the number of such groups in β-1,2,3,4,6-pentagalloyl-D-glucose. It should nevertheless also be noted that whilst β-1,2,3,4,6-pentagalloyl-D-glucose is a conformationally mobile molecule both vescalagin and castalagin are conformationally inflexible. These features will be commented upon later.

Besides the straightforward effects of polyphenol solubilty in water the influence of additives on that solubility should also be borne in mind .Thus the solubility of β-1,2,3,4,6-pentagalloyl-D-glucose in glass distilled water at 20°C is 1.0 mM. In hexanoic acid / sodium hexanoate (buffer ratio 1 : 1, 0.1 M, pH 4.7) the solubility is 40 mM, indicative of some form of hydrophobic micellar type of association between the aliphatic carbon chain of the organic acid and the polyphenol. This solubilising effect diminishes in magnitude as the number of carbon atoms in the buffer acid is decreased from seven to one (formic acid). Such solubilising factors generally decrease the effectiveness of a particular polyphenol as a 'tannin'.

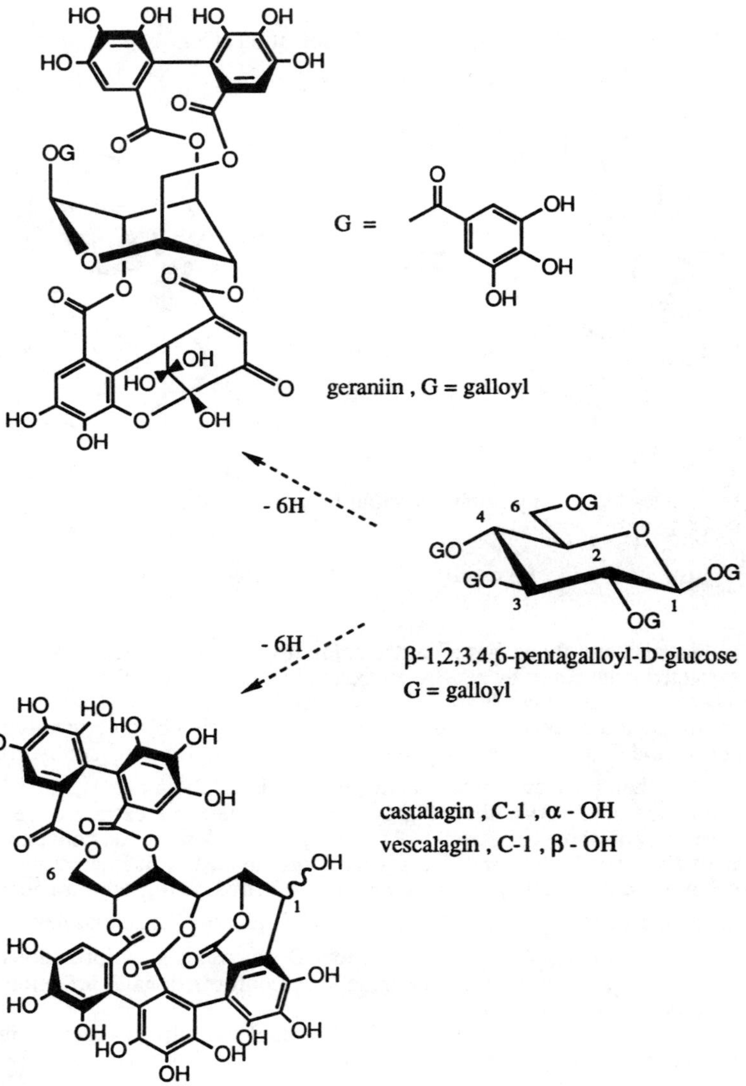

geraniin , G = galloyl

β-1,2,3,4,6-pentagalloyl-D-glucose
G = galloyl

- 6H

- 6H

castalagin , C-1 , α - OH
vescalagin , C-1 , β - OH

Inorganic Salts and Metal Ions. The association and the ultimate precipitation of polyphenols by proteins is remarkably sensitive to general salt effects and also to the presence of specific metal ions, such as Ca^{2+} and Al^{3+}. Table I shows the values for the maximum precipitation of β-1,2,3,4,6-pentagalloyl-D-glucose using a range of polypeptides, polymers and proteins in glass distilled water at 20°C (42).

Table I. Maximum precipitation values for the β-1,2,3,4,6-pentagalloyl-D-glucose[a] - protein / polymer systems in glass distilled water at 20°C

Protein / Polymer	Maximum precipitation %	Protein / polymer concentration x 10^{-6} M
Poly-L-proline, ($M_R \sim 7,800$)	100	2.8
Poly (pro.gly.pro), ($M_R \sim 5,300$)	100	17.0
Gelatin (Collagen), ($M_R \sim 100,000$)	82	0.5
Polyvinylpyrrolidone, ($M_R \sim 10,000$)	24	2.5
Sodium caseinate, ($M_R \sim 24,000$)	18	1.3
α-Lactalbumin, ($M_R \sim 14,400$)	18 - 25	1.3
Ribonuclease, ($M_R \sim 13,700$)	9	2.0
β-Lactoglobulin, ($M_R \sim 36,000$)	0	0
B.S.A., ($M_R \sim 65,000$)	0	0

[a] β-1,2,3,4,6-pentagalloyl-D-glucose concentration - 0.001M

Table II shows some typical values for similar systems, but in which polyphenol precipitation (β-1,2,3,4,6-pentagalloyl-D-glucose) takes place in a medium of 1.0 M sodium chloride. It is clear from this and similar data that the processes of aggregation and precipitation of the polyphenol - protein / polymer complexes is aided by general salt effects. At constant ionic strength ($I = 1/2 \Sigma m_1.z_1^2$) salts may be ranked in terms of their ability to enhance precipitation in the β-1,2,3,4,6-pentagalloyl-D-glucose / polyvinylpyrrolidone system as indicated (42). The series shows a strong resemblance to the familiar Hofmeister series first observed over 90 years ago. There has been a strong tendency to correlate such salt effects with increased hydrophobic interactions and decreased solubility of proteins in water. Presumably similar rationalisations can be made in respect of the increased aptitude of polyphenol-protein complexes to precipitate from aqueous media in the presence of salts.

Salts
$NaH_2PO_4 > Na_2SO_4 > CaCl_2 = MgCl_2 > KCl > NaCl = NaSCN > NaBr > Me_4NBr$
Cations
$Ca^{2+} = Mg^{2+} > K^+ > Na^+$
Anions
$H_2PO_4^- > SO_4^{2-} > Cl^- > Br^-$

Table II. Maximum precipitation values for the β-1,2,3,4,6-pentagalloyl-D-glucose[a] - polyvinylpyrrolidone system in 1.0 M NaCl at 20°C

Protein / Polymer	Maximum Polyphenol precipitation %	Protein / Polymer concentration x 10^{-6} M
Polyvinylpyrrolidone	97	1.6
Sodium Caseinate	75	0.9
β-Lactoglobulin	65	5.5
B.S.A.	17	0.5
Ribonuclease	10	2.3

[a] β-1,2,3,4,6-pentagalloyl-D-glucose concentration - 0.001M

More specific effects may be observed with certain metal ions such as Al^{3+} and Ca^{2+}. Thus aluminium salts complex strongly with phenolic compounds particularly those containing *ortho*-dihydroxy phenolic groups (e.g. galloyl esters). Indeed a complex may ultimately be precipitated from solution as aluminium salts are added to β-1,2,3,4,6-pentagalloyl-D-glucose. Aluminium salts also complex strongly with carboxylate groups (e.g. acetate < malate < citrate) and if aluminium salts are used in conjunction with polyphenolic materials then they mediate and enhance the complexation of the polyphenol with protein substrates. This synergism forms the basis of the well known use of aluminium (and titanium) salts in the tanning of leather by vegetable tannins. The process presumably occurs with the aluminium ion acting as a 'cement' binding to both protein (via carboxylate groups) and to the polyphenol (via the *ortho*-dihydroxy phenolic groups).

Caseins.   Similar observations in type may be made with calcium and the caseins (42). Caseins are a major group of secretory proteins synthesised during mammalian lactation and are stored and secreted as stable calcium phosphate complexes. These occur in milk in more or less spherical micelle particles. The caseins are phosphoproteins and fall into two groups which differ in their sensitvity to precipitation by calcium ions. In the cow these comprise the calcium insensitive κ casein family and the calcium sensitive caseins - α_{S1}, α_{S2} and β caseins. The resemblance between the α_S and β caseins is striking. A highly negatively charged segment, including several carboxylate side-chains and a phosphoserine cluster, is present in an otherwise very hydrophobic protein. The putative calcium binding sites in these proteins are believed to be associated with this negatively charged region. The κ casein has a labile peptide bond which is susceptible to low quantities of protease (rennin) and its cleavage triggers the clotting of milk. In this process the soluble C - terminal fragment of the casein (casein macropeptide, CMP ~ $^{1}/_{3}$rd. of the molecule) is removed. This fragment is rich in soluble carbohydrate residues. All three types of casein - α_S, β and κ caseins - have an unusually high content of proline, Table III.

Table III. Composition of the Caseins

Casein Component	Proline residue %	Phosphoserine residues*	$M_R \times 10^4$
α_{S1}	8.5	8 -10	2.36
β	16.5	5	2.4
κ	11.8	1	1.9

* Also phosphothreonine residues

The various caseins show contrasting behaviour in their respective abilities to precipitate β-1,2,3,4,6-pentagalloyl-D-glucose in the presence of calcium ions, although in the absence of calcium the extents of precipitation of the polyphenol are very similar, Table IV. The data clearly indicate the extreme sensitivity of β casein, the moderate sensitivity of α_{S1} casein, and the relative insensitivity of κ casein and the composite sodium caseinate towards calcium ions. Very interestingly a mixture of α_{S1}, β and κ casein approximating in % composition to that of sodium caseinate shows very similar behaviour to that of sodium caseinate in the precipitation of β-1,2,3,4,6-pentagalloyl-D-glucose. The amount of polyphenol precipitated is approximately half the theoretical value of 58% calculated on the basis of summation of the component (α_{S1}, β, κ) contributions. This evidence strongly suggests that the water soublising carbohydrate fragments on the calcium insensitive κ casein function to maintain the three component 'casein': polyphenol complex in solution. The actions of calcium ions are probably twofold. Firstly the general salt effect referred to above. Secondly the calcium ions may well act in a similar specific fashion to aluminium and function as a 'cement' helping to bind the polyphenol to the protein by coordinating carboxylate and phosphate groups on the protein and phenolic functionalities on the polyphenol. Spectroscopic evidence for this latter proposition comes from the observation that addition of calcium ions to a solution of methyl gallate produces a UV spectral shift analogous to that with aluminium ions, (42), Figure 11. In the case of methyl gallate this is indicative of the formation of a 1 : 1 coordination complex between the phenolic substrate and a calcium ion (42).

The importance of metal ions in the facilitation of the association of polyphenolic molecules with proteins and other molecules has not been subject to any systematic investigation but a number of observations point to its probable importance. Thus galloyl esters such as β-1,2,3,4,6-pentagalloyl-D-glucose, at concentrations of 2 - 5 mg./ml., readily form gels when aqueous solutions are cooled from 50 - 60°C to room temperature. Gel formation presumably arises from the ability of natural polyphenols to form extensive three dimensional lattices by stacking and intermolecular hydrogen bonding between galloyl ester groups, with the exclusion and localisation of solvent. Polyphenol gels may be disrupted by shaking or by the addition of solutes such as urea and β-octylglucoside. Gel formation is alternatively promoted by solutes such as caffeine and the peptide sweetener aspartame and by *inorganic salts,* which act presumably by facilitating formation of mixed three-dimensional lattices.

Table IV. Precipitation of β-1,2,3,4,6-pentagalloyl-D-glucose by Caseins.
The effects of calcium ions

Casein component	Maximum precipitation % in water	Maximum precipitation % in 0.012 M CaCl₂
α_{S1}	15	39
β	18	91
κ	13	16
α_{S1} (43%) β (42%) κ (15%) mixture	8	27*
Sodium caseinate	18	25

*Theoretical value by summation - 58%

Model Studies - Caffeine and Cyclodextrins. In addition to studies of reversible polyphenol complexation with macromolecules work has also been carried out on the association with small molecules including caffeine and related heterocycles, anthocyanins, anthracyclinones, methylene blue and α, β, and γ cyclodextrins, (43, 44, 45) .These studies are not only of intrinsic value - e.g. caffeine and 'tea creaming' and anthocyanin co-pigmentation - but they are of significance as models for polyphenol complexation generally.

Caffeine. In aqueous media polyphenols readily associate with several alkaloids including caffeine leading to the precipitation of intermolecular complexes. The complexes formed in black teas between caffeine and polyphenols are thought to alleviate many of the known undesirable physiological actions of caffeine. When precipitated, these particular complexes constitute the 'tea cream' which gives a measure of certain desirable attributes (strength and briskness) of the beverage. Structurally caffeine has features which are, in a sense, reminiscent of peptides derived from imino acids such as proline [*viz.*, the two -CO-N(Me)- groups] and studies of its association with polyphenols are also therefore of relevance to the important question of the role of proline in promoting the analogous complexation with proteins. Of the various purine and pyrimidine heterocycles which have been examined the affinity of caffeine for polyphenols is the strongest , suggesting the particular significance of the tertiary amide groups.

Whilst it is not possible to extrapolate directly to behaviour in solution, X-ray crystallographic analysis of various caffeine - phenol complexes (2, 46) confirm the importance of (a) apolar hydrophobic interactions, (b) hydrogen bonding, and (c) in certain situations coordination around a metal ion as primary intermolecular forces in caffeine polyphenol complexation. The crystal structure of the 1 : 1 methyl gallate - caffeine complex thus typically shows a layer lattice structure (46). In this array caffeine and methyl gallate molecules are arranged in alternating layers, approximately parallel, with an interplanar separation of 3.3 to 3.4 Å, Figure 12. This 'stacking' structure is complemented by an extensive in-plane system of hydrogen bonding between the three phenolic groups of methyl gallate (as proton donors) and the two ketoamide groups and the basic N-9 of caffeine, Figure 13.

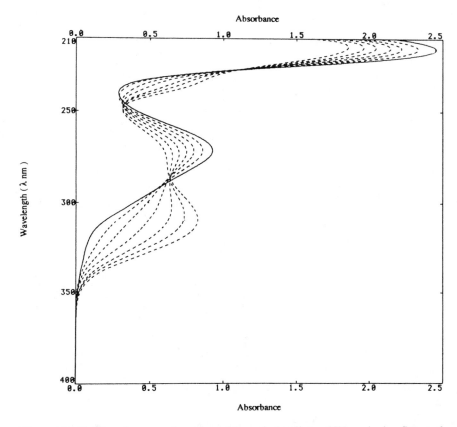

Figure 11. Calcium ion complexation with methyl gallate - UV analysis. Spectral changes produced by the addition of aliquots of CaCl$_2$ to methyl gallate in glass distilled water (with Bi Shi).

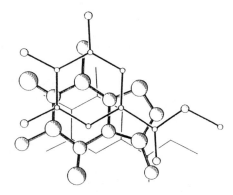

Figure 12. The Caffeine - Methyl gallate complex. Vertical 'π π' stacking.

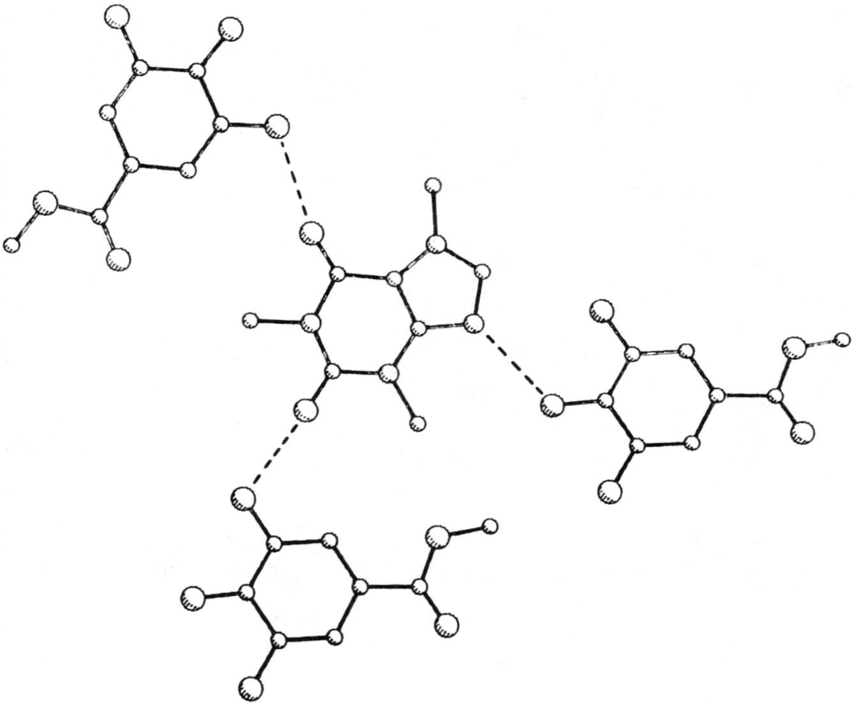

Figure 13. The Caffeine - Methyl gallate complex. In plane hydrogen bonding.

Caffeine

The crystal structure of the potassium chlorogenate - caffeine complex shows similar features (a, b) to those noted in the methyl gallate complex. However an additional critical stabilising factor (c) is the coordination of seven oxygen atoms in an irregular polyhedral arrangement around the central potassium ion (2).

This tendency of caffeine to form apolar 'π - π' stacking arrangements with phenolic substrates also occurs in solution and may be observed by the anisotropic desheilding of the aromatic protons of the phenol and correspondingly the three (-N-Me) groups and H-8 of the caffeine. This property has been exploited to determine the association constants for the formation of 1 : 1 complexes between caffeine and various polyphenols in solution, Table V.

Table V. Association constants (K, dm^3 mol^{-1}) for the formation of 1 : 1 complexes between caffeine and natural polyphenols in deuterium oxide. Determined by ^{1}H nmr spectroscopy using the chemical shift changes for H-8 of caffeine

Polyphenol	M$_R$	K (45°C)	K (60°C)
(-) Epicatechin	290	34.5	
(-) Epigallocatechin	306	35.6	
(-) Epigallocatechin-3-O-gallate	458	52.8	
(+) Catechin	290	26.1	
(+) Catechin-3-O-gallate	442	38.2	
β-1,2,3,4,6-Pentagalloyl-D-glucose	940		81.6
β-1,2,4,6-Tetragalloyl-D-glucose	788		53.4
β-1,3,6-Trigalloyl-D-glucose	636		37.1
Tellimagrandin-2 (Eugeniin)	938		61.9
Davidiin	938		24.7
Casuarictin	936		19.9
Rugosin-D	1874		138.0
Sanguin H-6	1870		63.0

Particularly striking in these results is the strong dependence on 'free' galloyl ester group content of the polyphenol and the tendency of the caffeine molecule to associate at or near one or more galloyl ester groups. Also of significance are the depressive effects on the association constant K which result from the restriction of conformational mobility in the polyphenol by the successive biosynthetic introduction of hexahydroxydiphenoyl groups formed from two galloyl ester functionalities (cf., β-1,2,3,4,6-pentagalloyl-D-glucose, tellimagrandin-2, casuarictin, Table V). Parenthetically it may be noted that these changes exactly mirror those found in polyphenol - protein complexation.

    Frequently the affinity of a particular substrate for different polyphenols has been measured by its ability to precipitate them from solution. Indeed this is the most convenient experimental technique to employ with proteins. The two processes of (i) association and (ii) aggregation and precipitation are physically and chemically quite different, Figure 1. It is therefore of some interest to be able to compare values of K for different polyphenols in relation to precipitation. In competitive experiments it may be seen, Table VI, that the extent of precipitation of a polyphenol from a mixture by caffeine is broadly, *but not directly*, related to the association constant K, Table V.

Table VI. Selective precipitation of Polyphenols by Caffeine

Polyphenol mixture (A)	% Polyphenol precipitated	
	i	ii
β-1,2,3,4,6-Pentagalloyl-D-glucose (K = 81.6)[b]	82	91
β-1,3,6-Trigalloyl-D-glucose (K = 37.1)[b]	28	45
Tellimagrandin-2 (Eugeniin) (K = 61.9)[b]	74	88

Polyphenol mixture (B)	% Polyphenol precipitated	
	i	ii
β-1,2,3,4,6-Pentagalloyl-D-glucose (K = 81.6)[b]	83	92
(+) Catechin-3-O-gallate (K = 38.2)[a]	30	46
(-) Epigallocatechin-3-O-gallate (K = 52.8)[a]	17	31

All initial polyphenol concentrations = 1.5×10^{-3} mol. dm^{-3}.
Caffeine concentrations :- (i) - 4.5 mmol dm^{-3}, (ii) - 9.0 mmol dm^{-3}
K (dm^3 mol^{-1}) for the formation of a 1 : 1 complex with caffeine, [a] 45°C, [b] 60°C

    Clearly caffeine possesses a number of features which optimise its effectiveness as a small molecule for complexation with polyphenolic substrates. The phenolic groups of polyphenols are good proton donors in hydrogen bonding systems. Likewise the tertiary amide carbonyl groups of caffeine are good proton acceptors. It thus seems probable that hydrogen bonding may ultimately make specific contributions to the stability of caffeine - polyphenol complexes through the entropy gain as bound water molecules of solvation are released .The 'hydrophobic effect' is nevertheless probably the most dominant influence on the interactions of polyphenols with caffeine in water. It is also of interest to note that in the various phenol - caffeine complexes (X-ray) the orientation of the two species in the 'vertical stacking' arrangement in the crystal are generally very similar, suggesting that the interaction may be strongly influenced by the development of complementary interacting dipoles within the two substrates.

Cyclodextrins. It has been known for some time that polysaccharides can adopt a variety of shapes in the condensed phase and in several instances such conformations persist in solution. Polyphenols bind strongly to some polysaccharides of this type

(e.g. starch) and the general significance of his type of cavity sequestration has been investigated in model studies with the cyclodextrins. One of the most distinctive properties of the cyclodextrins is their ability to include substrates, including aromatic molecules, as 'guests' in the 'host' cavity formed by the six (α) or seven (β) 1-α-4 cyclically linked D-glucopyranose residues. In solution, complexation with aromatic substrates can be readily monitored by following the ^{1}H nmr chemical-shift changes induced by the magnetic anisotropy of the aromatic guest as it penetrates the cavity, on the two concentric rings of C-H groups (at C-3 and C-5) which line the cavity. The nature of the driving forces which lead to inclusion within the cavity remain a matter for some speculation; suggestions include (a) hydrogen bonding, (b) van der Waals interactions, (c) release of strain energy in the cyclodextrin and (d) release of water molecule(s) fom the cavity. Determination of the association constants for the formation of 1 : 1 complexes between polyphenols and β-cyclodextrin has been carried out using ^{1}H nmr, Table VII. A number of features command attention. Most evident is the enhanced binding of phenolic flavan-3-ols compared to that of galloyl esters. It is dependent on the stereochemistry at C-3 of flavan-3-ols and is depressed by further hydroxylation in ring B. Complexation is facilitated by galloylation at the C-3 hydroxyl group, substantially in the case of (-) epigallocatechin. The expectation is that several modes of association are probably operative, involving the independent insertion of both rings A and B of the flavan substrates into the cyclodextrin cavity. These observations support the view that natural polyphenols have a peculiar affinity for pores or crevices in polysaccharide structures. In the case of cyclodextrins complexation is very much of the " lock and key " variety and is dependent on the size of the cavity and the structure and stereochemistry of the polyphenolic substrate. In the case of polysaccharides the association is probably much more dynamic in character and dependent on the fitting of polyphenols into developing polysaccharide cavities.

Table VII. Association constants for the formation of 1 : 1 complexes between Polyphenols and β-cyclodextrin (K, dm^3 mol^{-1}) in deuterium oxide

Flavan-3-ols[a]	K
(+) Catechin	2908
(+) Catechin-3-O-gallate	6232
(+) Gallocatechin	948
(-) Epiafzelechin	792
(-) Epicatechin	464
(-) Epigallocatechin	208
(-) Epigallocatechin-3-O-gallate	1889
Procyanidin B-2	63

Galloyl ester[b]	
β-1,2,3,4,6-Pentagalloyl-D-glucose	340

Determined at :- [a] 45°C; [b] 60°C

Proteins, Polypeptides and Polyamides. There can be very little doubt that from the earliest of times the interactions of polyphenols with proteins and related macromolecules, because of their perceived importance and practical significance, have commanded the greatest interest and attention. There is also no doubt that, after decades of semi-empirical work, the observations of Hagerman and Butler, made first in 1981 (10), are the most important. They are seminal and provide the basis for future work in this field. In contrast to the assumed non-specific nature of polyphenol binding to protein Butler found that the relative affinity of different proteins for the condensed proanthocyanidin from sorghum varied by over four orders of magnitude. Butler and his colleagues have thus shown (10, 48, 49) that flexible 'open' proteins, those rich in the amino acid proline (e.g. gelatin, certain mucalpolysaccharides and polyproline), bind polyphenols much more effectively than compact globular proteins. Hagerman and Butler attribute these differences to two major features. The first is the much more open and flexible conformations of the proline rich polypeptides and the second is their enhanced ability to form strong hydrogen bonds with the polyphenol, brought about by the increased accessibilty of the prolyl peptide bonds and the bis-alkyl substitution of the prolyl amide nitrogen (c.f., caffeine).

Hagerman and Butler (10) employed a competitive precipitation assay in which the precipitation of a standard radioisotopically labeled protein by polyphenol is compared to the precipitation of the same radioisotopically labeled protein when mixed with another protein of unknown affinity for 'tannin'. In this way these authors established a series showing the relative affinities of different proteins and polymers for a given polyphenol. Alternative approaches to these and related problems have been utilised in the authors' laboratory. Particular questions which have been addressed include :-

(i) How is the structure of a polyphenol related to its capacity to bind to a given protein ?

(ii) Can evidence be obtained to demonstrate the presence of soluble polyphenol - protein complexes ?

(iii) Is there a direct relationship between the ability of a polyphenol to reversibly complex with a given protein (K, Figure 1), and its ability to precipitate that protein ?

(iv) What are the other principal factors which influence polyphenol - protein complexation and precipitation ?

(v) How can the ability of a given polyphenol to bind to protein be enhanced or diminished ?

Answers to some of these questions have already been alluded to earlier.

Polyphenols - Structure and activity. Initial work concentrated on attempts to determine the equilibrium constant K for the reversible formation of soluble polyphenol - protein complexes (Figure 1), using equilibrium dialysis, (50). The protein BSA was employed and ultimately the data were analysed by the determination of the free energy of transfer ($\Delta G^{\theta,tr}$) of the protein from an aqueous medium to an aqueous medium containing the polyphenol ligand (concentration - m_f). Typical values are shown in Table VIII.

Molecular size and *conformational flexibility* are critically important as determinants of the ability of a polyphenol to bind to protein. Thus in the galloyl-D-glucose series the efficacy of association with protein BSA is enhanced with the addition of each galloyl ester group, (Table VIII, tri < tetra < penta) and reaches a maximum in the flexible 'disc-like' molecular structure of β-1,2,3,4,6-pentagalloyl-D-glucose. Likewise the substantially increased affinity of the 'dimeric' polyphenol

Rugosin D (Appendix I) compared to β-1,2,3,4,6-pentagalloyl-D-glucose is simply a reflection of the almost doubling of relative molecular mass (M_R). In the galloyl-D-glucose series where vicinal galloyl ester groups are constrained by the biosynthetic formation of a hexahydroxydiphenoyl group the loss of conformational freedom is reflected in a reduced capacity to bind to protein (c.f. Table VIII, β-1,2,3,4,6-pentagalloyl-D-glucose, tellimagrandin-2, vescalagin / castalagin) .The apotheosis of this effect is seen in the case of the two unique open chain D-glucose derivatives - vescalagin and castalagin - metabolites of *Quercus* and *Castanea* species. These rigid virtually inflexible molecules are, in a sense, analogues of β-1,2,3,4,6-pentagalloyl-D-glucose, but on a molar basis they are both very weakly bound to protein. In this context the observed 'relatively lower astringency' of the proanthocyanidins, compared to the galloyl ester derivatives, may be, in part, explicable in terms of the conformational restraints imposed by restricted rotation about the interflavan bonds (36). Although there are broad parallels between the values for the association constants K for the formation of 1 : 1 complexes of various galloyl esters with caffeine, (Table V), and values of $\Delta G^{\theta,tr}$, (Table VIII) for the corresponding esters, the figures suggest a much greater degree of discrimination in the affinity for protein.

Table VIII. Protein - Polyphenol Complexation. Values of the free energy of transfer ($\Delta G^{\theta,tr}$) of a protein, BSA, from an aqueous medium to an aqueous medium containing a polyphenol ligand (concentration, m_f)

Polyphenol	M_R	$-\Delta G^{\theta,tr}$ / Kj mole^{-1}
β-1,2,3,4,6-Pentagalloyl-D-glucose	940	26.9
β-1,2,3,6-Tetragalloyl-D-glucose	788	9.1
β-1,3,6-Trigalloyl-D-glucose	636	0.9
Tellimagrandin-2 (Eugeniin)	938	19.7
Vescalagin / Castalagin	934	1.0
Rugosin-D	1874	58.7

Determined in aqueous solution at 25°C, pH 2.2 and at a free ligand (polyphenol) concentration (m_f) of 4.5 mole Kg^{-1}.

Protein precipitation. The question of how the ability of a given polyphenol to form soluble complexes with a protein (as measured by $\Delta G^{\theta,tr}$) correlates with the ability to precipitate a protein from solution is similar to that posed earlier in the case of caffeine. Data in Table IX indicate that there is once again a broad but not direct correlation between these figures. The Table records figures, *under a fixed set of experimental conditions,* for the maximum extent of precipitation of a polyphenol and the gelatin concentration at which this occurs. Thus 80% of the β-1,2,3,4,6-pentagalloyl-D-glucose is precipitated whilst none of the β-1,3,6-trigalloyl-D-glucose is removed as an insoluble complex.

Also included in the table are figures relating to the ability of the black tea polyphenols, Figure 9, to precipitate gelatin under identical conditions. As compared to the galloyl-D-glucose esters it is clear that the various theaflavins are slightly more effective on a molar basis, (42). This is the first quantitative measure of the affinity of theaflavins and their gallate esters for proteins, and hence of their astringency.

Table IX. Precipitation of polyphenols by Gelatin

Polyphenol	$-\Delta G^{\theta,tr}$ Kj / mole^{-1}	% Ppt^n at maximum	Gelatin conc^n at maximum
β-1,2,3,4,6-Pentagalloyl-D-glucose (M_R = 940)	26.9	80	18
β-1,2,3,6-Tetragalloyl-D-glucose (M_R = 788)	9.1	44	36
β-1,3,6-Trigalloyl-D-glucose (M_R = 636)	0.9	-	-
Theaflavin (M_R = 564)	-	33	36
Theaflavin Monogallate (M_R = 716)	-	61	21
Theaflavin Digallate (M_R = 870)	-	85	15

Polyphenol concentration - 10μM; Gelatin concentration (μg / ml) at point of maximum polyphenol precipitation.

The results shown above and in Tables I - IV were obtained employing a novel technique developed (42) to measure polyphenol precipitation by proteins. The procedure is best suited to galloyl esters and their derivatives (e.g. β-1,2,3,4,6-pentagalloyl-D-glucose) and depends on the accurate, simultaneous, measurement of changes in the two principal absorption maxima of the polyphenol ($\lambda \sim$ 214 and 280 nm). This method has revealed many further subtleties in polyphenol - protein complexation. Addition of aliquots of gelatin to solutions of β-1,2,3,4,6-pentagalloyl-D-glucose leads, in the first instance, to *precipitation* of the polyphenol as a complex, and this is observed as a *decrease* in both absorbtion maxima. This is followed by a process of *resolubilisation* and a corresponding *increase* in both absorbtion maxima. These two phases - *precipitation* and *resolubilisation* - are characteristically portrayed in a plot of the absorbtion at λ_{max} 280nm versus protein concentration, Figure 14. It may be noted that the absorbtion at λ_{max} 280nm generally does not return to the original value upon resolubilisation. It is not yet clear whether this is due in some cases to incomplete resolubilisation or to the fact that the solubilised polyphenol - protein complex has a different extinction coefficient (ε) from that of the 'naked' polyphenol.

Figure 15 shows, in an entirely analogous way, the precipitation of β-1,2,3,4,6-pentagalloyl-D-glucose by polyvinylpyrrolidone in water alone and in 1.0M NaCl. These graphs illustrate the data shown in Tables I and II, and the effect of salts upon precipitation. A further phenomenon noted in all these experiments but most distinctly in the case of polyvinylpyrrolidone is the 'red-shift'. This is the change towards higher wavelength of the λ 280 nm absorbtion maximum, attendant upon resolubilisation of the polyphenol. In the case of polyvinylpyrrolidone this shift is typically 6 - 7 nm. For various proteins a generaly smaller (3 - 4 nm) but similar shift is observed. The 'red-shift' changes for the β-1,2,3,4,6-pentagalloyl-D-glucose - polyvinylpyrrolidone system are also shown in Figure 16. In water there is a steady change; in 1.0M NaCl the wavelength of absorbtion (~ 280 nm) does not change during precipitation but

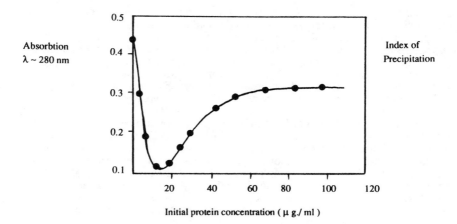

Figure 14. Precipitation of β-1,2,3,4,6-pentagalloyl-D-glucose by gelatin in glass distilled water. Plot of absorbtion at λ ~ 280 nm versus protein concentration. Precipitation and resolubilisation.

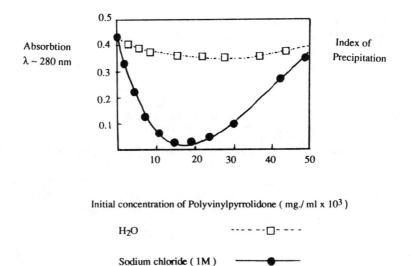

Figure 15. Precipitation of β-1,2,3,4,6-pentagalloyl-D-glucose by polyvinylpyrrolidone in glass distilled water and 1.0 M NaCl. Plot of absorbtion at λ ~ 280 nm versus polyvinylpyrrolidone concentration. Precipitation and resolubilisation.

does so at the onset of resolubilisation. The origins of this spectral shift are debatable but it has been assumed, because of its generality in association with the resolubilisation phase with a range of proteins and in situations where the polyphenol - protein complex is not precipitated and remains in solution, that it is characteristic of the formation of *soluble* polyphenol - protein (polymer) complexes.

Figure 17 shows a comparison of the behaviour in this precipitation assay of gelatin and a typical proline rich protein preparation from human saliva. The pattern is characteristic also of poly-L-proline and poly (pro.gly.pro) - namely the inability of these proteins and polypeptides to resolubilise the polyphenol.

Modulating Polyphenol Precipitation. Attention has been devoted to some of the ways in which the precipitation of polyphenol - protein complexes may be influenced (enhanced or inhibited). Some of the additives which act in one or other of these capacities are shown below. Reference has already been made to general salt effects and specific metal ion complexation, (Tables II and IV). Natural saponins and related steroids such as digitonin also enhance precipitation. The experimental data suggest that the saponins generally act by association with the protein, reinforcing the development of a hydrophobic layer on the protein surface, stimulating aggregation and then precipitation.

Enhancement of Precipitation
(i) - General salt effects.
(ii) - Specific metal ion complexation, e.g. Al^{3+}, Ca^{2+} with both substrates. The effects of Ca^{2+} may often be reversed by citrate.
(iii) - Natural saponins, digitonin.

Inhibition of Precipitation
(i) - Polysaccharide gums, Galactomannans.
(ii) - Synthetic bile acid 'dimer' salts.
(iii) - Sodium caseinate and α, β and γ caseins.

Conversely other steroids such as synthetic bile acid 'dimer' salts inhibit precipitation of polyphenol - protein complexes. These molecules act, it is thought, by the formation of a water soluble 'hydrophobic cleft' which can preferentially accomodate the polyphenolic substrate. The bis-steroidal salt thus acts as a competitor with the protein to complex the polyphenol.

Sodium caseinate and the various caseins similarly act as competitors with other proteins for polyphenolic substrates in water, in the absence of salts. Figure 18 shows the effect of adding sodium caseinate to gelatin on the precipitation of β-1,2,3,4,6-penta - galloyl-D-glucose; the onset of precipitation is delayed, and the protein (gelatin) concentration at which maximum precipitation is acheived is increased. The underlying effect is presumably that the sodium caseinate competes with the gelatin for the polyphenolic substrate and in doing so it forms a *water soluble complex* with the polyphenol.

Extensive studies have been carried out into the ways in which polysaccharide gums, and in particular the galactomannans, may inhibit polyphenol association and precipitation. The effect of ι-carrageenan on the precipitation of β-1,2,3,4,6-penta - galloyl-D-glucose by sodium caseinate in the presence of calcium ions is shown in Figure 19. The inhibitory pattern here (and with other polysaccharides) is quite distinct and leads to an overall reduction in the extent of precipitation at *all* concentrations of sodium caseinate. Table X shows the comparative inhibitory effects of ι and λ carrageenan and agarose in the same system at a given polysaccharide protein ratio.

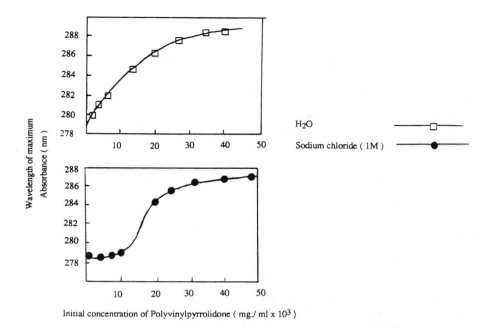

Figure 16. Bathochromic ('red') shift in absorbtion maximum at λ ~ 280 nm. Precipitation of β-1,2,3,4,6-pentagalloyl-D-glucose by polyvinylpyrrolidone in glass distilled water and 1.0 M NaCl. Plot of λ versus polyvinylpyrrolidone concentration.

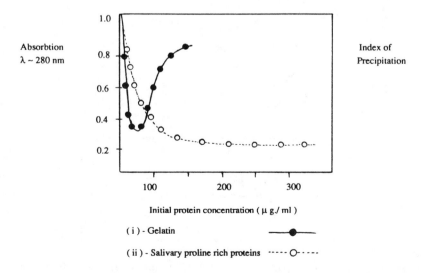

Figure 17. Comparison of precipitation of β-1,2,3,4,6-pentagalloyl-D-glucose by gelatin and salivary proline rich proteins in glass distilled water. Plot of absorbtion at λ ~ 280 nm versus protein concentration.

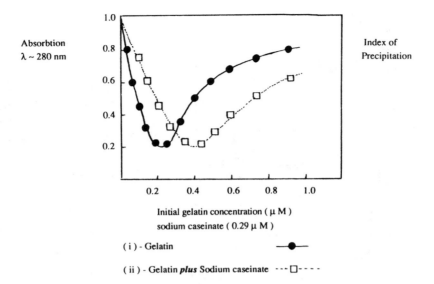

Figure 18. Inhibition of precipitation of β-1,2,3,4,6-pentagalloyl-D-glucose by gelatin using sodium caseinate in glass distilled water. Plot of absorbtion at λ ~ 280 nm versus gelatin concentration (with Ms.Heidi Grimmer) .

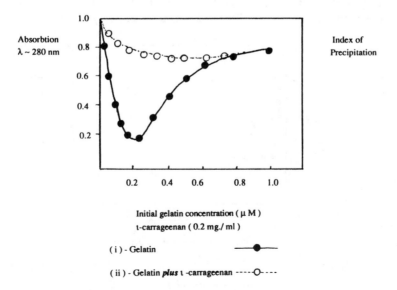

Figure 19.   Inhibition of precipitation of β-1,2,3,4,6-pentagalloyl-D-glucose by sodium caseinate in glass distilled water in the presence of Ca^{2+} using ι-carrageenan. Plot of absorbtion at λ ~ 280 nm versus protein concentration.

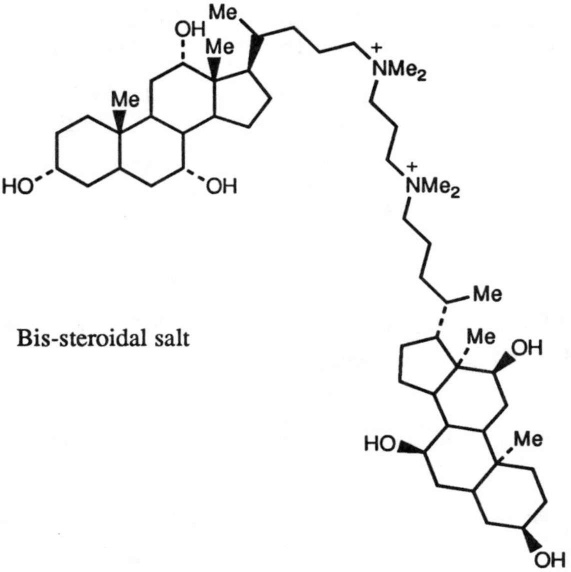

Bis-steroidal salt

'hydrophobic cleft'

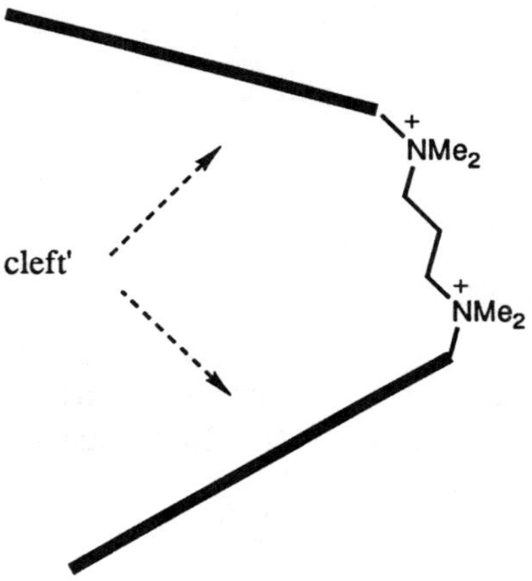

In Table XI similar data is presented for the effects of a range of polysaccharide gums in the gelatin - β-1,2,3,4,6-pentagalloyl-D-glucose system. It is interesting to note the inhibition of polyphenol precipitation by both pectin and polygalacturonic acid (as its K+ salt) - an observation which corroborates the earlier suggestion concerning the loss of astringency in the ripening of particular fruits.

Table X. Precipitation of β-1,2,3,4,6-pentagalloyl-D-glucose by Sodium caseinate :
the inhibitory effects of polysaccharides

Polysaccharide	% Reduction in precipitation of Polyphenol
	0
ι-Carrageenan	57
λ-Carrageenan	46
Agarose	7

β-1,2,3,4,6-pentagalloyl-D-glucose - 10.0 µM; sodium caseinate - 0.8 µM;
polysaccharide - 10.0 µg / ml.; $CaCl_2$ - 1.5 mM.

Table XI. Precipitation of β-1,2,3,4,6-pentagalloyl-D-glucose by Gelatin : the
inhibitory effects of polysaccharides

Polysaccharide	% Reduction in Precipitation of Polyphenol
	0
Carob (Locust bean) gum, Carubin	2
Tara gum	0
Guar gum, Guarin	0
Xanthan	22
Gellan	56
Gum Arabic	61
Polygalacturonic acid (K+)	62
Pectin	40
Dextran (M_R = 40,000)**	11
Glucose*	2

*Glucose molar concentration equivalent to the sub-unit concentration in Dextran**
β-1,2,3,4,6-pentagalloyl-D-glucose - 10.0 µM; Gelatin - 0.2 µM; Polysaccharide -
20.0 µg / ml.
Gellan and Xanthan are bacterial exopolysaccharides from *Auromonas elodea* and
Xanthomonas sp. respectively.

Carob (Locust bean, *Ceratonia siliqua*), Guar (*Cyamopsis tetragonolobus*) and Tara
(*Caesalpinia spinosa*) gums are plant galactomannans with differing ratios of
mannose and galactose in the polysaccharide structure, *viz* : Carob (mannose :
galactose ~ 4 :1), Tara (mannose : galactose ~ 3 :1) and Guar (mannose : galactose
~ 2 :1). The principal difference between these three hydrocolloids is the degree of
branching (galactose substitution) of the polymannan chain. These polysaccharides
form gels with the bacterial exopolysaccharide Xanthan. They do so, however, to
differing degrees. The extent of interaction between the galactomannans and the
Xanthomonas polysaccharide decreases with increase in the D-galactose content of the
plant galactomannan.   Thus more Tara than Carob gum is required to effect gelation.
Xanthan - Guar gum mixtures do not form gels. Table XII shows data which illustrate

the synergistic effect of these plant galactomannans upon the inhibition exerted by Xanthan gum on the precipitation of β-1,2,3,4,6-pentagalloyl-D-glucose by gelatin. Carob and Tara gum display substantial effects; Guar gum is without effect.

Table XII. Precipitation of β-1,2,3,4,6-pentagalloyl-D-glucose by gelatin : Synergistic effect of plant galactomannans upon Xanthan gum

Polysaccharide *added to* Xanthan	% Reduction in Precipitation of Polyphenol
	32
Carob (Locust bean) gum	55 - 91*
Tara gum	30 - 95*
Guar gum	26 - 30*

*Range of observed reductions in the precipitation of β-1,2,3,4,6-pentagalloyl-D-glucose by gelatin, corresponding to different ratios of galactomannan to Xanthan. β-1,2,3,4,6-pentagalloyl-D-glucose - 10.0 μM; Gelatin 0.2 μM; **total** polysaccharide - 98.0 μg / ml.

The behaviour of the various polysaccharides as inhibitors of polyphenol - protein precipitation may have several possible explanations. Increase in medium viscosity is one, but other experimental observations appear to rule this out as a direct primary cause. Rather it is suggested that the principal reason may lie in the fact that individual polysaccharides, or in the case of the plant galactomannans, combinations of particular polysaccharides are able to generate loose three-dimensional stuctured networks in aqueous solution (51). These, it is believed, may be able to encapsulate, either wholly or in part, the polyphenolic substrate (*vide supra,* the cyclodextrins). In this way the polyphenol is either fully prevented from association with the protein or, and this seems more probable, that the partial sequestration of the polyphenol in the polysaccharide 'cages' leads to the formation of tertiary protein - polyphenol - polysaccharide complexes in which the polyphenol acts as a link between protein and polysaccharide. These complexes would be expected to be much more hydrophilic in character and therefore not so readily precipitated from solution.

Summary and Conclusions

Early work on the questions surrounding polyphenol complexation were almost exclusively concerned with attempts to elucidate the identity of the 'active sites' on protein molecules and the nature of the reaction between these sites and the polyphenolic substrate which led to strong association. These investigations, and particularly those of two of the leading workers Grassman (52) and Gustavson (53), strongly favoured the peptide bond of proteins as the principal 'active sites' and led to the view that complexation was mediated by the formation of intermolecular hydrogen bonds between the polyphenol, with its abundance of phenolic hydroxyl groups, and the peptide functionality.

More recent studies have highlighted the intrinsic complexity of the molecular recognition processes which occur between polyphenols and other molecules in aqueous media. Whilst not denying the possibility in certain circumstances of there being specific sites which may be preferred ones to which the polyphenol may be finally 'anchored' by hydrogen bonding, the emphasis is now more clearly upon the whole process of molecular recognition beginning with the initially solvated species. From this point a whole range of factors may then influence the recognition process and the strength of the intermolecular bonds which are formed. One might be excused if, from a cursory inspection of a molecular model of a typical polyphenol, one assumed that the phenolic hydroxyl groups, which dominate the appearance of the exterior surface of the polyphenol, also dominated its complexation behaviour. Increasingly however the association of polyphenols with other molecules - proteins, polysaccharides, caffeine, cyclodextrins, anthocyanins, anthracyclinones - whatever their size and complexity are seen to be driven by *'hydrophobic effects'*. Nevertheless the multiplicity of phenolic hydroxyl groups in the polyphenol cannot be ignored. Intuitively the presence in polyphenols of a plethora of phenolic hydroxyl groups points to some role for hydrogen bonding in the associative processes.

Water acts as both hydrogen bond donor and acceptor in the formation of hydrogen bonds. Therefore, in an aqueous medium, the development of hydrogen bonds as a means of molecular recognition between host and guest molecules also requires that as a precondition hydrogen bonds involving each of the substrates to water are themselves first broken. The energetics of interaction thus depend on the stability of the hydrogen bonds both *made* and *broken*. Interactions of this type may nevertheless be marginally energetically favourable processes since there may be a net gain in entropy due to the release of substrate bound water into the bulk medium. Although isolated monofunctional hydrogen bonds may be expected therefore to have a minimal influence on molecular recognition processes, *polyfunctional hydrogen bonding networks,* formed cooperatively and radiating from the polyphenol substrate, probably contribute significantly as part of a second phase in polyphenol complexation. This concept seems particularly apposite when seeking to rationalise the very strong binding of polyphenols to the loose, flexible structures of proline rich proteins , first noted by Hagerman and Butler (10). There is thus ample evidence, particularly in aprotic systems, that the carbonyl function in tertiary amides (e.g. prolyl peptides) is itself a very good hydrogen bond acceptor (54).

The effectiveness of plant polyphenols thus derives from the fact that they are polydentate ligands with a multiplicity of potential secondary binding sites (the phenolic nuclei) on the periphery of the molecule. Because of the propinquity of these groups they exert cooperativity in the secondary processes of association; because of the molecular size of polyphenols they are also able to function to cross-link separate polypeptide chains in the same or different protein molecules.

Molecular recognition of macromolecules, such as proteins, by polyphenols is best described by the "*hand in glove*" metaphor. In the initial stages the process is driven strongly by 'hydrophobic effects'. There is not a static matching of binding groups in the host and the guest, rather the association is in its second phase time dependent and dynamic (c.f., the importance of conformational flexibility in both polyphenol and protein noted earlier). The required segmental mobilities of host and guest to bring appropriate tertiary (prolyl) amide and phenolic hydroxyl groups in close juxtaposition to form hydrogen bonds are probably small, entropically favourable, and therefore not energetically prohibitive. The binding energy in this model thus derives from the summation of a relatively large number of contacts developed in a time dependent and dynamic manner.

Acknowledgements

The authors thank the Nestle Company Ltd.(Vevey, Switzerland), Mars-Four Square (G.B), Unilever plc, The Government of the Peoples Republic of China, and the Agricultural and Food Research Council (U.K.) for financial support to undertake this work; the contributions of Ms.Heidi Grimmer , Mr.Zhang Dunxin and Mr.Bi Shi to parts of this work are also recognised as are the intellectually stimulating and informative discussions with Mme.Daniele Magnolato (Nestec Co.Ltd.) and Mr.Michael Saltmarsh (Mars Four-Square G.B.) . Mr Michael Manterfield is warmly acknowledged for his patience, skill and expertise in setting up and printing this document.

Literature Cited

1. Bate-Smith, E.C. *J.Linnaen Soc.(Bot.)* **1962,** *58,* 95.
2. Martin, R.; Lilley, T.H.; Bailey, N.A.; Falshaw, C.P.; Haslam, E.; Magnolato, D.; Begley M.J. *Phytochemistry,* **1987,** *26* ,2733.
3. Davy, H. *Phil.Trans.*1803.
4. Mayer, A.; Harel, E. *Phytochemistry* **1979,** *18* ,193.
5. Cheynier, V.; Osse, C.; Rigaud, J. *J.Food Sci.* **1988,** *53,* 1729, 1760.
6. Brunet, P.J.C.; Coles, B.C. *Proc.Roy.Soc.* **1974,** *198B,* 133.
7. Sugumaran, M. *Bio-org.Chem.* **1987,** *15,* 194.
8. Rich, A.C.; Crick, F.H.C. *J.Mol.Biol.* **1961,** *3,* 483.
9. Bear, R.S. *Adv.Protein Chem.* **1952,** *7,* 69.
10. Hagerman, A.E.; Butler, L.G. *J.Biol.Chem.* **1981,** *256,* 4494.
11. Bate-Smith, E.C. *Food* **1954,** *23,* 124.
12. Bate-Smith, E.C. *Adv.Food Res.* **1954,** *5,* 262.
13. Goldstein, J.L.; Swain, T. *Phytochemistry* **1963,** *2,* 371.
14. Matsuo, T.; Ito, S. *Agric.Biol.Chem.* **1982,** *46,* 683.
15. Haslam, E.; Lilley, T.H. *C.R.C. Critical Rev.Food Sci.Nutr.* **1988,** *27,* 1.
16. Somers, T.C. *Phytochemistry* **1971,** *10,* 2175.
17. Somers, T.C.; Evans, M. *J.Sci.Food Agric.* **1979,** *30,* 623.
18. Haslam, E. *Phytochemistry* **1980,** *19,* 2577.
19. Haslam, E. *J.Chem.Soc.Chem.Communications* **1974,** 594.
20. Willstatter, R.; Zollinger, E.H. *Liebig's Annalen* **1916,** *412,* 195.
21. Robinson, G.M.; Robinson, R. *Biochem.J.* **1931,** *25,* 1687.
22. Robinson, G.M. *J.Amer.Chem.Soc.* **1939,** *61,* 1606.
23. Goto, T. *Progress Chem. Org.Nat.Prod.* **1987,** *52,* 113.
24. Brouillard, R.; Mazza ,G.; Saad, Z.; Albrecht-Gary, A.M.; Cheminat, A. *J.Amer.Chem.Soc.* **1989,** *111,* 2604.
25. Mistry, T.V., Cai, Y.; Lilley, T.H.; Haslam, E. *J.Chem.Soc. (Perkin Trans.2)* **1991,** in press.
26. Brouillard, R.; Mazza ,G. *Phytochemistry* **1990,** *29,* 1097.
27. Brouillard, R. In *The Flavonoids - Advances in Research;* Ed. Harborne, J.B.; Chapman and Hall : London, **1988,** 525.
28. Haslam, E., Gupta, R.K. *J.Chem.Soc.(Perkin Trans.1)* **1978,** 892.
29. Mehanso, H.; Butler, L.G.; Carlson, D. *Ann.Rev.Nutr.* **1987,** *7,* 423.
30. Millin, D.J.; Rustidge, D.W. *Process Biochem.* **1967,** *2,* 9.
31. Lunder, T. *Farmaceutisch.Tijds.voor Belgie* **1989,** *66,* 34.
32. Sanderson, G.W. *Recent Adv.Phytochemistry* **1972,** *5,* 247.
33. Nonaka, G.I.; Kawahara, O.; Nishioka, I. *Chem. Pharm.Bull.* **1983,** *31* , 3906.
34. Roberts, E.A.H. *J.Sci.Food Agric.* **1958,** *9,* 212.
35. Takino Y.; Imagawa, H.; Horikawa, H.; Tanaka, A. *Agric.Biol.Chem.* **1964,** *28,* 64.
36. Haslam, E. *Plant Polyphenols - Vegetable Tannins Revisited;* Cambridge University Press : Cambridge, **1989.**
37. Jencks, W.P. *Catalysis in Chemistry and Engineering;* McGraw-Hill : New York, **1969** ,393.
38. Fersht, A.R. *Trends Biochem.Sci.* **1984,** *9,* 145; **1987,** *12,* 301.

39. Okuda, T.; Yoshida, T.; Hatano, T. *J.Chem.Soc.(Perkin Trans.1)* **1982, 9.**
40. Haslam, E.; Haddock, E.A.; Gupta, R.K. *J.Chem.Soc.(Perkin Trans.1)* **1982,** 2535.
41. Okuda, T.; Mori, K.; Hatano ,T. *Chem.Pharm.Bull.* **1985,** *33*, 1424.
42. Haslam, E.; Lilley, T.H.; Warminski, E.; Hua, L.; Cai, Y. ;Martin, R.; Gaffney, S.H.; Goulding, P.N.; Luck, G. Unpublished work.
43. Cai Y.; Martin, R.; Lilley, T.H.; Magnolato, D.; Haslam, E. *Planta Med.* **1989,** *55*, 1.
44. Gaffney, S.H.; Martin, R.; Lilley, T.H.; Magnolato, D.; Haslam, E. *J.Chem.Soc.Chem.Communications* **1986,** 107.
45. Cai, Y.; Gaffney, S.H.; Martin, R.; Lilley, T.H.; Magnolato, D.; Spencer, C.M.; Haslam, E. *J.Chem.Soc.(Perkin Trans.2)* **1990,** 2197.
46. Martin, R.; Lilley, T.H.; Bailey, N.A.; Falshaw, C.P.; Magnolato, D.; Haslam, E.; Begley, M.J. *J.Chem.Soc.Chem.Communications* **1986,** 105.
47. Bender, M.; Komiyama, K. *Cyclodextrin Chemistry;* Springer Verlag :Basle, **1978.**
48. Butler, L.G. In *Toxicants of Plant Origin;* Ed. Cheeke, P.R.; C.R.C. Press Inc. : Boca Raton, Florida, **1989,** 95 - 121.
49. Butler, L.G.; Riedl, D.J.; Lebrk, D.G.; Blytt, H.J. *J.Amer.Oil Chemists' Soc.* **1984,** *61*, 916.
50. McManus, J.P.; Davis, K.G.; Beart, J.E.; Gaffney ,S.H.; Lilley, T.H.; Haslam, E. *J.Chem.Soc.(Perkin Trans.2)* **1985,** 1429.
51. Rees, D.A. *Pure Appl.Chem.* **1981,** *53*, 1.
52. Grassmann, W. *Collegium* **1937,** *809*, 530.
53. Gustavson, K.H. *J.Polymer Sci.* **1954,** *12*, 317.
54. Wolfenden, R. *Science* **1983,** *222*, 1087.

RECEIVED December 17, 1991

Appendix 1

Natural galloyl and hexahydroxydiphenoyl esters

β-1,2,3,4,6-pentagalloyl-D-glucose , $M_R = 940$

β-1,2,3,6-tetragalloyl-D-glucose , $M_R = 788$

β-1,3,6-trigalloyl-D-glucose , $M_R = 636$

Tellimagrandin 2 , M_R = 938

Rugosin D , M_R = 1874

Chapter 3

Production of Phenolic Compounds by Cultured Plant Cells

Chee-Kok Chin[1] and Henrik Pederson[2]

[1]Department of Horticulture, Rutgers, The State University of New Jersey,
New Brunswick, NJ 08903
[2]Department of Chemical and Biochemical Engineering, Rutgers, The State
University of New Jersey, Piscataway, NJ 08855

Plants produce a variety of important chemicals including
phenolic compounds. Plant cell cultures offer an
alternative to whole plants as a source of these
compounds. Its main advantage is that it allows for
controllable and reliable manufacture of the
phytochemicals in any location.

The main constraint of using plant cell cultures to
produce phytochemicals is low productivity. Fortunately,
plant cell culture is quite amenable to manipulation.
Several approaches including optimization of culture
media and culture conditions, elicitation,
biotransformation, two stage culture, two phase culture,
immobilization, genetic selection, and genetic
manipulation have been separately found to be able to
increase productivity. Strategies comprising of a
combination of these approaches together with optimal
fermentor design, and downstream processing should one
day permit phytochemicals to be produced from cultured
cells in industrial scale at reasonable costs.

Plants are a source of a variety of materials including food, fiber
and chemicals. Many phytochemicals are used as medicine,
pesticides, fungicides, pigments, fragrances and flavor compounds.
Some of these are phenolic compounds.

Phytochemicals are usually extracted from whole plants, usually
cultivated. Many of these plants have regional and climatological
requirements. Consequently, the supply of chemicals from these
plants is susceptible to geographical and climatological
restrictions, and political manipulations.

0097–6156/92/0506–0051$06.00/0
© 1992 American Chemical Society

While some phytochemicals can be synthesized, others are impossible to synthesize or can only be synthesized with difficulty and at high costs.

In recent years with the advancement of plant cell culture techniques it is possible to grow plant cells as a suspension in large quantity similar to growing of microorganisms. It is an appealing alternative to use plant cells cultured in fermentors or bioreactors to produce phytochemicals. However, often culture plant cells do not produce many chemicals found in whole plants or only produce them in low levels. Several factors may be responsible for the failure of the cultured cells to produce the chemicals. These are: (a) the enzymes for the biosynthetic pathways are not being produced; (b) the precursors for the pathways are not available; and (c) the biosynthetic enzymes and the precursors are located in different compartments. The problems found in using cultured cells to produce phenolic compounds are similar to that of using the cells to produce phytochemicals in general. In the following we will discuss several strategies to overcome these constraints for higher production. These strategies should apply to all secondary metabolites including phenolic compounds.

Plant Cell Culture

Plant cell culture could be initiated from cells and tissues of different organs such as leaves, stems, roots and flowers. Unlike animal cells which maintain their differentiated cell types, plant cells normally go through a dedifferentiation stage when they are cultured. For example, mesophyll cells from leaf tissues will convert to unorganized and dedifferentiated cells and lose their chloroplasts in culture. The dedifferentiation poses a problem for the production of chemicals since product formation is typically associated with particular cell types resulted from morphological differentiation. Fortunately, cultured plant cells are amenable to manipulation through which the production of chemicals can be induced or enhanced.

Culture Medium

The major factor affecting the growth and differentiation of plant cell culture is the culture medium which provides the nutritional need to the cells. Plant cell culture medium normally contains mineral salts, a carbon source, vitamins and plant growth substances (22). The cells can synthesize all other essential compounds from this rather simple list of constituents. Among the constituents, the growth substances auxin and cytokinin have the most dramatic effects on differentiation and production of phytochemicals (23,30). These growth substances may affect the production of chemicals through their effects on differentiation or may produce their effects independent of differentiation. Other constituents such as mineral salts and carbon source will also affect growth, differentiation, and production of chemicals. For example, production of berberine by cultured <u>Thalictrum</u> <u>rugosum</u> cells is inhibited by high phosphate but promoted by high sucrose concentrations (<u>7</u>). Another example to illustrate the importance of mineral salts is the production of shikonin by cultured <u>Lithospermum</u> <u>erythrorhizon</u> cells (<u>1</u>,<u>15</u>). Shikonin pigments are found only when the cells are grown on media resembling White's medium (<u>29</u>), whereas growth of the cells is favored on Linsmaier and Skoog's medium (<u>20</u>). One difference between the two media is that the latter contains ammonia whereas the former does not. The presence of ammonia is responsible for the inhibition of shikonin synthesis. Further improvements in productivity could be realized by adjustments in nitrate and copper ion concentrations. The final productivity of the system was improved 12-fold by 'tuning' the media and staging the system to separate growth and production phases.

Culture Conditions

Plant cell cultures normally consist of single cells and cell aggregates of various sizes. Constant agitation is required to ensure adequate aeration and to facilitate cell separation. Suspension cell cultures are commonly cultured in Erlenmyer flasks placed on gyrotory shakers set at 50 to 250 rpm. Other devices for agitation include magnetic stirrers, roller bottles and spinning cultures. Certain fermentors used for microorganisms can be modified for plant cell cultures by reducing shear stress (<u>19</u>). The agitation force plays a role in cell growth, aggregate size and chemical production (<u>19</u>).

Other culture conditions such as temperature, light intensity and light quality may also affect chemical production. Light is known to be required for chloroplast development and chloroplasts is the site of a number of enzymes involved in metabolism of phenolic compounds, e.g., p-coumarate hydroxylase and diphenol oxidase. Phenylalanine ammonia lyase, the enzyme which catalyses the conversion of phenylalanine to trans-cinnamic acid is activated by light. Most often light stimulates production of phytochemicals by cultured plant cells. For example, production of polyphenolic compounds by tea cell cultures (<u>14</u>), synthesis of polyphenolic compounds and anthocyanin by Paul's scarlet rose cell cultures (<u>9</u>), synthesis of catharanthine by <u>Catharanthus</u> <u>roseus</u> cell cultures (<u>11</u>), and synthesis of flavone glycoside by <u>Petroselinum</u> <u>hortense</u> cell cultures (<u>16</u>,<u>17</u>), are all stimulated by light. However, light stimulation of phytochemical production is not universal. In some cases light inhibits production of phytochemicals. For example, tannin synthesis in juniper cultures is reduced by light (<u>8</u>).

In addition to light intensity, photoperiod, i.e., the relative lengths of light and dark periods, also affects the synthesis of phytochemicals. Van Den Berg et al. reported that in suspension culture of <u>Rhomnus</u> <u>purshiana</u> the accumulation of anthracene was relatively high at a photoperiod of 12 hours (<u>28</u>). Increasing the light periods beyond 12 hours suppressed the production.

Elicitation

Production of certain phytochemicals can be increased with compounds called elicitors (2,10). Elicitors can be biotic or abiotic. Whole plants can respond to microbial challenge to produce pathogen related proteins which in turn lead to production of phytoalexins which are antimicrobial compounds (12). Elicitors can be prepared from cell walls of various phytopathogenic or even non-phytopathogenic fungi. These elicitors when applied to cell cultures can increase the production of certain phytochemicals. For example, accumulation of sanguinarine, chelerythrine, chelirubine and macarpine in suspension cultures of Eschscholtzia californica increases rapidly and dramatically when the cells were treated with cell wall extracts prepared from yeast, Collectotrichum lindemuthianum or Verticillium dahliae (2,6). Other elicitors of biological origin include ethylene (2,3), salicylic acid (24), and chitosan (4). Abiotic elicitors include certain heavy metals, surfactants, and metabolic inhibitors (10).

Biotransformation

If low productivity is due to lack of a precursor in the biosynthetic pathway, it may be possible to add the precursor to the culture medium and let the cells to convert it to the product, a process called biotransformation. For example, tryptamine or tryptophan when added to Catharanthus roseus cultures increased the indole alkaloid biosynthesis (13,18). Yeoman et al. reported that supplying immediate precursors of capsaicin, the pungent and hot flavor compound of hot pepper to Capsicum frutescens cultures increased the production of capsaicin (32). Earlier and more general precursors such as amino acids also increased the production capsaicin, but to a lesser extent. Other substances that have been successfully obtained through biotransformation include cardenolides, steroids, terpenoids, alkaloids and glycosides (1).

Not all precursors added to the medium could be biotransformed into the products of interest. A number of factors, such as penetraton, availability of the enzymes in the biosynthetic pathway, and feedback or toxic effects of product accumulation could restrain the production.

Two Phase Culture

Accumulation of chemicals in plant cells is physiologically regulated. Typically, the synthesis rate declines as the intracellular concentration of the product increases. When a certain concentration is reached the synthesis ceases. The plant secondary metabolites are usually stored within the vacuolar compartments of the cell. In cell culture these metabolites may be released into the medium. Such release will reduce the intracellular concentration and, therefore, facilitate further synthesis. The release of products can be increased by permeabilization of the membranes by treatments with compounds such as dimethyl sulfoxide, ethyl acetate and isopropanol (4,5). Another way to reduce intracellular concentration is to facilitate the accumulation of products extracellularly. Maisch et al. found that the addition of XAD-4 resin to Nicotiana tabacum cultures enhanced the production of phenolic compounds several times compared to adsorbent-free control (21).   Payne et al. reported that when XAD-7 was added to the medium, the accumulation of total indole alkaloids increased, and the production of the specific alkaloid aymalicine and serpentine was stimulated (25). Strategically, it is desirable that a second phase be added to the medium to specifically absorb the product of interest.   Indeed, Byun et al. reported that production of benzophenanthridine alkaloids sanguinarine, chelerythrine, chelirubine and macarpine in suspension cultures of Eschscholtzia californica cell cultures was significantly enhanced in a two phase culture system with a dimethyl siloxane polymer as the second phase (6). A good second phase material should facilitate accumulation of the product and should not be toxic to the cells. The use of an effective second phase not only enhances yield but could also simplify the downstream processing and purification.

Genetic Selection and Manipulation

Ability to produce a particular chemical by plant varies with species and even within individuals of a species. Cell cultures derived from individual plant which has a high accumulation of secondary products tends to contain high amounts of the same product (27). The comparisons have been shown in harmane alkaloid and serotonin production by Peganum hormala (26).   Thus, it is important to initiate cell culture from a high producing individual. Sometimes cells in cell culture undergo changes to produce a mixture of cells called variants with different capacities to produce chemicals. This could be due to genetic or epigenetic changes.   Although variants are produced at low frequency it is possible to select from a population of cells high producing variants for chemical production. This is especially true if the products have recognizable characteristics such as fluorescence, UV absorbance or color so that simple visual selection can be made. An example of this is the selection of Euphorbia milli cell lines via consecutive collecting of high producers based on color (31).

Synthesis of chemicals by plants depends on the enzymes in the biosynthetic pathway. The production of these enzymes is often associated with particular cell types and a particular development stage. For example, many flavanoids are only produced in petal cells during flowering stage. In other organs and cell types the enzymes in the pathway are not produced because the genes for the enzymes are not expressed. Gene expression is dependent on the regulatory sequences of the gene consisting of promotor and terminating sequence. Some promotors are active constitutively but others are active only in certain cell types and developmental stages.   Although expression of some nonconstitutive genes can be brought about by manipulation of the media and culture conditions, these approaches are not always effective.   An alternative is to identify the limiting step and the enzyme involved in the pathway. Using molecular biology techniques the gene for the enzyme can be isolated.   The promotor of the gene can be substituted with a constitutive or an inducible promotor.   When the engineered gene is introduced back to the cell, this would allow the gene to function constitutively or to be induced at an appropriate time.   With such genetically engineered cells the constraints of production would be removed and these cells could be used in large scale production of the chemicals of interest in a fermentor or bioreactor.

Conclusion

Steady advances in plant cell culture technology in recent years are giving confidence to researchers in this field that the goal of using cultured plant cells to produce phytochemicals can be realized in the near future.   The advances include: (a) now it is possible to grow cells of most plants in suspension culture;   (b) it is possible to grow plant cells in large quantities in fermentors and bioreactors.   Nevertheless, for most plant cells a bottleneck for production, via., low productivity, remains.   The approaches examined in this report should be useful in easing this bottleneck.   With the rapid advances in molecular biology, the approach of using genetic manipulation to produce high yielding cell lines is particularly attractive.   Generally excessive accumulation of a product is harmful to cells and, therefore, would reduce growth.   This problem may be solved by fusing the gene involved in product formation to an inducible promotor.   This not only would allow the production levels to increase but also would allow easy separation of growth and production stages.   Clearly for the goal of commercial production of phytochemicals by cultured plant cells to be realized it is very desirable to have the cooperation of tissue and cell culturists, chemists, molecular biologists and chemical engineers.

Literature Cited

1.   Alfermann, A.W.; Reinhard, E. Bull. Soc. Chim. Fr. 1980, 1-2, 35-45.
2.   Bollert, T.; In Primary Signals and Secondary Messengers in the Reaction of Plants to Pathogens; Boss, W.F.; Morre, D.J.; Ed. Alan R. Liss, New York; 1989, 227-255.
3.   Boller, T.; Vogli, V. Plant Physiol. 1984, 74, 442-444.
4.   Brodelius, P.; Funk, C.; Haner, A.; Villegas, M. Phytochem. 1989, 28, 2651-2654.
5.   Brodelius, P.; Nilsson, K. Eur. J. Appl. Microbiol. Biotechnol. 1983, 17, 275-280.
6.   Byun, S.Y.; Pedersen, H.; Chin, C.K. Phytochemistry, 1990, 29, 3135-3139.
7.   Choi, J.W. Ph.D. Thesis, Rutgers University, New Brunswick, New Jersey. 1987.

8. Constabel, F.; <u>Proc. Int. Conf. Plant Tissue Culture</u>. Penn State University. 1963, 183.
9. Davies, M.S.; <u>Planta</u>. 1972, <u>104</u>, 50-65.
10. Dicosmo, F.; Misawa, M. <u>Trends Biotechnol.</u> 1985. <u>3</u>, 318-322.
11. Drapeau, D.; Blanch, H.W.; Wilke, C.R. <u>Planta Medica</u>, 1987, 373-376.
12. Ebel, J. <u>Ann. Rev. Phytopath.</u> 1986, <u>24</u>, 235-264.
13. Fontanol, A.; Tabata, M. <u>Nestle Research News</u>, 1987, 93-103.
14. Forrest, G.I.; <u>Biochem. J.</u> 1969, <u>113</u>, 765-772.
15. Fujita, Y.; Hara, Y.; Suga, C.; Morimoto, T. <u>Plant Cell Report</u>. 1981,<u>1</u>, 61-63.
16. Hahlbrock, K.; Wellman, E. <u>Planta</u>. 1970, <u>94</u>, 236-239.
17. Kreuzaler, F.; Hahlbrock, K. <u>Phytochem.</u> <u>12</u>, 1149-1152.
18. Kutnay, J.P.; Aweryn, B.; Choi, L.N.G.; Kolodziejezyk, P.; Kurz, W.G.W.; Chatson, D.B.; Constabel, F. <u>Heterocycles</u>. 1981, 1169.
19. Leckie, F.; Scragg, A.H.; Cliffe, K.C. <u>Enzyme Microb. Technol.</u> 1991, <u>13</u>,296-305.
20. Linsmaier, E.M.; Skoog, F. <u>Physiol. Plant</u>. 1965, <u>18</u>, 100-127.
21. Maisch, R.; Knoopo, B.; Beiderbeck, R. <u>Z. Naturforsch, C. Biosci</u>. 1986, <u>41C</u>. 1040-1044.
22. Murashige, T. <u>Tissue Culture Methods and Applications</u>; Kruss, P.F., Jr.; Patterson, M.D. ed.; Academic Press, New York 1973, 698-703.
23. Nakagawa, K.; Fukui, H.; Tabata, M. <u>Plant Cell Rep.</u>, 1986, <u>5</u>, 69-71.
24. Ohshima, M.; Itoh, H.; Matsuska, M.; Murakami, T.; Ohashi, T. <u>Plant Cel.</u> 1990, <u>2</u>, 95-106.
25. Payne, G.F.; Payne, N.N.; Shuler, M.L.; Asada, M. <u>Biotech. Letters</u>, 1988, <u>88</u>, 187-192.
26. Sasse, F.; Heckengerg, U.; Berlin, J. <u>Plant Physiol.</u> 1982. <u>62</u>, 400.
27. Shuler, M.L. <u>Annals</u>, New York <u>Acad. Sci.</u> 1981, <u>369</u>, 65.
28. Van Den Berg, A.J.J.; Radema, M.H.; Labadie, R.P. <u>Phytochem.</u> <u>27</u>, 415-417.
29. White, P.R. A Handbook of Plant Tissue Culture. Jaques Cattell Press, Lancaster, PA. 1943.
30. Yamada, Y.; Fujita, Y. Handbook of Plant Cell Culture. Macmillan Co., New York, 1983.
31. Yamamoto, Y.; Mizuguchi, R. <u>Theor. Appl. Genet.</u> 1982. <u>61</u>, 113-116.
32. Yeoman, M.M.; Miedzybrodzka, M.K.; Linsey, K.; McLauchlan, W.R.; In <u>Plant Cell Cultures: Results and Perspectives</u>, Saha, E.; Parisi, B.; Cella, R.; Ciferio, O.; Eds., Elsevier, Amsterdam, 1980, 327-343.

RECEIVED December 2, 1991

ANALYTICAL METHODOLOGY

Chapter 4

Determination of Vanillin, Other Phenolic Compounds, and Flavors in Vanilla Beans

Direct Thermal Desorption—Gas Chromatography and —Gas Chromatography—Mass Spectrometric Analysis

Thomas G. Hartman[1], Karl Karmas[1], Judy Chen[2], Aparna Shevade[2], Maria Deagro[2], and Hui-Ing Hwang[2]

[1]Center for Advanced Food Technology and [2]Department of Food Science, Cook College, Rutgers, The State University of New Jersey, New Brunswick, NJ 08903

Dried and cured vanilla (*Vanilla planifolia*, Madagascar Bourbon variety) beans were analyzed using direct thermal desorption - gas chromatography (TD-GC) and gas chromatography-mass spectrometry (TD-GC-MS) methodologies. Samples of vanilla bean cross sections, seeds and tissue homogenates were subjected to thermal desorption using a Short Path thermal desorber accessory directly interfaced to the injection port of a GC and GC-MS system. The technique provides for outgassing of the volatile components of the vanilla beans via ballistic heating and delivery of the desorbed volatiles into the GC or GC-MS for subsequent analysis. Using this methodology over 60 flavor compounds were identified, 18 of which were phenolic species. Several of these compounds have not been previously reported as occurring in vanilla. The technique was demonstrated to be quantitative and reproducible. Quantification of vanillin and other compounds were achieved using 2,6-di-methoxyphenol as a spiked surrogate internal standard. An observed vanillin content of 1.94 % was found to be consistent with the value obtained using conventional analysis. Furthermore, other phenolic compounds characteristic of Bourbon vanilla were also detected.

Vanilla is the world's most popular flavor. The production of natural vanilla and its synthetic counterparts greatly exceed that of any other flavoring compound. Estimates of worldwide production of vanilla beans are in the range of 900 to 1500 tons annually (1) with Bourbon vanilla constituting approximately 75 %. Other production regions include Mexico, Tahiti and Indonesia.

Analytical investigations are routinely carried out on vanilla beans to determine the vanillin content and for several other important reasons. Quantitatively vanillin is the major flavor compound present in the beans. It is found along with many other compounds which contribute to overall flavor character. Pioneering investigations into the nature of volatiles identified over

0097–6156/92/0506–0060$06.00/0
© 1992 American Chemical Society

170 flavor compounds (2,3). However, these studies involved exhaustive and time consuming extraction and isolation studies. According to a recent review, over 250 individual flavor compounds have been reported to be present in vanilla (4). The identification of these additional compounds is therefore of importance to flavor chemists attempting to understand or duplicate its delicate aroma bouquet. A second analytical goal is to determine the chemical composition of vanilla beans and use the data to pinpoint regions of origin. For instance, using this approach it may be possible to differentiate Bourbon vanilla produced in Madagascar, Reunion and Comores which is considered the highest quality from lesser quality beans produced in other regions such as Mexico and Indonesia. These determinations are made possible by the fact that regional and agricultural variations effect the resultant chemistry of the vanilla beans produced. For instance, Bourbon vanilla's are found to contain the highest vanillin concentrations and Tahitian vanilla's which consitutes another botanical variety are unique in that they contain relatively high levels of anisyl alcohol and related compounds which are less predominant in other varieties (4). The most difficult analytical goal is the detection of synthetic vanillin adulteration in natural vanilla. Various approaches to this problem have been proposed. Several involve detailed analysis and accurate determination of the relative ratios of vanillin and additional phenolic species such as vanillic acid, p-hydroxybenzaldehyde and other characteristic compounds (5). This approach necessitates the analysis of a large number of samples to define a data base which is still subject to geographical, seasonal, species and agricultural variances and is therefore limited. The most definitive evidence for adulteration is obtainable by ^{13}C isotope ratio mass spectrometry (6) or deuterium NMR analysis of the vanillin component (7). However, the problem of adulteration is mainly confined to the analysis of vanilla extracts and not the vanilla beans themselves.

Various methodologies have been described for the analysis of vanilla beans. Most methods involve an extraction step combined with some type of separation and determination. Some analytical strategies involve separation using high performance liquid chromatography (HPLC) with ultra-violet detection (5). Others utilize gas chromatographic separations with flame ionization or mass spectrometric detection (8) and some involve derivatization followed by spectrometric determinations (9).

In this manuscript we describe a novel analysis protocol for vanilla beans involving direct thermal desorption - gas chromatography using flame ionization and or mass spectrometric detection (TD-GC-FID, TD-GC-MS). The analysis is rapid, quantitative, and reproducible. A typical analysis is completed in 35-45 minutes and provides both qualitative and quantitative data on a plethora of volatile and semivolatile flavor compounds. This methodology contrasts with the extremely time consuming and laborious extraction and isolation protocols used in previous flavor investigations. Since no organic solvent extraction step is required it reduces worker exposure to potentially hazardous vapors and eliminates disposal costs associated with these solvents. Considerable time savings are achieved and artifact production associated with complex extraction and isolation protocols are minimized. The methodology described may be useful for flavor studies, general QA/QC testing or for screening batches of vanilla beans from different regions and vendors.

Materials

Samples of dried and cured Bourbon vanilla beans (*Vanilla planifolia*) from the island of Madagascar were generously donated by Dr. Daphna Havkin-Frenkel of David Michael & Co., Inc.. The beans supplied were described as "first" quality with a vanillin content of 2.0 %. The compound used as internal standard, 2,6-dimethoxyphenol and vanillin analytical standard were obtained from Aldrich Chemical Co., Milwaukee WI. Both compounds had a stated purity of 99 %. Tenax TA adsorbent, 60-80 mesh was obtained from Alltech Associates, Deerfield, IL. Silanized glass wool was from Supelco Inc., Belefonte, PA. The model TD-1 Short Path Thermal Desorber and all desorption accessories were obtained from Scientific Instrument Services Inc., Ringoes, NJ.

Experimental

Silanized glass lined stainless steel desorption tubes (3.0 mm i.d. X 10 cm length) from Scientific Instrument Services Inc. (SIS), Ringoes NJ, were packed with a 2 cm bed volume of Tenax-TA adsorbent between plugs of silanized glass wool. The tubes were conditioned by passing helium through them at a rate of 40 ml per minute while heating in a SIS desorption tube conditioning oven. The tubes were temperature programmed from 50°C to 320°C at rate of 10°C per minute with a 1 hour minimum hold at the upper limit. The tubes were then cooled to room temperature and capped prior to addition of sample. This conditioning step is performed to insure that no volatile compounds are present in the adsorbent resin or desorption tube prior to sample loading. Cross sections or homogenates of the vanilla bean samples (10 mg) were then weighed into the desorption tubes, being placed above the Tenax adsorbent bed. Vanilla bean homogenates were prepared by grinding in a small mortar and pestle. Cross sections of the beans were prepared using a clean razor blade. Figure # 1 shows the desorption tube containing the Tenax bed and vanilla sample. The loaded desorption tube was then spiked with 10 µg of 2,6-dimethoxyphenol internal standard by injecting 1.0 µl of a methanol stock solution (10 mg/ml) using a solvent flush technique to insure quantitative delivery. This represents a spike level of 1000 parts per million (ppm) w/w. In several of the GC-MS experiments the internal standard was spiked at a reduced level of 100 ppm in an attempt to better estimate the concentrations of flavor volatiles present at lower levels. The spiked desorption tube was then promptly connected to an SIS model TD-1 Short Path Thermal Desorber module interfaced directly to the injection port of the GC or GC-MS. A Helium flow (20 ml/min.) was then established through the desorber for 1-5 minutes to purge air and methanol from the desorption tube. The Tenax adsorbent bed is incorporated in the desorption tube so that no loss of sample or internal standard occurs during this initial purge step conducted at ambient temperature. The methanol solvent used to deliver the internal standard passes directly through the Tenax as this adsorbent does not retain very low molecular weight polar species. Experimental comparisons of desorption experiments conducted with and without the Tenax bed indicated slightly higher (and more reproducible) recoveries of internal standard were obtained when adsorbent was incorporated. The desorption tube was then injected into the GC and thermally desorbed for 5 minutes. Figures 2-5 illustrate the desorption apparatus, injection and analysis procedure. Three different desorption temperatures were investigated. Experiments were conducted using desorption temperatures of 150, 220 and 250°C. The gas chromatograph was a Varian 3400 equipped with a 30 meter X 0.32 mm i.d. DB-1 capillary column (J&W Scientific Co.) containing a 0.25 µm film thickness. The injector

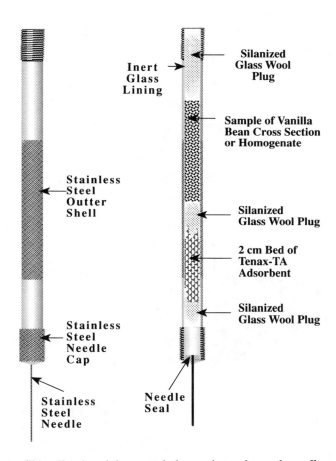

Figure 1. Glass lined stainless steel desorption tube and needle assembly used for direct thermal desorption - gas chromatography or gas chromatography - mass spectrometric (DTD-GC, DTD-GC-MS) analysis of vanilla bean samples.

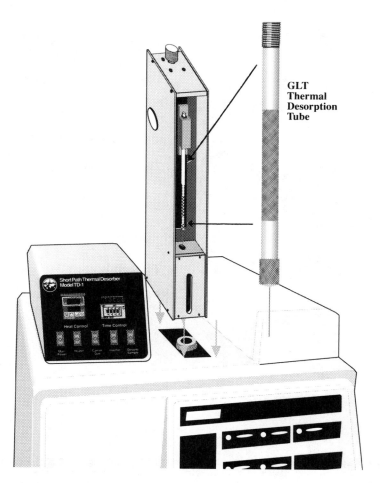

GLT
Thermal
Desorption
Tube

Figure 2. Connection of a loaded desorption tube to the Short Path Thermal
Desorber apparatus which is interfaced directly to the GC injection port.

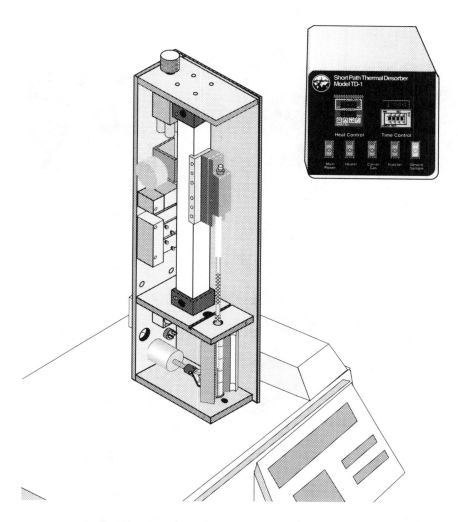

Figure 3. Short Path Thermal Desorber module shown injecting the desorption tube into the GC injector and positioning the tube in front of a pair of bilateral actuated heating blocks (open position).

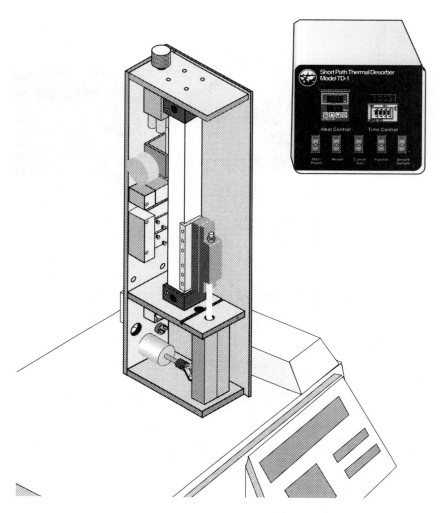

Figure 4. Short Path Thermal Desorber shown in heat and desorb mode of operation. Heating blocks are engaged around the desorption tube and carrier gas is flowing through the system thus purging the desorbed volatile and semi-volatile sample components directly into the GC injection port.

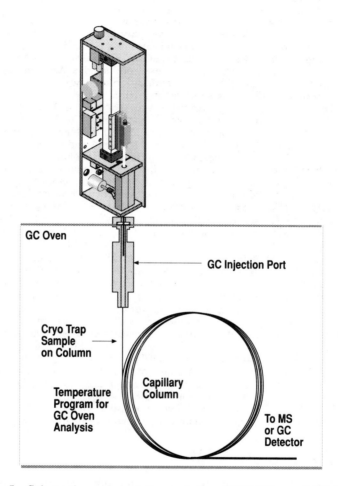

Figure 5. Schematic representation of the DTD-GC or DTD-GC-MS analysis technique. Volatile and semi-volatile compounds thermally desorbed from the sample are purged with carrier gas directly into the GC injector. The desorbed components are cryofocused at subambient temperatures to preserve chromatographic resolution. Following the thermal desorption interval the GC column is temperature programmed and the separated components are delivered to the GC or MS detector.

temperature was 250°C and a 100:1 split ratio was employed. The column was temperature programmed from -20°C (hold 5 minutes during thermal desorption interval to achieve cryofocusing) to 40°C at a rate of 10°C per minute then to 280°C at a rate of 4°C per minute with a 30 minute hold at the upper limit. Helium was used as carrier gas with a flow rate of 1.0 ml per minute (20 cm per second linear carrier velocity). The GC experiments utilized flame ionization detection (FID) and chromatograms were recorded and processed using a Varian 4290 integrator and a VG Multichrom chromatography data system. GC-MS experiments were conducted exactly as described above except that the end of the GC capillary column was inserted directly into the ion source of the mass spectrometer via a heated transfer line maintained at 280°C. The mass spectrometer was a Finnigan MAT 8230 high resolution double focusing magnetic sector instrument. The MS was operated in electron ionization mode scanning masses 35-350 once each second with a 0.8 second interscan time. The MS data was acquired and processed using a Finnigan MAT SS300 data system. All mass spectra obtained were background subtracted and library searched against the National Institute of Standards and Technology (NIST) mass spectral reference collection (formerly the National Bureau of Standards, NBS). Further searches were conducted using the Wiley, Eight Peak Index and our own custom library of mass spectra. For quantification purposes, a five point vanillin calibration curve was established by spiking desorption tubes with a constant (10 μg) amount of internal standard and a varying amount of vanillin (1.0, 5.0, 10.0, 50.0, 100.0 and 500.0 μg). The tubes were then analyzed exactly as described for the actual samples using GC-FID. To check analytical precision, five replicate determinations were made using the dilution from the midpoint of the calibration curve (10 μg vanillin & 10 μg internal standard). The dynamic range of the calibration curve was sufficient to encompass the vanillin determinations in all samples analyzed. Control experiments included periodic analysis of conditioned desorption tubes containing Tenax adsorbent and internal standard only. Numerous analyses were conducted on the vanilla beans. Some experiments included dissecting the beans and analyzing the seeds and pod materials in separate experiments. Other analyses compared vanilla bean cross sections to homogenates. In general TD-GC-MS was used primarily to identify the components present and for quantification of the low level flavor components. The TD-GC-FID experiments were used to quantify vanillin and other high level compounds since greater dynamic range was attainable using GC-FID integration.

Results and Discussion

In addition to vanillin, over 60 volatile compounds were identified in the vanilla beans, 18 of which were phenolic species. Other compounds detected included acids, alcohols, esters, aldehydes, ketones, heterocylclic compounds and a series of aliphatic hydrocarbons. Several of the components detected remain unidentified. The data is summarized in tables 1 and 2. An ion current chromatogram of a typical TD-GC-MS analysis of vanilla bean is shown in figure 6. An expanded view of this chromatogram highlighting the region in which the phenolic compounds elute is shown in figure 7. The peak assignments for the phenolic components in the chromatogram are listed in table 1. Please note that the concentrations listed in tables 1 and 2 for all compounds

Table   1 Phenolic Compounds Identified in Vanilla Beans

[1]Peak #	Assignment	[2]Estimated Conc. PPM
1	phenol	28
2	2-methoxyphenol (guaiacol)	19
3	1,2-benzenediol (catechol)	29
4	* 2-hydroxy-5-methylacetophenone	24
5	p-hydroxybenzylmethyl ether	17
6	4-ethyl-2-methoxyphenol (p-ethylguaiacol)	33
7	2,6-dimethoxyphenol (internal standard)	100
8	p-hydroxybenzyl alcohol	29
9	p-hydroxybenzaldehyde	1040
10	4-hydroxy-3-methoxybenzaldehyde (vanillin)	19400
11	p-hydroxymethylbenzoate	36
12	* 4-hydroxy-3-methoxyacetophenone (acetovanillone)	37
13	p-hydroxybenzoic acid	28
14	* 1-(4-hydroxy-3-methoxyphenyl)-2-propanone	31
15	* 4-(4-hydroxyphenyl)-2-butanone	31
16	4-hydroxy-3-methoxybenzoic acid (vanillic acid)	440
17	* 4-(4-hydroxy-3-methoxyphenyl)-2-butanone (zingerone)	46
18	4-hydroxy-3,5-dimethoxybenzaldehyde (syringaldehyde)	10
19	vanillin-2,3-butyleneglycol acetal	10

[1] peak assignments for chromatogram shown in figure 7
[2] Estimated concentrations for all compounds other than vanillin were made by peak area comparisons to that of 2,6-dimethoxyphenol (internal standard) with no correction for individual detector response factors. Vanillin content was accurately determined from internal standard response curve data.
* Compounds not previously reported as occurring in vanilla

Table 2 Other Volatile Compounds Detected in Vanilla Beans

[1]R.T.	Assignment (by class)	[2]Estimated Conc. PPM
	Acids	
8:54	formic	62
9:43	acetic	151
18:34	benzoic	38
19:47	nonanoic	31
22:17	p-methoxybenzoic (anisic)	49
27:22	tetradecanoic (myristic)	12
27:40	pentadecanoic	24
28:48	hexadecanoic (palmitic)	258
29:47	heptadecanoic	30
30:22	octadecenoic (oleic)	13
30:32	octadecadienoic (linoleic)	135
30:47	octadecanoic (stearic)	39
	Esters	
11:15	methyl acetate	84
28:28	methyl palmitate	13
30:09	methyl linoleate	21
	Ketones	
14:21	* 3-methylcyclopentanone	25
17:20	* 2,4,6-heptanetrione (diacetyl acetone)	57
27:09	* 2,4,6-trimethylacetophenone	69
27:40	6,10,14-trimethyl-2-pentadecanone	23
	Aldehydes	
7:59	* 3-methylbutanal	27
17:30	* nonanal	53
20:08	* 2,4-decadienal (2 isomers)	25
23:04	* 4-acetyloxy-3-methoxybenzaldehyde (vanillylacetate)	27
	Alcohols	
12:00	2-pentanol	95
12:44	1,3-butanediol (1,3-butyleneglycol)	47
19:39	2-octen-4-ol	25
20:12	p-methoxybenzyl (anisyl alcohol)	9
27:59	* 1-tetradecanol	13
28:09	* 1-hexadecanol	38

Table 2 Other Volatile Compounds Detected in Vanilla Beans (Cont'd)

Heterocyclics

12:12	2-furancarboxaldehyde (furfural)	54
13:06	2-furanmethanol (furfuryl alcohol)	15
13:50	* 5-methy-tetrahydro-2-furanmethanol (tetrahydromethylfurfuryl alcohol)	17
14:27	5-methyl-2-furancarboxaldehyde (methyl furfural)	24
16:05	* 4,4-dimethyl-3-hydroxy-dihydro-2-(3H)-furanone (pantolactone)	51
16:57	* 2-furancarboxylic acid methyl ester	14
18:15	* 6-methyl-3,5-dihydroxy-2,3-dihydro-4(H)--pyran-4-one (hydroxydihydromaltol)	335
18:47	* 2,3-dihydrobenzofuran	114
18:52	* 2-methyl-3,5-dihyroxy-4H-pyran-4-one (hydroxymaltol)	12
19:32	5-(hydroxymethyl)-2-furancarboxaldehyde (hydroxymethylfurfural, HMF)	460

Hydrocarbons

29:00	eicosane	39
30:30	10-methyleicosane	25
31:00	heneicosane	23
31:20	docosane	22
32:16	tricosane	29
33:06	tricosene	38
33:12	tetracosane	13
34:15	pentacosene	39
34:21	hexacosane	24

[1] GC retention times correspond to chromatogram shown in figure 6.
[2] Estimated concentrations for all compounds other than vanillin were made by peak area comparisons to that of 2,6-dimethoxyphenol (internal standard) with no correction for individual detector response factors.
* Compounds not previously reported as occurring in vanilla

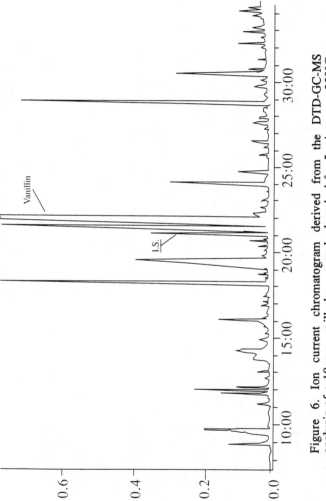

Figure 6. Ion current chromatogram derived from the DTD-GC-MS analysis of a 10 mg vanilla bean sample desorbed for 5 minutes at 250°C. Peak assignments are given in Table 2.

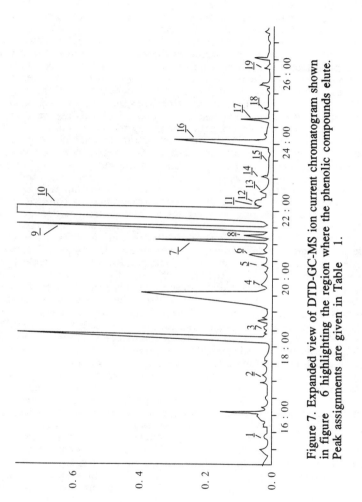

Figure 7. Expanded view of DTD-GC-MS ion current chromatogram shown in figure 6 highlighting the region where the phenolic compounds elute. Peak assignments are given in Table 1.

other than vanillin represent semi-quantitative estimates based on peak area integration comparisons to that of the internal standard. No corrections were made to account for differences in individual detector response factors toward the internal standard aside from vanillin.

Most of the volatile flavor compounds observed in this study have been previously identified (2,3). However, several newly identified species were found in our investigation. The heterocyclic compounds 5-methyldihydrofurfuryl alcohol, pantolactone, 2-furancarboxylic acid methyl ester, hydroxymaltol, hydroxydihydromaltol and 2,3-dihydrobenzofuran to the best of our knowledge are all new identifications. The furanoid and pyranone species most certainly arise from dehydration or thermal degradation of 6-carbon sugars. The pantolactone is a known thermal decomposition product of the B-complex vitamin, pantothenic acid which occurs in all living tissues but is especially abundant in flower blossoms (10). These observations suggest the possibility that some of these compounds may be thermal degradation artifacts of compounds originally present in the vanilla beans as sugar or vitamin precursors. Another noteworthy group of newly identified compounds include several phenolic species with alkyl ketone moieties substituted primarily in the para position. These include 2-hydroxy-5-methylacetophenone, 4-hydroxy-3-methoxyacetophenone (acetovanillone), 1-(4-hydroxy-3-methoxyphenyl)-2-propanone, 4-(4-hydroxyphenyl)-2-butanone and 4-(4-hydroxy-3-methoxyphenyl)-2-butanone (zingerone). The compound zingerone is a known constituent of ginger (11). It is also known to form as a thermal degradation product of gingerols via retro aldol decomposition (11). Newly identified ketone species include 3-methycyclopentanone, diacetylacetone and 2,4,6-trimethylacetophenone. Several aldehydes and alcohols newly identified include 2,4-decadienal (2-isomers), 3-methylbutanal, nonanal, 1-tetradecanol and 1-hexadecanol. A series of long chain paraffinic and olefinic hydrocarbons in the range of C-20 (eicosane) through C-26 (hexacosane) not previously observed were found to be present in this study especially when the desorption temperature was raised to 250°C. The flavor attributes of such compounds are probably very low. However, the waxes, resins and other high boiling components of the vanilla bean matrix are known to serve as fixatives which time control the release of volatile flavor components. Control experiments conducted by analyzing blank desorption tubes containing only adsorbent and internal standard were free of artifacts and interferences. On occasion several trace level peaks were observed in the control blanks. These peaks were identified by mass spectrometry to be dimethylpolysiloxane oligomers which result from GC septum bleed and were ignored in the data treatment. No sample to sample cross contamination was observed and no deterioration of capillary column performance occured despite several weeks of constant usage. This is due to the novel design features of the Short Path thermal desorption system utilized. A completely independent pneumatic circuit is established for each sample analyzed thus preventing cross contamination of the instrument. Furthermore, the desorption tube serves as a sacrificial injection port delivering outgassed volatiles into the GC while preventing the non-volatile residues from entering and contaminating the injection port.

The experiments conducted using various desorption temperatures (150, 220 and 250°C) showed that slightly higher recoveries of higher boiling phenolics such as vanillic acid were achieved at the two higher temperatures. However, the chromatograms obtained using 250°C desorption were found to contain much higher levels of long chain aliphatic hydrocarbons, waxes and other resinous substances with little or no gain in recovery of vanillin or related phenolics. Therefore, all subsequent experiments were conducted using a maximum desorption temperature of 220°C for five minutes. The data obtained indicates that our analysis protocol causes volatility discrimination with regard to some of the higher boiling phenolic compounds. For instance vanillin and p-hydroxybenzaldehyde are detected at concentrations and in relative ratios which are consistent with published values for Bourbon vanilla. However, the recovery of vanillic acid appears slightly lower than literature values and syringic acid which is reportedly present in Bourbon vanilla at levels as high as 0.3 % remained undetected in our analyses. This discrepancy is surely related to the inherent low volatility of these higher boiling phenolics. Other volatile flavor investigations have also failed to detect some of these species (2,3). Analyses involving HPLC determinations are better suited for such applications since they do not discriminate on the basis of volatility.

A vanillin calibration curve was generated by plotting the ratio of vanillin/internal standard peak area's versus vanillin concentration using the experimentaly derived data. The curve was linear throughout the range of concentrations tested and had a correlation coefficient of 0.999. All quantitative vanillin measurements made on the vanilla beans were within the dynamic range of the calibration curve. The five replicate analyses of vanillin and internal standard from the midpoint of the calibration curve (10 ?g each compound) yielded a mean ratio of 0.81 with a standard deviation of 0.02 and a 2.47 coefficient of variation. This experiment establishes that the analytical precision of the methodology employed is excellent. Indeed, this level of precision is equal to that obtained by conventional direct injection into the GC using a syringe. A vanillin concentration determination made on a vanilla bean homogenate yielded a value of 1.94 % which correlates well with the 2.0 % value stated by the provider of the beans. The latter concentration value was reportedly obtained by conventional extraction and HPLC analysis. Analysis of seeds removed from vanilla beans showed them to be generally lower in vanillin and other flavor compounds as compared to the homogenates, pod material and bean cross sections. The mean vanillin concentration for five replicate determinations of seeds was found to be only 0.82 %. Five analyses of vanilla bean cross sections gave vanillin concentrations of 1.36, 2.90, 2.22, 1.64 and 1.57 %. The mean of these five determinations is 1.94 %, the same value obtained by analyzing a homogenized composite sample. This data most certainly is a reflection of the heterogeneous nature of the beans with respect to vanillin content. This data is not at all surprising considering the small (10 mg) sample size analyzed. For this reason it is imperative to composite samples by homogenization prior to analysis in order to obtain representative vanillin concentrations in the beans. The ability to analyze larger samples would also be desirable in this regard since it is cumbersome and difficult to accurately weigh such small samples. However, the sample load permitted by our present methodology is limited to a maximum of 10 mg. The reason for this is that despite the fact that a 100:1 split ratio is employed we are still limited to the loading capacity of the capillary column utilized. Theoretically, we should be able to increase the sample size to 100 mg by using a 0.53 or 0.75 mm i.d. wide bore capillary column with a thicker phase coating. However, the move to a larger diameter column may decrease resolution. Perhaps if combined with a slower temperature program an ideal medium can be achieved. Experiments are currently under way to investigate these possibilities.

Acknowledgments

We acknowledge the Center for Advanced Food Technology (CAFT) Mass Spectrometry Lab facility for providing instrumentation support for this project. CAFT is an initiative of the New Jersey Commission of Science and Technology. We also thank Scientific Instrument Services Inc. of Ringoes, New Jersey for donation of the Short Path Thermal Desorption instrument used in this study and Dr. Daphna Havkin-Frenkel of David Michael & Co. Inc. for her generous donation of the vanilla beans. This work was the culmination of three semesters of undergraduate research projects designed to provide food science students with special high technology laboratory skills. This is NJAES publication # F 10569-4-91.

Literature Cited

1) Report on Vanilla Bean Global Production Quotas for the Years 1970-1990, obtained from Dammann & Co., Inc., 20 Potash Road, Oakland, New Jersey

2) Klimes, I.; Lamparsky, D. Int. Flavours Food Additives **1976** 7, 272

3) Galetto, W. G.; Hoffman, P. G. J. Agric. Food Chem. **1978** 26, 1, 195

4) Riley, K. A.; Kleyn, D. H. Food Technol. **1989** 43, (10), 64

5) Fayet, B.; Tisse, C.; Guerere, M.; Estienne, J. Analusis **1987** 15, 217

6) Bricout, J. J.A.O.A.C. **1974** 57, 713

7) Toulemonde, B.; Horman, I.; Egli, H.; Derbesy, M. Hele. Chim. Acta **1983** 66, 2342

8) Fraisse, D.; Maquin, F.; Stahl, D.; Suon, K.; Tabet, J. C. Analusis **1984** 12, 63

9) Carnero, R. C.; Heredia, B. A.; Garcia, S. F. J. Agric. Food Chem. **1990** 38, 178

10) Williams, R. J.; Major, R. T. Science, **1940** 91, 246

11) Chen, C.-C.; Rosen, R. T.; Ho, C.-T. J. Chromatogr. **1986** 360, 175

RECEIVED November 7, 1991

Chapter 5

High-Performance Liquid Chromatographic Analysis of Phenolic Compounds in Foods

Amrik L. Khurana

Whatman Specialty Products, Inc., 341 Kaplan Drive, Fairfield, NJ 07004

HPLC analysis of various food phenolic materials is discussed. Several mechanisms such as reverse phase, normal phase and anion-exchange to separate phenolic compounds on C-8, C-18, weak cation exchange and amino phases are elaborated. HPLC-resolution of the natural phenolic components such as anthocyanins, flavones, carotenoids, beet pigments, curcumins, mangiferin and gingerolsis reviewed. Isolation and analysis of phenolic components produced from degradation of natural products during food processing is also discussed.

The phenolic components are widely distributed in nature, have been identified and quantified in water *(1-5)*. The natural phenolic components from various food sources are distributed as colorants like flavones, anthocyanins, beet pigments, curcumins from Curcuma Longa L., gingerols from ginger root and mangiferin from mango fruit *(6-12)*. Gas chromatographic *(13-16)* and HPLC *(6-12)* chromatographic techniques have been generally applied to analyze various phenolic components. Most GC methods require derivatization of phenolic compounds *(13-15)*. In the case of HPLC, resolution of food phenolics has been reported on C-18 reverse phases *(6-9)*. The present paper describes a summarized review of HPLC of various naturally occurring food phenolic components along with other possible mechanisms of their separation beside the reported resolution on C-18 reverse phased.

Several mechanisms such as reverse phase, normal phase and anion-exchange to resolve phenolic compounds on C-8, C-18, weak cation-exchange and amino phases will be discussed. These HPLC bonded phased were obtained from Whatman Specialty Products Inc., Fairfield, N.J. and can also be prepared by reacting silane materials such as octydimethyl chlorosilane, octadecyldimethyl chlorosilane, carboxypropyldimethyl methoxysilane and aminopropyl dimethyl chlorosilane with silica gel. Various HPLC columns were packed by slurring materials in methanol and applying 5000 psi pressure.

Figure 1 represents reverse phase resolutions of simple phenolic compounds such as phenol and 1-naphthol on C-8 and C-18 columns by methanol-water system as mobile phase. Weak-cation exchanger (WCX) with carboxylic acid functionality is generally used to resolve various amino components by cation-exchange mechanism with a counter-ionic salt incorporated in the mobile phase. The same bonded phase can also be used to perform reverse phase or normal phase separation

0097–6156/92/0506–0077$06.00/0

© 1992 American Chemical Society

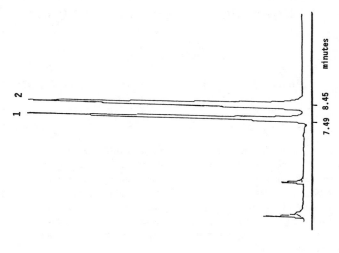

Figure 1. Resolution of phenol and 1-naphthol on C-18 and C-8 columns (11cm x 4.6mm, I.D.) Mobile phase: water: methanol (40:60, v/v) at 0.9ml/min; λ_{max}: 254nm; 1. phenol and 2. 1-naphthol. A= C-18 and B= C-8 columns

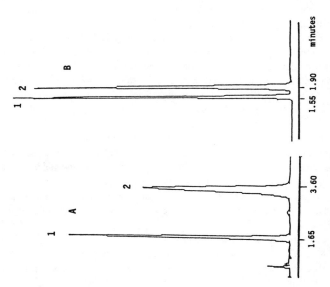

Figure 2. Reverse phase resolution of phenol and 1-naphthol on WCX column (11cm x 4.6mm, I.D.) Mobile phase: water: ethanol (80:20, v/v) at 0.2ml/min; λ_{max}: 254nm; 1. phenol and 2. 1-naphthol

when used in nonionic form. Figure 2 shows reverse phase separation of phenol and 1-naphthol on WCX column by using water-methanol mixture as mobile phase. The normal phase separation on the same column is shown in Fig. 3. In this case, hexane-ethanol mixture is used as a mobile phase. Figure 4 shows anion-exchange resolution of the components on amino column. In this case, ammonium phosphate has been used as a counter ion in the mobile phase.

Flavonoids

A group of naturally occurring substances derived from flavone (phenyl-γ-pyrone) are called flavonoids. The citrus fruit exclusively contains the methoxylated flavones *(17)* which can be detected in leaves as well as in all fruit parts: flavedo, albedo, membranes, juice and seeds. RP C-18 and normal phase silica gel columns have been used to analyze flavones from various foods *(8)*.

Cellulose triacetate phase has been used to achieve enantiomeric resolution of racemic flavonones such as naringenin, hesperetin, eriodicytiol, homoeriodicytiol, pinocembrin, isosakuranetin, pinostorbin and sakuranetic *(18)* with methanol as mobile phase. The mechanism of chiral recognition on cellulose triacetate has not been understood in detail. Presence of two hydroxyl groups in positions 5 and 7 (Fig. 5) gave the best results. The 6-methoxylated and 6-hydroxy flavanone were not resolved. Substitution in the ring A was considered to be important for chiral recognition, although these substitutions were not in the vicinity of chiral center. It may be due to absorption of a molecule in such a way that discrimination between the two optical antipodes was easier. It is noteworthy that resolution of enantiomeric hydroxy flavones and methoxylated flavones may also be achieved on commercially available γ-cyclodextrin chiral phase which separates the racemic isomers with three benzene or cyclohexane rings in the system.

Resolution of polymethoxylated and acetylated flavones has been reported *(8,19)* on RP-18 and silica gel phases using aceonitrile-water (40:60, v/v) and ethanol (100%) as mobile phases respectively. The polymethoxylated flavones are found in orange juice while acetlyated flavones occur in soy flakes.

Anthocyanins

Red, blue and purple colors exhibited by flowers, fruits and other plant tissues are due to anthocyanins. The major anthocyanins are derived from substitution on the flavylium cation (Fig. 6).

There has been increased interest in the use of water soluble plant extracts as food colorants due to official delisting of artificial food colors. RP-18 columns have been used to analyze these compounds *(20-21)*. Resolution of the anthocyanin molecules can also be accomplished by normal phase or cation-exchange chromatography.

HPLC resolution of anthocyanin glucosides and galactosides form cowberry and chokeberry fruits on C-18 columns *(20-21)* requires the use of formic acid in the mobile phase to deactivate residual silanols on the silica surface. The analysis of these anthocyanin glucosides and galactosides under acid conditions may lead to hydrolysis which can interfere with their results.

β-Cyanins and Amaranthin

The phenolic coloring matters such as β-cyanins and amaranthin occur in red beetroot *(22-25)* and in the leaf and stems of the plant amaranth. The structures of pigments are given in Figure 7. These differ from amaranthin with respect on the substituent on position 5.

Betanine in solution is known to degrade to botanic acid (BA) and cyclo-dopa-5-O-glucoside (CDG). The regeneration process involves a Schiff's base condensation of the nucleophilic amine of CDG with aldehyde of BA *(26)*. Reverse phase HPLC on C-18 columns has been used to analyze β-cyanins, the degradation products and amaranthin from red beetroot and amaranth *(9, 27)*. Modifiers

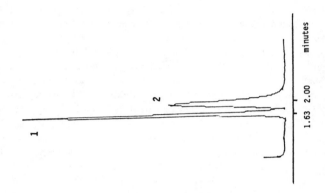

Figure 4. Anion-exchange resolution of phenol and 1-naphthol on Amino column (11cm x 4.6mm, I.D.) Mobile phase: 0.1M ammonium phosphate: ethanol (80:20, v/v) at 1ml/min. λ_{max}: 254nm. 1. phenol and 2. 1-naphthol

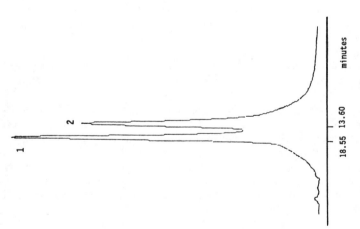

Figure 3. Normal phase separation of phenol and 1-naphthol on WCX column (11cm x 4.6mm, I.D.) Mobile phase: hexane: ethanol: water (300:38:0.75, v/v) at 0.2ml/min; 1. 1-naphthol and 2. phenol

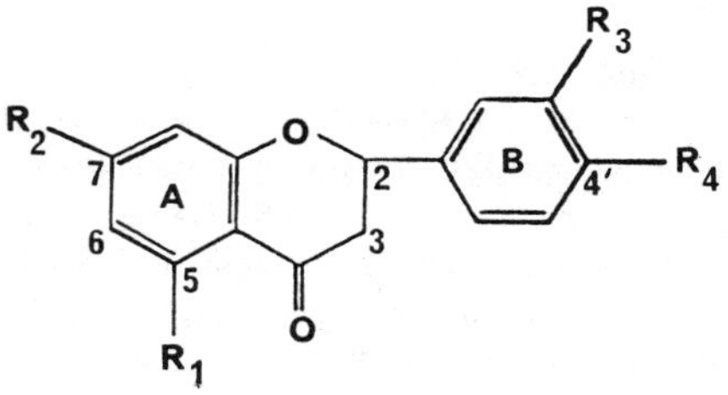

Figure 5.   Structure of Flavonoids

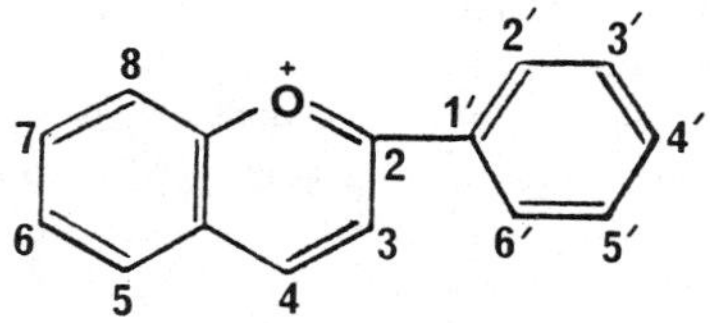

Figure 6.   Structure of Anthocyanins (Flavylium cation)

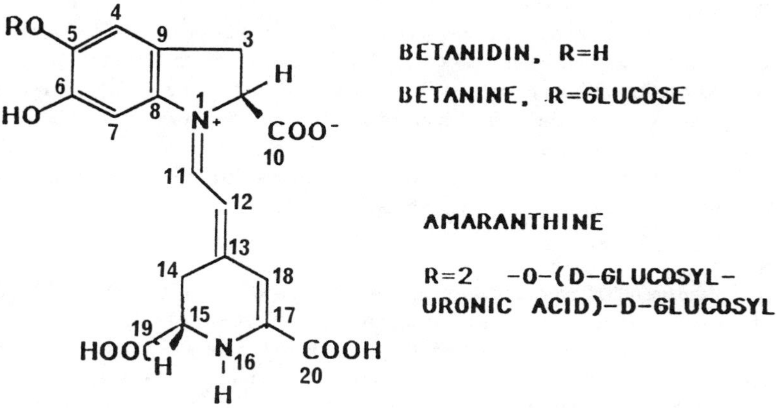

BETANIDIN, R=H

BETANINE, R=GLUCOSE

AMARANTHINE

R=2   -O-(D-GLUCOSYL-
URONIC ACID)-D-GLUCOSYL

Figure 7.   Structure of β-Cyanins and Amaranthine

such as picric acid and tetrahydroammonium phosphate were used in the mobile phase. These pigments have carboxylic acid groups in their skeleton and it will be quite possible to achieve resolution of such phenolic compounds by using an anion-exchanger with a computer ionic agent in the mobile phase.

Curcumins and Related Compounds

The phenolic coloring pigments, curcumin(1,7-bis-4-hydroxy-3-methoxy-phenyl)-1,6-heptadiene-3,5-dione) and its isomers demethoxy curcumin (1-(4-hydroxy-3-methoxy-phenyl)-7-(4-hydroxyphenyl)-1,6-heptadiene-3,5-dione)andbisdemethoxycurcumin(1,7-bis-(4-hydroxyphenyl)-1,6-heptadiene-3,5 dione) are found in the rhizomes of the plant Curcumin Longa L. (Zingiberaceae). These curcuminoids have found their use as coloring agents for food and drugs and are regarded as drug/drug model and antiheptotoxic agents (12). The photo-oxidation of curcuminoids produce vanillin, 4-hydroxybenzaldehyde, ferulic aldehyde, p-hydroxybenzoic acid, vanillic acid and ferulic acid (10). Normal phase HPLC on amino columns have been used to analyze curcumin and its isomers (11, 12). The photo-oxidation phenolic compounds of curcumin ant its isomers have also been analyzed by normal phase HPLC on Whatman PartiSphere-5 WCX - column (11). The reverse phase analysis of curcumins and their isomers has been reported on C-18 HPLC Column (28). A normal phase silica gel column has been used in an attempt to resolve curcuminoids. It has been reported that 1,3-diketone groups of the curcumin system interact with the active sites on silica surface. It has been further observed that the possibility of the hydroxy groups on the same skeleton interacting with the surface of the stationary phase to provide a normal phase separation cannot be ruled out (11). The amino columns can also be used to resolve the curcuminoids by an anion exchange mechanism which will require a counter ion like ammonium formate in the mobile phase.

Mangiferin

Mangiferin (Fig. 8) is a well known member of xanthones and occurs as a glucoside in mangoes. It can be easily distinguished from flavones and quinones by spectral data. Mangiferin is known for its cardiotonic, spasmolic, diuretic, choliretic and antiphlogistic actions (29). HPLC resolution of mangiferin from two terpinoid coumarins, colladin and colladonin which are found in C. Triquetra along with mangiferin, has been reported on RP-C18 column. Sulfuric acid (0.2%) has been used in the mobile phase to suppress the ionization of residual silanols (29). Other commercially available amino cyano (PAC form Whatman, Inc.) or cyano columns which are used for resolution of sugars may also be used to resolve mangiferin form other materials. It is also possible to resolve the mangiferin by an anion exchange mechanism on PAC column.

Gingerols

Gingerols are the phenolic flavor components derived from ginger root. The gingerols undergo retro-aldol degradation on injection into gas chromatography (7). C6, C8, C10 aldehydes and gingerone are formed as a result of retro-aldol reaction. These components have been analyzed on an RP-C18 column (6). A Whatman - WCX (weak-cation exchange) column has also been used to analyze the same components by a reverse phase mechanism. A gradient starting with 100% water to 100% methanol was used. The unionized carboxylic acid group of the weak - cation exchanger exhibits a reverse phase mechanism under neutral conditions (10).

Capsaicins

Capsaicins such as nordihydrovapsaicin, capsaicin, dihydrocapsaicin, homocapsaicin and homodihydryocapsaicin are heterocyclic phenolic flavor components. Reverse phase HPLC on a C-18 column has been used to analyze these components *(30)*. Water :acetonitrile (54.7:45.3, v/v) has been used as a mobile phase. The electochemical detection of capsaicin can provide a higher degree of sensitivity and specificity as these are very electroactive due to the easily oxidized phenolic functional group. It is easy to identify these phenolic compounds on the basis of hydrophobicities and retention times *(31-33)*. An improvement in resolution can be achieved by decreasing the acetonitrile with methanol as modifier.

In short, the phenolic components in food can be resolved on various HPLC phases like C-8 and C-18 by reverse phase mechanism/ weak cation-exchanger both by reverse and normal phase mechanisms and dialkyl amiopropyl phase by anion-exchange mechanism. Various optical isomeric phenols may be separated on types of commercially available chiro-phases like cellulose triacetate and cyclodextrins.

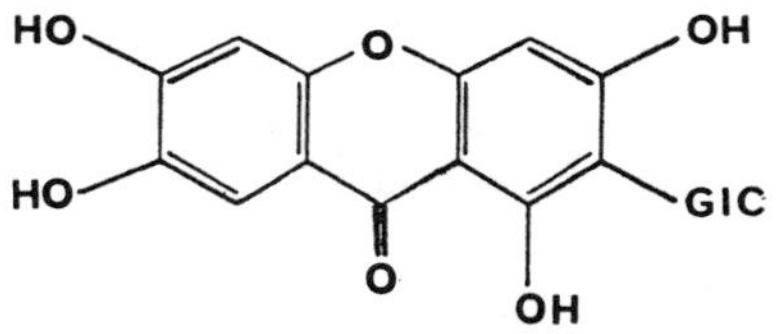

Figure 8.   Structure of Mangiferine

Literature Cited

1. Armmentrout, J.D.; McLean, J.D.; Long, M.W. *Anal. Chem. 1979, 51, 1039-45.*
2. Boryx, A. *J. Chromator. 1981, 261, 361-66.*
3. Chao, G.K.; Suatoni, J.C. J. Chromatogr. Sci. 1982, 20, 430-40.
4. Kuwata, K.; Uebori, M.; Yamazaki, Y. *Anal. Chem. 1980, 52, 857-62.*
5. Kuwata, K.; Uebori, M.; Yamazaki, Y. *Anal. Chem. 1981, 53, 1531-34.*
6. Chen, C.C.; Rosen, R.T.; Ho, C.-T. *J. Chromatogr. 1986, 360, 175.*
7. Chen, C.C.; Rosen, R.T.; Ho, C.-T. *J. Chromotogr. 1986, 360, 163.*
8. Heimhuber, B.; Gallensa, R.; Herrman, K. *J. Chromatogr. 1988, 439, 481.*
9. Huang, A.A.; Von Elbe, J.H. *J. Food Sci, 1980, 51, 670.*
10. Khurana, A.L.; Butts, E.T.; Ho, C.-T. *J. Liq. Chromatogr. 1988, 11, 1615.*
11. Khurana, A.L.; Ho, C.-T. *J. Liq. Chromatogr. 1988, 11, 2295.*
12. Tonnesen, H.; Karlsen, J. *J. Chromatogr. 1983, 259, 367.*
13. Guerin, M.R.; Olerich, G.; Horton, A.D. *J. Chromatogr. Sci. 1974, 12,385-91.*
14. Cuerin, M.R.; Olerich, G. *Tob. Sci., 1976, 19, 44-48.*
15. Sakuma, H.; Kusama, M.; Munakata, S.; Ohsimi, T.; Sugawara, S. Beitr. *Tabakforsch. 1963, 12, 63-71.*
16. Spears, A.W. *Anal. Chem. 1963, 35, 320-22.*
17. Hortowitz, R.M. In the Orange, its chemistry and physiology. Sinclair, W.B. Ed. University of California Press, Berkley , Ca. 1961,11.
18. Krause, M.; Galensa, R. *J. Chromatogr. 1988, 441, 417.*
19. Farmakalidis, E.; Murphy, P.A. *J. Agric. Food Chem. 1985, 33, 385.*
20. Kallio, H.; Pallasaho, S.; Karppa, J.; Linko, R.R. *J. Food Sci. 1986, 51, 408-410.*
21. Osmianski, J.; Snapis, J.C. *J. Food Sci., 1988, 53, 1241.*

22. Stark, D.; Reznik, H. Z. *Pfanzenphysiol. 1979, 94, 183.*
23. Stark, D.; Engel, U.; Rezmik, H. Z. *Phanzenphysiol. 1979, 101, 215.*
24. Vincent, K.T.; Scholt, R.G. *J. Agric. Food Chem. 1978, 26, 812.*
25. Von Elbe, J.H.; Huang, A.S.; Attoe, E.L.; Nank, W.K. *J. Agric. Food Chem. 1986, 34, 512.*
26. Schwartz, S.J.; Von Elbe, J.H. Z. Levensm Unters Forsch. 1983, 176, 448.
27. Schwartz, S.J.; Von Elbe, J.H. *J. Agric. Food Chem. 1980, 28, 580.*
28. Amakawa, K.E.; Hirata, K.; Ogiwara, T.; Ohnishi, K. *Jpn. Anal. 1986, 33, 586.*
29. Simova, M.; Tomov, E.; Pangarova, T.; Palvlova, N. *J. Chromatogr. 1986, 351, 379-382.*
30. Chiang, G.H. *J. Food Sci. 1986, 51, 499.*
31. Hoffman, P.G.; Lego, M.C.; Galetto, W.G. *J. Agric. Food Chem. 1983, 31, 1326.*
32. Rouseff, R. In Liquid chromatographic analysis of food and beverages. Charalambous, G. Ed. Academic Press, Orlando, Fl. 1979, 1, 16.

RECEIVED July 16, 1992

Chapter 6

Glycosidically Bound Phenolic and Other Compounds in an *Umbelliferous* Vegetable Beverage

Tarik H. Roshdy[1], Robert T. Rosen[1], Thomas G. Hartman[1], Joseph Lech[1], Linda B. Clark[2], Elaine Fukuda[1], and Chi-Tang Ho[2]

[1]Center for Advanced Food Technology and [2]Department of Food Science, Cook College, Rutgers, The State University of New Jersey, New Brunswick, NJ 08903

An *Umbelliferous* vegetable beverage to be used by the National Cancer Institute (NCI) for chemical and biological studies directed towards the possible prevention of cancer was analyzed. This beverage consisted of carrot, celery, and tomato juices, with pepper, basil, paprika, garlic and rosemary added as spices and natural antioxidants.

This paper reports on the extraction and direct mass spectrometric evidence of bound phenolic and other compounds from an *Umbelliferous* vegetable beverage. Phenolics observed included ferulic acid, hydroxycinnamic acid, hydroxymethoxycoumarin, homovanillic acid, acetosyringone, coniferol, vanillin and analogs.

It is recognized that compounds in fruits and vegetables may exist at a greater concentration as conjugates than as aglycones. The hydrolysis, for example, during digestion or ripening, of glycosides in vegetables and fruits, is a major pathway leading to the enrichment of flavors. Hydroxy containing species such as phenolics are directly bound to sugars. Compounds such as coumarins, which are lactones, exist as hydroxy acids when bound to sugars. Dihydroxy benzene derivatives yield the corresponding phenolics. Recent work on the bound fractions of pineapple, peach, celery and hog plum has been published (1-5).

The NCI under the auspices of Dr. Herbert Pierson, and in cooperation with Dr. Phillip Crandall of the University of Arkansas Food Science Department, has formulated an Umbelliferous vegetable beverage to be used for the chemical and biological studies directed towards the possible prevention of cancer. This beverage consisted of carrot, celery, and tomato juices, with pepper, basil, paprika, garlic and rosemary added as spices and natural antioxidants.

Phenolics observed included ferulic acid, hydroxycinnamic acid, hydroxymethoxycoumarin, homovanillic acid, acetosyringone, coniferol, and vanillin. These results were obtained by isolation and extraction techniques using Amberlite XAD-2 resin followed by enzyme hydrolysis with subsesequent determination by gas chromatography (GC) and GC-mass spectrometry (GC-MS).

0097–6156/92/0506–0085$06.00/0
© 1992 American Chemical Society

Experimental

Materials

β-Glucosidase 50,000 units containing approx. 100% protein from Sigma Chemical Co.

Supelpak adsorbent, purified form of Amberlite XAD-2 resin from Supelco Inc.

Glass column 1 cm I.D. with a 50 cm length from Fisher Scientific Company (for column chromatography).

HPLC grade Dichloromethane, acetone, methanol, n-pentane, and water (Fischer Scientific).

Gas Chromatographic & Mass Spectrometry Operating Conditions:

a. GC Conditions (splitless mode):

GC Type:	Varian 3400
Column Type:	60 meter DB-1 Fused Silica Capillary (J&W Scientific).
Column Film Thickness:	0.25 micron
Column ID:	0.32 mm
Carrier Gas:	40 psi High Purity Grade Helium for 1ml/min.
Make-up Gas for FID:	40 psi High Purity Grade Helium 30ml/min
Air:	60 psi High Purity Grade 300ml/min
Initial Column Temp.:	50°C
Initial Hold:	5 min
Rate:	2°/min
Attenuation Range:	1×10^{-12}
Final Temp.:	290°C
Final Hold:	90 min
Total Time:	215.00 min

Data Acquisition: The output from the GC was integrated as well as recorded and logged on a VG Multichrom chromatography data acquisition system as well as a strip chart recorder output of Varian 4290 integrator.

b. Mass Spectrometry Conditions:

Instrument: Finnigan MAT 8230 high resolution mass spectrometer directly interfaced to a Varian 3400 gas chromatograph.

Mode:	electron ionization.
filament emission current.	70 eV. 1 mA
Accelerating Voltage:	3 kV
Multiplier Voltage:	1800 V corresponding to a gain of 10^6
Ion Source Temperature:	250°C.
GC Conditions:	Identical to those used on the off-line GC.
Data Acquisition:	Finnigan MAT SS300 data system.

Methodology

One hundred milliliters of the *Umbelliferous* vegetable beverage was divided and placed into four 50mL capacity heavy-duty centrifuge glass tubes with screw caps. The tubes were then centrifuged for 20 min at 2500 RPM. Forty grams of dry Amberlite XAD-2 resin was placed into a glass column [50 cm x 1 cm]. Glass wool was packed tightly at the top and bottom of the column to prevent the loss of the stationary phase and to prevent the column bed from rising with the water level and producing air bubbles. The XAD-2 column was sequentially conditioned with water, methylene chloride/n-pentane [1:1, v/v], methanol, and water. The second water elution was used to clear the methanol from the column prior to adding the sample. The supernatant from the centrifuge tubes was then loaded on the top of the glass column. HPLC grade distilled water was used to elute the water soluble (yellow-orange) fraction until the eluted water was clear. HPLC grade n-pentane / methylene chloride [1:1, v/v] was used to elute the second (yellow-red) fraction until the eluted solvent was clear of color. Finally, 100 mL of HPLC grade methanol was used to elute the most polar organic compounds (yellow fraction).

The methanol fraction (bound fraction) was concentrated to 1.0 mL using a rotary evaporator with the assistance of a water bath at 50°C. The concentrate was transferred into a 250 mL flask and 100 mL of citrate-phosphate buffer (pH 5) was added. Nitrogen gas was bubbled into the solution at a slow rate to eliminate traces of solvent which would inhibit the enzyme. The sample was then hydrolyzed by adding 60 mg of β-glucosidase (5.5 units/mg) and warmed to 37°C for 72 hrs in a shaking incubator. The liberated aglycones were extracted with three 150 mL portions of methylene chloride using a separatory funnel. The organic phase was dried over anhydrous sodium sulfate. This fraction was concentrated to 0.3 mL under a gentle stream of nitrogen gas. Triacontane was added as an internal standard (I.S.). This fraction was flushed with nitrogen and kept in a freezer until analysis.

Results and Discussion

The methodology for the determination of glycosidically bound compounds in celery has been published (4). This method described the identification and the quantitation of phthalides and coumarins in celery. In our experiments this methodology was applied to the *Umbelliferous* beverage. The technique involves applying a filtered or centrifuged aliquot of the sample on the top of an XAD-2 resin column. The column is then washed with water to elute very polar interferences (yellow-orange fraction). This eluent represents the sugars, acids and other very water-soluble substances which were adsorbed onto the XAD-2 column. Subsequent elution with methylene chloride/n-pentane eliminated the free organics (not covalently bound to sugars) (yellow fraction). Lastly, methanol was used to elute the glycosidically bound phenolics and alcohols.

Twelve liberated phenolic compounds have been identified in the polar fraction (bound fraction) of the *Umbelliferous* beverage. These compounds are eugenol, vanillin, 4-hydroxyacetophenone, acetovanillone, 3-hydroxyphenyl acetic acid, homovanillic acid, acetosyringone, coniferyl alcohol, homovanillic acid methyl ester, hydroxycinnamic acid, ferulic acid and 7-hydroxy-6-methoxy-1(2H)-benzopyran-2-one. The summary of the twelve phenolic compounds identified in the bound fraction of the *Umbelliferous* beverage with their molecular weights and their structures are given in Table I.

The internal standard used for quantitative purposes, triacontane, was chosen because it was clearly separable from all the components in the methanol extract.

Figure 1 shows the GC-MS profile of the liberated compounds from the bound fraction. The phenolic compounds were identified by matching

**TABLE I PHENOLIC COMPOUNDS IDENTIFIED
IN THE BOUND FRACTION**

COMPOUNDS IDENTIFIED	MW	ppm	STRUCTURE
Eugenol or 2-methoxy-4-(2-propenyl)-phenol ($C_{10}H_{12}O_2$)	164	0.51	
Eugenol isomer ($C_{10}H_{12}O_2$)	164	0.12	
Vanillin or 3-hydroxy-4-methoxy-benzaldehyde ($C_8H_8O_3$)	152	0.24	
4-hydroxy-acetophenone ($C_8H_8O_2$)	136	0.35	
Acetovanillone or 1-(4-hydroxy-3-methoxyphenyl)-ethanone ($C_9H_{10}O_3$)	166	0.89	
3-hydroxyphenyl acetic acid ($C_8H_8O_3$)	152	0.62	
Homovanillic acid or (4-hydroxy-3-methoxyphenyl) acetic acid ($C_9H_{10}O_4$)	182	0.74	

TABLE I Cont.

COMPOUNDS IDENTIFIED	mw	ppm	STRUCTURE
Acetosyringone or 1-(4-hydroxy-3,5-dimethoxy-phenyl)-ethanone ($C_{10}H_{12}O_4$)	196	0.61	
Coniferyl alcohol or γ-hydroxyisoeugenol or 3-(4-hydroxy-3-methoxyphenyl)-2-propen-1-ol ($C_{10}H_{12}O_3$)	180	0.61	
Homovanillic acid methyl-ester ($C_{10}H_{12}O_4$)	196	0.68	
hydroxycinnamic acid or 3-phenyl-2-propenoic acid ($C_9H_8O_3$)	164	0.09	
Ferulic acid or 3-(4-hydroxy-3-methoxyphenyl)-2-propenoic acid ($C_{10}H_{10}O_4$)	194	0.49	
7-hydroxy-6-methoxy-2H-1-benzopyran-2-one ($C_{10}H_8O_4$)	192	0.61	

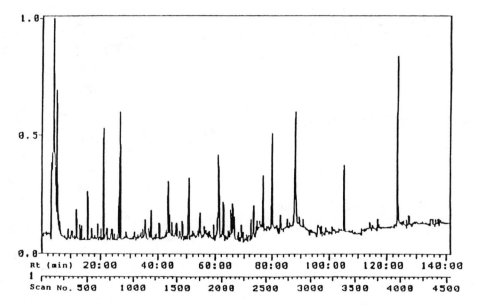

Figure 1. GC-MS profile of aglycones liberated from bound fraction of the *Umbelliferous* vegetable beverage.

their electron ionization mass spectra with a NBS (NIST) computer library. Other isomers than those presented are possible. Most of these species were not observed in the free fraction, that is, the fraction which was eluted with dichloromethane/n-pentane.

Other bound compounds which were identified in the bound fraction of the *Umbelliferous* beverage are listed in Table II. Eight liberated nonphenolic compounds have been identified in the polar fraction (bound fraction) of the Umbelliferous beverage. These compounds are benzaldehyde, methionol, benzene methanol, maltol, 2,3-dihydroxy-2-methoxy- 6-methyl-4H-pyran-4-one, 4-(3-hydroxy-2,6,6-trimethyl-1- cyclohexen-1-yl)-3-buten-2-one, 1H-indole-3-ethanol, and 7-hydroxy-6-methoxy-2H-1-benzopyran-2-one.

This work was initiated as the use of enzymes for the liberation of conjugated species is important in human biochemistry and in flavor chemistry. The human intestine has enzymes generally classified as "fecalase" which liberate aglycones from glycosides, subsequently allowing adsorption of potentially important phytochemicals through the colon. Some of the phenolic and other compounds identified in this study have important antimutagenic activity. Ferulic acid has activity inhibiting formation of nitrosamines *in vivo* (6) and protecting DNA from electrophilic attack (7). Other phenolics inhibit formation of certain prostaglandins implicated in tumor growth (8). Indole carbinols and some homologs increase the rate of metabolism with subsequent reduction of estradiol, a compound thought important in induction of breast cancer (9).

In addition, enrichment of these previously bound conjugates through enzymatic digestion has the potential of increasing yields of naturally occurring flavorants in which the public so greatly demands.

Table II Other Compounds Identified in the Bound Fraction of the *Umbelliferous* Beverage

Compounds Identified	MW	ppm
benzaldehyde	106	0.97
methionol	108	0.12
benzene methanol	108	1.76
maltol or 3-Hydroxy-2-methyl-4H-pyran-4-one	126	0.94
2,3-dihydro-3,5-dihydroxy-6-methyl-4H-pyran-4-one	144	0.07
4-(5-hydroxy-2,6,6-trimethyl-1-cyclohexen-1-yl)-3-buten-2-one	208	0.09
1H-indole-3-ethanol	161	0.83
7-hydroxy-6-methoxy-2H-1-benzopyran-2-one	192	0.61

Acknowledgment: The Center For Advanced Food Technology is a New Jersey Commission on Science and Technology Center. The authors thank the NCI for the funding of the project, and Dr. Herbert Pierson of NCI for his comments and encouragement during the period in which the work was done.

Literature Cited

1. Adedeji, J.; Hartman, T.G.; Rosen, R.T.; Ho; C.-T. *J. Agric. Food Chem.*, 1991, 39, 1494-1497 .
2. Wu, P.; Kuo, M.-C.; Hartman, T.G.; Rosen, R.T.; Ho, C.-T. *J. Agric. Food Chem.*, 1991, 39, 170-172.
3. Ho, C.-T.; Sheen, L.-Y.; Wu, P.; Kuo, M.-C.; Hartman, T. G.; Rosen, R. T. *In Flavour Science and Technology*. Thomas, A. F.; Bessier, Y. Eds.; 1990, 77-80.
4. Tang, J.; Zhang, Y.; Hartman, T.G.; Rosen, R.T.; Ho, C.-T. *J. Agric. Food Chem.*, 1990, 38, 1937- 1940.
5. Wu, P.; Kuo, M.-C.; Hartman, T.G.; Rosen, R.T.; Ho, C.-T. *Perfumer and Flavorist*, 1990, 15, 51-54.
6. Kuenzig, W., Chan, J., Norkus, E. Holowaschenko, H. Newmark, H., Mergens, W. and Conney, A.H., *Carcinogenisis* 1984, 5, 309-313.
7. Newmark, H.L., *Nutr. Cancer*, 1984, 6, 58-70.
8. Dehirst, F.E. *Prostaglandin,* 1980, 20, 209-214
9. Bradlow. L.H.; Michnovicz, J.J., *J. National Cancer Institute*, 1990, 82, 613-615.

RECEIVED July 16, 1992

Chapter 7

Evaluation of Total Tannins and Relative Astringency in Teas

P. J. Rider[1,3], A. Der Marderosian[1,2], and J. R. Porter[1,2]

[1]Departments of Chemistry and [2]Biology, Philadelphia College of Pharmacy and Science, 43rd Street and Woodland Avenue, Philadelphia, PA 19104

The results of our analysis of total tannin levels and relative astringency of various commercial and traditional teas is reported herewith. Our method involves complexation with BSA (bovine serum albumin), centrifugation and the assay for protein remaining in supernatant with Coomassie Blue. Tannic acid equivalence (TAE) is calculated by dividing the concentration of tannic acid required to completely precipitate the BSA by the concentration of tea required to do the same. Complete absence of supernatant BSA is verified by capillary electrophoresis (CE) using a novel buffer system. While less sensitive than more sophisticated procedures of tannin analysis, this method is simple, rapid and relatively free of opportunities for experimental error.

Morton and others have long believed that ingestion of highly astringent (high tannin) foods and beverages may be strongly associated with the development of esophageal cancer (*1*). Other researchers contend that there are health benefits to be gained from dietary tannins (*2*). It may be that the deleterious effects of high tannin diets only apply when, inadvertently, carcinogens like nitrosamines are included in the diet. The purpose of this research has been to develop an assay for astringency in teas so that future studies might determine the correlation between the strength of astringency in various teas and their potential health effects.

Explanation of Terminology

Astringency. Astringency is defined in Medicine as the drawing together or constriction of tissue (*3*). Some foods and beverages taste astringent, having a "puckering" taste. According to Bate-Smith, the primary reaction whereby this astringency develops in the palate is by precipitation of glycoproteins in the mucous secretions of salivary glands (*4*). In the field of Phytochemistry, astringency has often been used as if synonymous with protein precipitation (*5*).

---

[3]Current address: Department of Biology, LIF 136, University of South Florida, Tampa, FL 33620

0097–6156/92/0506–0093$06.00/0
© 1992 American Chemical Society

Tannins. Tannins are a group of polyphenolic phytochemicals which are believed to be responsible for dietary astringency. Indeed, by definition, naturally occurring polyphenolic compounds must have the ability to form stable water-insoluble complexes with protein to be classified as tannins (6). Phytochemists restrict their definition of tannins to polyphenolics of 500 daltons or more (7).

Astringency assays. Astringency assays are often called "total tannin" assays because no other kind of assay is believed to be better in estimating the overall amount of tannins (6). A standard protein is usually mixed with a test solution containing an unknown amount of tannin. After centrifugation, the amount of precipitation is calculated. The most popular protein sources are bovine serum albumin (BSA) and fresh human blood (5). BSA is the preferred choice because it is readily available, inexpensive and less likely to absorb light in regions where common plant constituents may interfere (8). Astringency assay results are usually expressed in tannic acid equivalence (TAE).

Tannic Acid. Tannic acid is an amorphous mixture of compounds prepared by fermentation and extraction of galls from the young twigs of various oak trees (9). The primary contents of tannic acid are hydrolysable tannins, which are mainly polyphenolic combinations of gallic acid and glucose. High performance liquid chromatography (HPLC) of tannic acid from various sources has revealed differences in composition and tanning capacity (10). Hagerman and others have stressed the need for more highly uniform standards for astringency and tannin assays (11).

Methods

No method for the assay of tannin in teas is currently recognized by the Association of Official Analytical Chemists (A.O.A.C.) (12). Of the many procedures for total tannin estimation described in the literature, we chose to investigate those utilizing protein precipitation because of their increasing promotion in recent years (6). These procedures generally begin by mixing the astringent liquid with a protein solution like BSA (1 mg/ml in 0.20 M acetate, 0.17 M NaCl buffer at pH of 5) followed by centrifugation and analysis of the protein in either the supernatant or pellet. One method, recently promoted in the literature, uses ninhydrin to estimate astringency via the amino acids released following a partial hydrolysis of the tannin-protein pellet (8). We examined this method but found it much more time-consuming and less reliable than measuring the residual supernatant protein with Coomassie Blue (13). Thus the Coumassie Blue or Bradford Method was integrated into the procedure outlined in Figure 1.

Bovine albumin was obtained from Sigma (A-7030; lot 10H0261). Tannic acid was also obtained from Sigma (T-0125; lot 87F-0745). Coomassie Blue (Brilliant Blue G) was obtained from Aldrich (20,139-1; lot KT02325JP).

Teas were prepared by pouring 150 ml of boiling water over dry, weighed leaf material and allowing the mixture to steep for 5 minutes thus approximating the extraction efficiency of a standard cup of tea, before removing 5 ml. This was allowed to cool to room temperature before 1 ml aliquots were used to complex with BSA (Fig. 1). Tannic acid solutions were prepared in distilled water. Sephadex columns were equilibrated prior to addition of the supernatant with a 50% aqueous dilution of the same acetate buffer prepared for the BSA solution. This diluted buffer was used to wash pellets, elute BSA and provide blanks for capillary electrophoresis (CE).

The Bradford assay was accomplished by mixing 100 ul of cleaned supernatant with 4 ml of Bradford reagent (*14*). After waiting 5 minutes for color stabilization, the mixtures were poured into disposable plastic cuvettes and measured at 595 nm. on a LKB Biochrome Novaspec II Model 4040-011 Spectrophotometer against blanks prepared from 100 ul of the diluted buffer instead of supernatant.

Considerable time was saved in the following replication protocol. Typically, ten different concentrations of a tea were prepared for each run. Two aliquots were taken from each and, after complexation, centrifuged. Half of the samples were analyzed by the Bradford method without Sephadex cleanup of the BSA to find approximate concentrations of minimal residual BSA. Tubes were then selected for Sephadex cleanup accordingly, and based on the Bradford results of these, a limited series of tubes were employed in the CE assay. This helped keep the protein in our capillary column to a minimum.

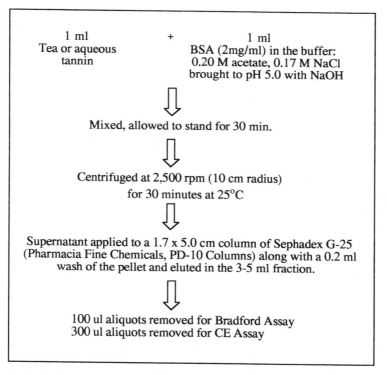

Figure 1. Outline of the procedure used to estimate the astringency or total tannin of various teas.

The tannic acid curve in Figure 2 was prepared without the Sephadex clean up. It illustrates the interference at 595 nm due to nonproteins, like phenolics, which are removed by the Sephadex columns. Capillary electrophoresis verification indicated that 1.05 mg/ml is the concentration of tannic acid required for complete BSA precipitation.

Capillary electrophoresis was performed on an ISCO instrument model 3140. The main objective which led to our adoption of a novel buffer system for BSA estimation was our reluctance to subject our fused silica column to the derivatization normally employed to deactivate silica and make it non-adsorptive toward proteins (*15*). A Borate 25 mMol/ Deoxycholate 25 mMol buffer system was developed and provided by Gupta (see acknowledgments). It proved ideal for our purposes and eliminated the need for derivatization. A representative pair of electropherographs is shown in Figure 3 along with details of the operation.

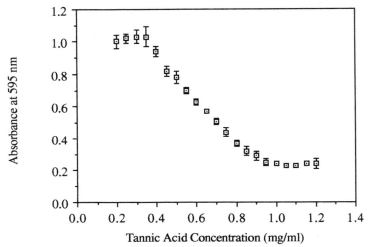

Figure 2. Curve of the Bradford absorbance at 595 nm due primarily to residual BSA in the supernatant after complexation with tannic acid and centrifugation. (Error bars represent standard deviations. n = 4).

Automatically following each run, a 10 minute capillary purge sequence was programmed to condition the instrument for the next run. This operation consisted of 2 minutes of distilled, deionized, degassed water, then 2.5 minutes of 0.1N NaOH, followed by 3 minutes of the water again and 2.5 minutes of the borate/deoxycholate buffer.

Results and Discussion

The teas studied and their relative astringencies are listed in Table I. All of the teas, except for *Eriodictyon californicum L.* (Yerba Santa), are basically various grades of *Camelia sinensis L.* (*16*). The mechanical crushing and fermentation of the leaves turns green tea into black and oxidizes the abundant but poorly soluble proanthocyanidins or condensed tannins into smaller and more soluble compounds like the theaflavins which give black tea its "brisk" flavor, color and high astringency (*17*). Cutting the tea leaf increases the overall surface area for faster extraction and a stronger cup of tea. These two manufacturing processes explain most of the major differences in astringency among those teas derived from *C. sinensis*. Green teas are not fermented, while Oolong and Black teas are. Black teas are fermented the longest and have the highest astringency. The partial fermentation of Oolong teas affects primarily the essential oil and catechin composition of the tea (and its aroma) while the amounts of tannins remain

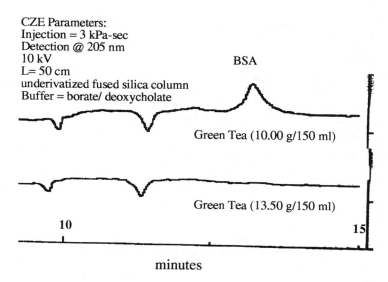

Figure 3. Capillary zone electrophoretic traces and operating parameters.

Table I. Results of Bradford/CE astringency assay

Type of Tea[a]	Brand Name[b]	Required Concentration (g/150 ml) and Range		TAE (mg TA/g Leaf)
Orange Pekoe/	Tetley	3.45	3.40 - 3.50 (n = 2)	45.80
Cut Black Tea	Lipton	3.72	3.70 - 3.75 (n = 2)	42.47
Black Tea	Lichee (Sunflower)	8.12	8.00 - 8.25 (n = 2)	19.46
from China	Keemun	7.25	6.50 - 8.00 (n = 2)	21.79
Oolung Tea from China	Ti Kuan Yin	15.17	15.00-15.50(n = 3)	10.42
Oolong Tea from Taiwan	Summit Imports	14.33	13.50-15.00(n = 3)	11.03
Jasmine Tea from Taiwan	Mong Lee Shang	14.50	14.00-15.00(n = 2)	10.90
Green Tea from China	Long Ching (Tin Mend Brand)	13.50	13.00-14.00(n = 4)	11.70
Eriodictyon californicum L.	"Yerba Santa" from Penick Co.	11.50	11.00-12.00(n = 2)	13.74

[a]All teas are various grades of *Camelia sinensis L.* unless noted otherwise.
[b]Exotic teas purchased from a local grocery store.

unchanged (*16*). Our results tend to support this, demonstrating no significant difference in the astringency of Green and Oolong teas. Jasmine teas are Green or Oolong teas which have been enhanced with the fragrant addition of jasmine blossoms. Based on organoeleptic evaluations of our teas, we believe that the Jasmine tea used in this study is an Oolong tea. The floral parts appear to have no significant effect on the astringency of the tea.

Yerba Santa has been implicated in the high risk for esophageal cancer among American Indians in California (Morton, J. F., University of Miami, personal communication, 1990). Its astringency potential falls between the unfermented teas and uncut black teas. Actual astringency would vary with the individual's preference for method of tea preparation.

In light of the need to establish a pure and readily available standard tannin for more meaningful representation of the results of total tannin or relative astringency assays we are attempting to isolate geraniin, the only pure crystalline tannin to be used for that purpose to our knowledge (7). Geraniin is produced in several species of *Geranium* (not *Pelargonium*) including-with particularly high yield-*Geranium carolinensis L.* (*7, 18*) which grows in the Southeastern United States (Wunderlin, R. P., University of South Florida, personal communication, 1991).

Acknowledgments   For the use of his CZE instrument, novel buffers and encouragement, we thank Dr. P. Gupta (Pharmaceutics Department, Philadelphia College of Pharmacy and Science). We would also like to thank Dr. J. Nikelly, Dr. A. Gennaro and Dr. L. Delisser-Mathews (Chemistry Department, Philadelphia College of Pharmacy and Science), Dr. R. F. Orzechowski, Pharmacology Department, Philadelphia College of Pharmacy and Science), Dr. J. F. Morton (Morton Collectanea, University of Miami), Dr. J. T. Mullins (Department of Botany, University of Florida), Dr. A. E. Hagerman (Department of Chemistry, Miami University, Ohio), Dr. T. Okuda (Faculty of Pharmaceutical Sciences, Okayama University), Dr. R. P. Wunderlin and Dr. J. T. Romeo (Department of Biology, University of South Florida).

Literature Cited

1. Kapadia, G. J.; Rao, G. S.; Morton J. F. In *Carcinogens and Mutagens in the Environment: Naturally Occurring Compounds;* Strich, H. F., Ed.; CRC Press: Boca Raton, FL, **1983,** Vol. 3; pp 3-12.
2. Wang, Z. Y.; Khan, W.A.; Bickers, D.R.; Mukhtar, H. *Carcinogenesis.* **1989,** 10, 411-415.
3. *Taber's Cyclopedic Medical Dictionary;* Thomas, C. L., Ed.; 14th ed.; F.A. Davis Co.: Philadelphia, PA, **1981.**
4. Haslam, E. *Plant Polyphenols: Vegetable tannins revisited;* Chemistry and Pharmacology of Natural Products; Cambridge University Press: New York, NY, **1989;** pp 154-214.
5. Bate-Smith, E.C. *Phytochem.* **1973,** 12, 907-912.
6. Harborne, J.B. *Phytochemical Methods: A Guide to Modern Techniques of Plant Analysis;* 2nd ed.; Chapman and Hall: New York, NY, **1988,** pp 87-88.
7. Okuda, T.; Mori, K.: Hatano, T. *Chem. Pharm. Bull.* **1985,** 33, 1424-1433.
8. Makkar, H. *J. Agric. Food Chem.* **1989,** 37, 1197-1202.
9. Trease, G. E.; Evans, W. C. *Pharmacognosy;* 12th ed.; Bailliere Tindall: Philadelphia, PA, **1983,** pp 378-380.
10. Verzele, M.; Delahaye, P.; Van Damme, F. *J. Chromatogr.* **1986,** 362, 363-374.
11. Hagerman, A. E.; Butler, L. G. *J. Chem. Ecology.* **1989,** 15, 1795-1810.

12. *Official Methods of Analysis;* Williams, S. Ed.; 14th ed.; Association of Official Analytical Chemists: Arlington, VA, **1984,** pp 187-188.
13. Martin, J. S.; Martin, M. M. *Oecologia.* **1982,** 54, 205-211.
14. Bradford, M. M. *Anal. Biochem.* **1976,** 72, 248-254.
15. Green, J.S.; Jorgenson, J.W. *J. Chromatogr.* **1989,** 478, 63-70.
16. Schapira, J.; Shapira, D.; Shapira, K.*The Book of Coffee & Tea: A guide to the appreciation of fine coffees, teas and herbal beverages;* St. Martin's Press: New York, NY, **1975;** pp 143-240.
17. Robertson, A.; Bendall, D. S. *Phytochem.* **1983,** 22, 883-887.
18. Okuda, T.; Mori, K.; Hatano, T. *Phytochem.* **1980,** 19, 547-551.

RECEIVED December 17, 1991

SELECTED OCCURRENCE
OF PHENOLIC COMPOUNDS

Chapter 8

Manufacturing and Chemistry of Tea

Douglas A. Balentine

Thomas J. Lipton Company, Englewood Cliffs, NJ 07632

Tea is the most widely consumed beverage in the world. The tea plant, *Camellia sinensis*, is a very important agricultural and commercial product with a unique horticulture and manufacturing process. There are numerous polyphenolic compounds produced in growing tea shoots. The pathways for *in vivo* biosynthesis of these phytochemicals have been generally elucidated and are reviewed. The chemistry of green and black tea is reviewed and includes a summary of the polyphenolic composition of tea. HPLC techniques are now available for rapidly measuring a number of important tea polyphenols including the catechins, flavonols, flavonol glycosides, and the theaflavins. These methods are briefly reviewed.

Tea beverages are prepared from the leaves of *Camellia sinensis*. Second only to water, tea is the most widely consumed beverage in the world today with annual per capita consumption exceeding 40 liters *(1)*.

Tea has an ancient and intriguing history. Chinese mythology teaches that in the year 2737 BC the Emperor Shen Nung discovered tea drinking while on an outing to the countryside. The Emperor noted the fragrant smell of tea when leaves from branches used to fuel his fire fell into a kettle of boiling water. A brew of the leaves was prepared and the merits of tea drinking were found. The first true accounts of the use and horticulture of tea were written by the Chinese scholar, Kou P'o, in 350 AD. The Chinese have historically used tea as a beverage valued for its pleasant flavor and medicinal qualities. Tea was introduced in Japan in 600 AD, in Europe in 1610, and in the American colonies in 1650 *(2,3)*.

0097–6156/92/0506–0102$06.00/0
© 1992 American Chemical Society

Botony and Agriculture

The tea plant is indigenous to China and Southeast Asia. It has been classified as *Camellia sinensis* and has two common varieties, *sinensis* and *assamica*. The variety *sinensis* is a shrub-like plant growing to a height of 4-6 meters with leaves 5-12 cm in length. This variety is more commonly used in agriculture. The variety *assamica* grows to 12-15 meters in height and has leaves of 15-20 cm in length *(2)*.

Through cultivation, tea has become an important agricultural product throughout the world, particularly in regions lying close to the equator. Geographical areas having annual rainfall of 50 inches or more per year and an average temperature of 30°C are most favorable for tea growth and agriculture *(2)*.

Traditionally, *Camellia sinensis* has been propagated, hybridized and bred through seeds. To maintain genetic purity and more rapidly establish new productive stands of tea, vegetative propagation has become common. Leaf cuttings are taken and planted in nurseries where they develop into seedlings within six months. The seedlings are transplanted to the fields and new stands of tea are ready for harvesting in three years.

In the fields, tea is maintained as a shrub-like bush through frequent harvesting. The emerging shoots and young leaves of the plant, called the flush, are harvested every 8-12 days. The flush is rich in polyphenols and enzymes which are important compounds in tea manufacturing and its subsequent quality. Harvesting is generally done by hand. Mechanical harvesting methods have been developed, but are limited to regions where labor is expensive and where tea is not grown on steep mountain slopes. Mechanical harvesting is most common in Australia, Argentina, and the nation formerly known as the USSR. Immediately after harvest, tea leaves are brought to factories at large tea estates for manufacturing. It is the manufacturing process which determines the type of tea to be produced. There are three general types of tea produced: green (unfermented), oolong (partially fermented), and black (fully fermented). The manufacturing process is designed to either prevent or allow tea polyphenols to be oxidized by catalytic enzymes. The green tea manufacturing process involves rapid steaming or pan firing of freshly harvested leaves to inactive enzymes, preventing fermentation, followed by rolling and firing (drying) the leaves. When oolong and black teas are produced, fresh leaves are allowed to wither (rest) until the moisture content of the leaves is reduced to 55-65% of the leaf weight, causing the concentration of polyphenols in the leaves and deterioration of the leaf structure. The withered leaves are rolled and crushed which initiates fermentation of the tea polyphenols. Oolong tea is prepared by firing the leaves shortly after rolling in order to terminate the oxidation process and dry the leaves. Black tea is prepared from crushed, withered leaves through a separate fermentation process in which warm air is circulated through the rolled and crushed leaves. The fermentation process results in the oxidation of simple polyphenols to more complex, condensed polyphenols which give black tea its bright red color and brisk, astringent flavor. The fermented tea leaves are fired to inactivate the enzymes and dry the leaves.

Manufactured tea is sized, graded and evaluated for flavor and infusion color by professional tea tasters. Tea is packaged into sacks or wooden chests and is sold at the world tea auctions.

Total world production of tea in 1989 was 2.45 million metric tons. From this total, green tea accounted for 490 thousand metric tons and black tea accounted for 1.96 million metric tons. The general distribution of tea production by region is shown in Table I (1).

Table I. 1990 World Tea Production in Metric Tons (MM)

Geographic Region	Total Tea	Black Tea	Green Tea
Asia	2.001	1.530	0.471
Africa	0.323	0.323	0
South America	0.039	0.039	0
Oceanic	0.009	0.009	0
USSR	0.115	0.092	0.023
Total	2.497	2.003	0.494

Adapted from ref. 1.

Composition of Tea and Biosynthesis of Tea Polyphenols

The leaves of *Camellia sinensis* are similar to most plants in general morphology and contain all the standard enzymes and structures associated with plant cell growth and photosynthesis. Uniquely, tea contains relatively large quantites of polyphenols and methylxanthines. The general composition of fresh tea flush is presented in Table II (*Graham H., personal communication, 1991*). Organic acids important to tea chemistry include gallic acid, cinnamic acids, and their quinnic acid esters (Figure 1). The composition of tea leaves will vary somewhat as a function of genetic strain, climate, and growing conditions.

Green Tea Polyphenols

There are numerous types of polyphenols in green tea leaves, the most important of which are the flavan-3-ols (catechins) and the flavonols, and their glycosides. The flavan-3-ols (catechins) in green tea leaves are listed in Table III and their structures are shown in Figure 2 (4).

Table II. Compostion of Fresh Green Tea Leaf
% of Dry Weight

Polyphenols	39	Protein	15
Methyl xanthines	3.5	Lignin	6.5
Amino acids	4.0	Lipids	2
Organic acids	1.5	Chlorophyll	0.5
Carbohydrates	25	Ash	5

Graham, H., personal communication.

Table III. Green Tea Flavan-3-ols (Catechins)
% of Dry Weight

Catechin	0.1
Epicatechin	0.9
Epicatechin gallate	0.8
Gallocatechin	3.5
Epigallocatechin gallate	4.4
Epigallocatechin	3.9
Total	10.1

Gallocatechin data from ref. 4.
Determined by HPLC anlysis: Hypersil ODS C-18 column
(200 x 2.1 mm); Solvent : acetonitrile, methanol, 0.15%
formic acid and 0.3% triethylamine buffer run in gradient
at 0.35 ml/minute at 40°C.

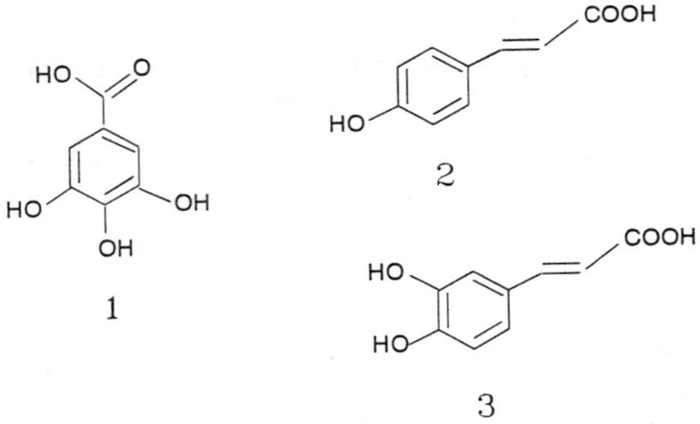

Figure 1. Organic Acids in Tea
1: Gallic acid
2: Coumaric acid
3: Caffeic acid

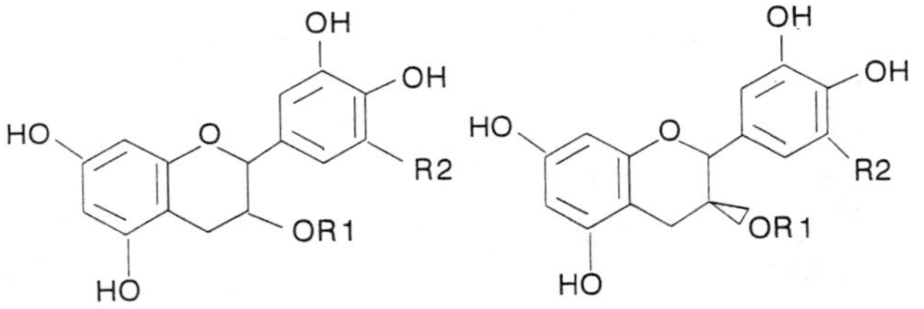

Figure 2. Green Tea Flavan-3-ols (Catechins)
Structure 1

Catechin	:R1, R2=H
Gallocatechin	:R1=H, R2=OH

Structure 2

Epicatechin	:R1, R2=H
Epigallocatecin	:R1=H, R2=OH
Epicatechin gallate	:R1=gallate, R2=H
Epigallocatechin gallate	:R1=gallate, R2=OH

These polyphenolic constituents are the key reactants invovled with the enzymatic fermentation of green tea to black tea. The flavonols and a number of flavonol glycosides in green tea leaves are listed in Table IV *(Collier, P. Unilever Research, Colworth House, personal communication) (5,6)*. The structures of the flavonol aglycones are shown in Figure 3.

Table IV. Flavonol and Flavonol Glycosides in Green and Black Tea (mg/g Dry Leaf)

Flavonol Group	Green Tea	Black Tea
Quercetin	0.40	-
Myricetin	0.34	-
Kaempferol	0.52	-
Rutin	1.58	1.32
Quercetin-G	1.00	0.76
Kaempferol-G	1.33	0.70
Quercetin-G2	3.17	-
Kaempferol-G1	2.30	-
Isoquercetrin	1.82	-
Kaempferol-G2	4.30	-
Kaempherol-G3	-	1.01

G = 3-glucosyl(1 -> 3)rhamnosyl(1 -> 6)galactoside
G1 = 3-rhamnoglucoside
G2 = 3-rhamnodiglucoside
G3 = 3-O-rutinoside
Adapted from ref. 7 and Collier, P., Unilever Research Colworth House, personal communication.

The pathways for the *de novo* biosynthesis of these green tea components have been generally elucidated and are outlined in Figures 4-7. The precursors to the polyphenols of *Camellia sinensis* are produced from carbohydrates through the shikimic acid pathway. This pathway produces the amino acid phenylalanine from glucose. Deamination of phenylalanine forms cinnamic acid, and hydroxylation of cinnamic acid forms coumaric acid. These two organic acids are key precursors of green tea polyphenols. Co-enzyme A esters of these organic acids react with 3 molecules of malonyl-CoA to form chalcones. Ring closure of chalcones results in formation of dihydroflavonols (Figure 5). The

R
OH
HO
O
R1
OH
HO
O

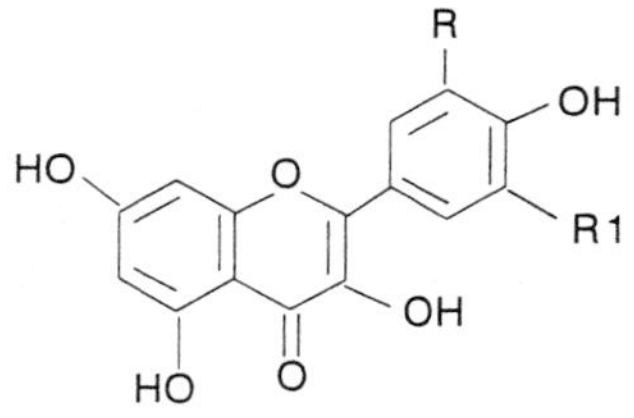

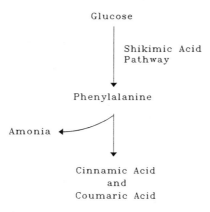

Figure 3. Green Tea Flavonols
Kaempferol :R1= R2=H
Quercetin :R1=OH, R2=H
Myricetin :R1, R2=OH

Glucose
Shikimic Acid
Pathway
Phenylalanine
Amonia
Cinnamic Acid
and
Coumaric Acid

Figure 4. Shikimic Acid Pathway

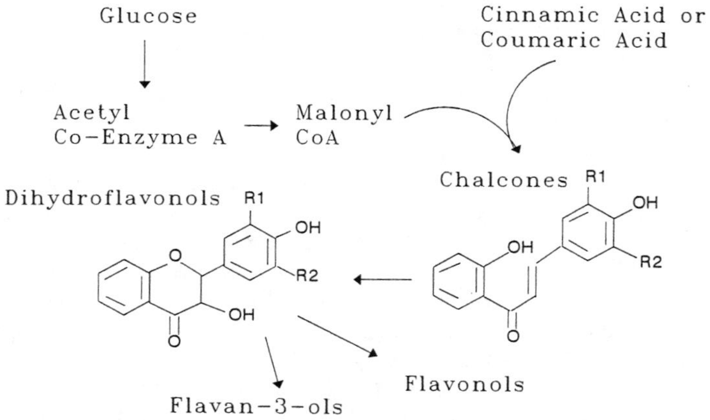

Glucose
Cinnamic Acid or
Coumaric Acid
Acetyl
Co-Enzyme A
Malonyl
CoA
Chalcones
Dihydroflavonols
R1
OH
R2
O
OH
O
OH
Flavonols
Flavan-3-ols

Figure 5. Dihydroflavonol Synthesis

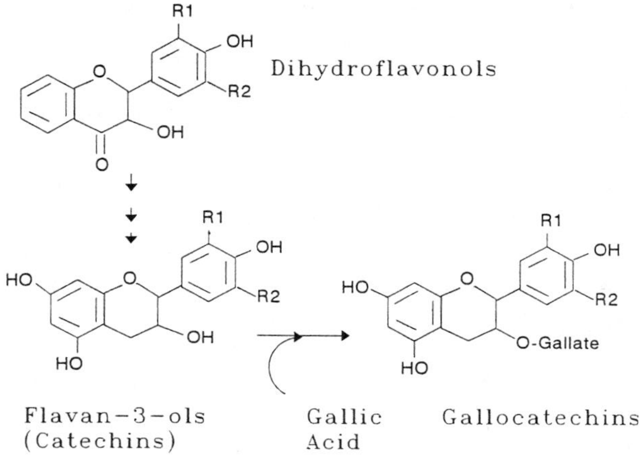

Figure 6. Flavanol Synthesis

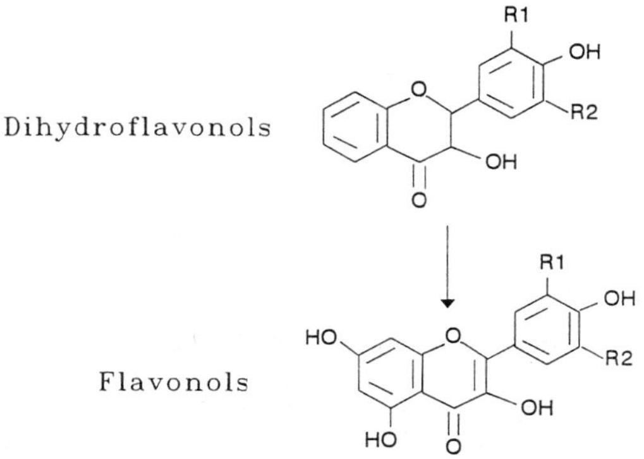

Figure 7. Flavonol Synthesis

dihydroflavonols react to form the flavonols through ketoreduction of ring B and hydration of ring A or flavan-3-ols through ring A hydration and ring B reduction *(7,8)*. The enzymes responsible for catalysis of these reactions are listed in Table V (7).

Table V. Enzymes Involved With Biosynthesis of Tea Polyphenols

Enzymes	E.C. Number
Acetyl-CoA carboxylase	6.4.1.2
Phenylalanine ammonia-lyase	4.3.1.5
Cinnamate 4-hydroxylase	1.12.12.11
4-Coumarate-CoA ligase	6.2.1.12
Chalcone synthase	2.3.1.74
Chalcone isomerase	5.5.1.6
2-Hydroxyisoflavanone synthase	
Flavone synthase	
(2S)-Flavanone 3-hydroxylase	1.14.11.9
Flavonol synthase	
Dihydroflavonol 4-reductase	
Flavan-3, 4-cis-diol 4-reductase	

Adapted from ref. 7.

The Chemistry of Tea Manufacturing

The manufacturing of oolong and black tea involves the enzymatic oxidation of green tea flavan-3-ols contained in the leaf cytoplasm to a number of condensed polyphenols: theaflavins, thearubigens, bisflavonols, and epitheaflavic acids. Polyphenol oxidase is the key enzyme responsible for the oxidation reaction. This enzyme is a metallo protein containing a copper co-factor with a molecular weight of 140,000 daltons. The enzyme is compartmentalized in the leaf microsomes and is released by the rolling and crushing of tea leaves after the withering process which allows the enzyme to react with the cytoplasmic catechins in green tea *(2-4)*.

Theaflavins are responsible for the bright red color of black tea. Theaflavins are formed from the green tea flavan-3-ols through the mechanism outlined in Figure 8. The reaction involves the oxidation of ring B of the catechin molecule

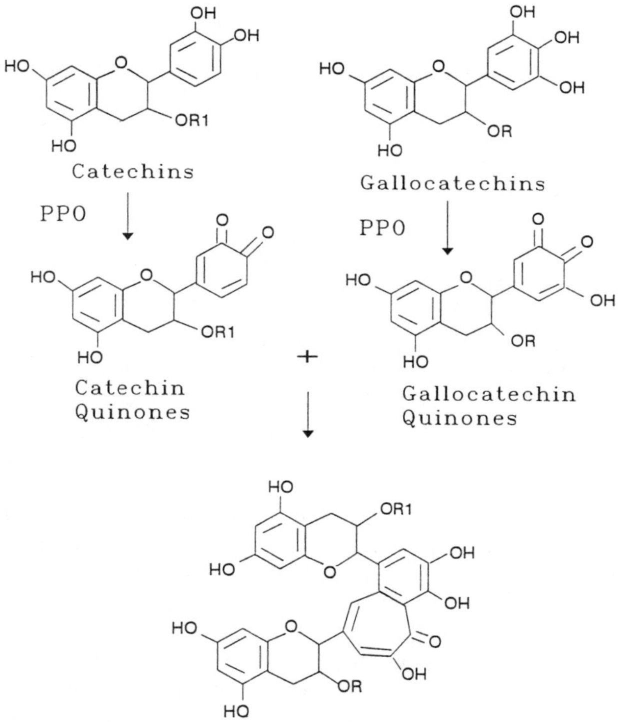

Figure 8. Formation of Theaflavins

Theaflavin	R=R1=H
Theaflavin gallate A	R=gallate and R1=H
Theaflavin gallate B	R=H and R1=gallate
Theaflavin digallate	R=R1=gallate

to the quinone by polyphenol oxidase. The quinone of a gallocatechin reacts with the quinone of a catechin through a Michael Addition followed by carbonyl addition across the ring and subsequent decaboxylation to form theaflavins. The theaflavin structure has a characteristic benzytropolone ring *(Collier, P., Confidential Unilever Report, 1970)*. Theaflavins can be readily determined by HPLC *(9)* and their content in black tea is presented in Table VI.

Table VI. Theaflavin Content of Black Tea Extract Solids

Component	mg/g Dry Solids
Theaflavin	6.8
Theaflavin gallate A	8.3
Theaflavin gallate B	2.5
Theaflavin digallate	0.7

Black Tea leaves (1 gram) were infused in 100 grams boiling water for 5 minutes and the leaves were filtered out. The infusion was directly analyzed for theaflavins by HPLC. The column, mobile phase, and conditions are described in the legend following Table III.

Epitheaflavic acids (Figure 9) are produced through the same basic mechanism as theaflavins from a reaction between a gallic acid quinone and the quinone of epicatechin or epicatechin gallate *(3)*. According to current knowledge, epitheaflavic acids are a minor component of black tea leaves. There are no reliable methods for the quantitation of these polyphenols.

Bisflavonols are another group of polyphenols which form during the manufacturing of black tea. These compounds are dimers of the catechins produced from quinones of flavan-3-ols by covalent linkage of ring B of the catechins (Figure 9)*(3)*. Currently, there are no useful methods for the quantitation of bisflavonols in tea.

The thearubigens are a group of complex polphenols which make up a large amount of the water soluble phenolic constituents of black tea and are the brown pigment in tea. The chemical nature of thearubigens has not been elucidated. They have been generally classified as polymeric proanthocyanidins *(3,4)*. Acid hydrolysis of thearubingens results in anthocyandins (Figure 10).

The general composition of black tea leaves is presented in Table VII *(Balentine, D. A., Thomas J. Lipton Company, unpublished data) (3,4)*. In comparison to green tea, black tea is low in catechins and has significant quantities of complex polyphenols. A comparison of the composition of beverages prepared from green and black tea is presented in Table VIII. Of significance, both green and black tea contain similar quantities of

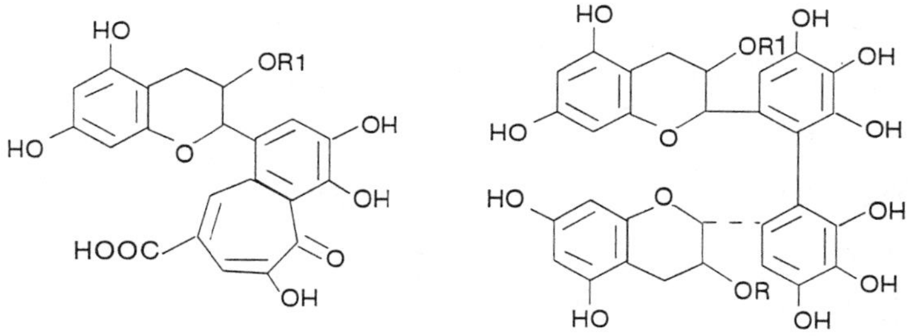

Figure 9. Epitheaflavic Acids and Bisflavanols

Epitheaflavic acid	R1=H	Bisflavanol A	R=R1=gallate
Epitheaflavic gallate	R1=gallate	Bisflavanol B	R=gallate and R1=H
		Bisflavanol C	R=H and R1=gallate

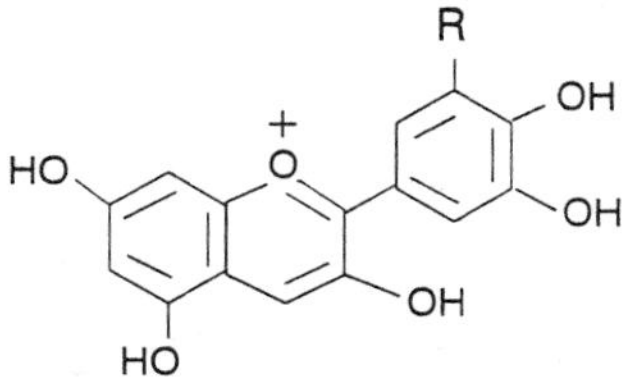

Figure 10. Anthocyanadins Derived From Thearubigens

methylxanthines (caffeine and theobromine). While green tea is an excellent source of simple polyphenols or catechins, black tea beverages contain complex polyphenols.

Table VII. Composition of Black Tea Leaves

Component	mg/g Dry Leaf	Component	mg/g Dry Leaf
Caffeine	4.0	Thearubigens	59.4
Catechin	2.3	Protein	150
Epicatechin	4.1	Fiber	300
Epicatechin gallate	8.0	Amino acids	40
Epigallocatechin	10.5	Ash	50
Epigallocatechin gallate	16.6	Carbohydrate	70
Flavonol glycosides	0.5		
Theaflavin	2.5		
Theaflavin gallate A	1.7		
Theaflavin gallate B	2.4		
Theaflavin digallate	2.5		

Adpated from refs. 2,3,4, and Balentine D. A., Thomas J. Lipton Company, unpublished data.

Analysis of Tea Polyphenols

Early work on tea chemistry and the analysis of polyphenols in tea empolyed thin layer chromatography techniques. Recently, HPLC methods have been developed for the analysis of both the simple and complex polyphenols in *Camellia sinensis* and tea beverages. The HPLC methods generally employ reverse phase C-18 chromatography columns using acetonitrile and acetic acid buffers as the mobile phase (10,11). The use of a PRP (Poly [styrene-divinylbenzene]) column with acetonitrile as the mobile phase has recently been reported for analysis of black tea polyphenols (10). The difficulty in obtaining purified compounds to use as standards has hindered developments in the analysis and quantitation of polyphenolic constituents in tea. The availability of HPLC/MS technology and capillary electrophoresis techniques should pave the way for development of improved analytical methods and for isolation of standards allowing for detailed analysis and quantitation of the polyphenols in tea. An HPLC chromatograph of black tea is shown in Figure 11. The use of a

Table VIII.   Composition of Green and Black Tea Beverages Percentage of Dry Extract Solids

Component	Green Tea	Black Tea
Catechins	34.0	4.2
Theaflavins	-	1.8
Thearubigens	-	17.0
Flavonols	0.4	-
Flavonol glycosides	4.4	1.4
Protein	7.6	10.7
Amino acids	5.3	4.8
Caffeine	6.9	7.1
Carbohydrates	12.5	13.5
Organic Acids	9.5	11.0

Adapted from ref. 2,3,4, Graham H., personal communication and Balentine D. A., Thomas J. Lipton Company, unpublished data.

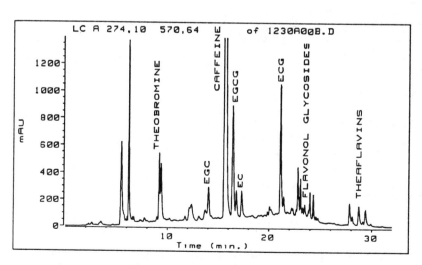

Figure 11. Black Tea HPLC Chromatogram at 274 NM
HPLC Analysis: Hypersil ODS C-18 column 200 mm x 2.1 mm.
Solvent: acetonitrile, methanol, and 0.15% formic acid and
triethylamine buffer run in gradient at 0.35 ml/minute at 40°C.

diode array detector to confirm the characteristic spectra of the theaflavins during HPLC analysis was employed and the UV spectra of these compounds are shown in Figure 12.

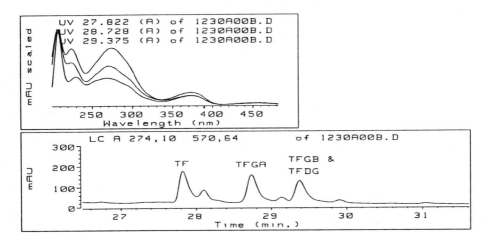

Figure 12. Black Tea Theaflavin UV Spectra and HPLC Chromatogram

TF   :Theaflavin
TFGA   :Theaflavin gallate A
TFGB   :Theaflavin gallate B
TFDG   :Theaflavin digallate

Acknowledgments

The author wishes to recognize Wendy Balentine and Josephine Toic for their assistance with proof reading and editing of the manuscript and Denise Marsh for her support of our computer system.

Literature Cited

1.   International Tea Committee Ltd. *Annual Bulletin of Statistics 1990*. London.
2.   Graham, H. In *Kirk-Othmer Encyclopedia of Chemical Technology*; Wiley: New York, N.Y., **1983**, 3rd Ed. 22, 628-644.
3.   Graham, H. In *The Methylxanthine Beverages and Foods; Chemistry, Consumption, and Health Effects*; Alan R. Liss Inc: New York, N.Y., **1984**, 29-74.

4.   Sanderson, G.W. In *Structural and Functional Aspects of Phytochemistry*; Academic Press Inc.: New York, **1972**, 247-316.
5.   Yamaguchi, Y.; Hayashi, M.; Yamazoe, H.; Kunitomo, M., *Fol Pharm J.* **1991**, *97*, 329-337.
6.   Finger, A.; Engelhardt, U.H.; Wray, V., *Phytochem.* **1991**, *30*, 2057-2060.
7.   Heller, W.; Forkmann, G. In *The Flavonoids*; Editor, J.B. Harborne; Advances in Research Since 1980; Chapman and Hall: London, **1988**, 399-422.
8.   Goodwin, T.W.; Mercer, E.I. *Introduction to Plant Biochemistry*; Pergamon Press: Elmsford, N.Y.,**1983**, 528-566.
9.   Steinhaus, B.; Englehardt, U.H. *Z. Lebensm. Unters. Forsch.* **1989**, *188*, 509-511.
10. Bailey, R.G.; Nursten, H.E. *J. of Chromatography.* **1991**, *542*, 115-128.
11. Bailey, R.G.; McDowell, Il; Nursten, H.E. *J. Sci. Food Agric.* **1990**, *52*, 509-525.

RECEIVED February 18, 1992

Chapter 9

Phenolic Compounds in Spices

Carolyn Fisher

Kalsec, Inc., P.O. Box 511, Kalamazoo, MI 49005

A general overview of the phenolic compounds in spices is presented, along with information about their chemistry, their attributes for the food industry (flavor, color, antioxidants), as well as what is known about their biological activities. Among the compounds and spices mentioned are capsaicinoids from chillies, curcuminoids from turmeric, piperine from black pepper, gingerols from ginger, eugenol from cloves, coumarin from cinnamon, myristicin and safrole from nutmeg and mace, and carnosol and rosmarinic acid from rosemary and sage.

Spices are an important part of the human diet. Not only do they enhance the taste and flavor of foods, spices also increase their shelf-life by being both antimicrobial and antioxidant. Spices also exhibit a wide range of physiological and pharmacological properties. The present paper reports the taste and flavor of these spice phenols and their uses in the food industry.

Pungent Principles

Capsicum (Red Pepper). Chillie peppers are used throughout the world and give us many dishes which are indigenous to certain regions such as India, tropical America and Northern China. Chillies are popular in countries to enhance the bland flavor of the staple food, such as rice in India and China, and corn & beans in Mexico. Dishes from many of these regions are catching on throughout the world.

The compounds responsible for the well-known hot sensation from chillie peppers are the family of phenolic amides known as the capsaicinoids (1) (Figure 1). They are vanillylamides of fatty acids of varying lengths. These interesting compounds consist of a phenol, an amide linkage, and a long chain fatty acid. All these functionalities are necessary for their physiological activity (2). Having a polar end and non-polar end, when applied to the mouth or skin, they dissolve through the natural skin oils, reacting with nerve endings to give a warming, burning sensation in the area of contact.

Large or frequent doses of capsaicinoids reduce the sensitivity of an individual to subsequent usage. Also thermoregulation is affected; pyrexia, the abnormal elevation of body temperature, occurs upon large or frequent doses of capsaicinoids (*2*). Capsaicinoids are the active ingredient in linaments to relieve local muscle or joint pains (e.g. Heet, Omega Oil, Sloan's, Stimurub). Capsaicinoids are also used therapeutically in the treatment of cutaneous disorders: psorisis, postherpetic neuralgia, postmastectomy pain syndrome, painful diabetic neuropathy and vulvar vestibulitis (*3*).

The quality of commercial extracts is currently controlled by the organoleptic Scoville method (*4,5*). In the past fifteen years, many HPLC and GC methods have been developed, (*6-13*) but none have yet been adopted by the Association of Official Analytical Chemists (*14*). In our laboratory, we have developed a more practical HPLC method utilizing an internal standard that can give good quantitation of the capsaicinoids. Using the compound, 4,5-dimethoxybenzyl-4-methyloctamide (DMBMO) (Figure 2), which is a non-pungent synthetic capsaicinoid as the internal standard, we have improved the HPLC assay (Figure 3). (T. Cooper, *J. Agric. Food Chem., submitted.*)

The capsaicinoids are quite stable to heat and light. Typical Capsicum fruits contain 0.01 % to 1 % capsaicinoids. Typical commercial extracts (Oleoresin Capsicum) contain 3 % to 6 % capsaicinoids. For different varieties of Capsicums, the amount of different capsaicinoids varies a lot (*15*). However, Capsicum Annum Annum is usually the variety grown commercially with the major capsaicinoids being in the following approximate ratio: 1:4:3 nordihydrocapsaicin : capsaicin : dihydrocapsaicin.

Black Pepper. The confusion between red and black pepper began when Columbus discovered America. He was looking for the spices of the Orient, but instead, he found a pungent fruit of a small annual shrub (Capsicum) rather than the white (ripe and fermented) or black (immature) berries of a vine (Piper nigrum, L.). Black pepper is used to add zest to many dishes. From eggs to steak, we enjoy the flavor and bite of this spice.

Piperine (*16*), which gives the 'bite' that is enjoyed in black pepper, has a structure somewhat different from the capsaicinoids (Figure 4). It is not strictly a phenol, but instead contains a methylene-dioxy group attached to the aromatic ring. Perhaps because of this, it is much less pungent than the capsaicinoids, but gives a similar type of warming sensation. The other piperinoids (chavicine, isochavicine and isopiperine) are minor components of black pepper extracts and impart little taste (*17*).

Piperine is also an effective anticonvulsant drug that antagonizes convulsions induced by both physical and chemical means. During the testing of piperine in the modulation of electroshock seizures, many derivatives of piperine were synthesized and studied (*18*). Six derivatives were found to have marked anticonvulsant effects. Since 1975, antiepilepsirine has been used clinically to effectively treat epilepsy (*19*).

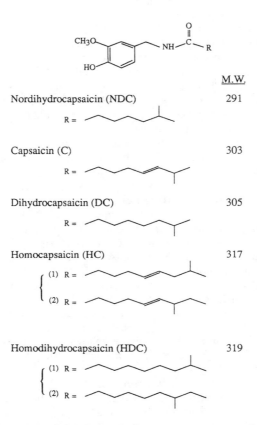

Figure 1.   Capsaicinoids: The pungent principles of Capsicum (Red Pepper).

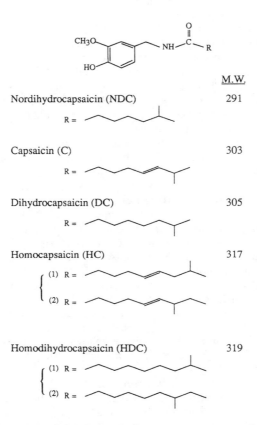

Figure 2.   Dimethoxybenzylmethyloctamide.   Internal standard for HPLC assay for capsacinoids.

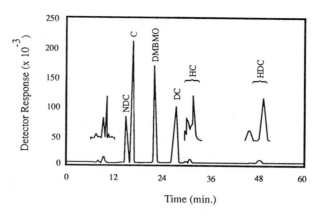

Figure 3.   Chromatogram of capsaicinoids with internal standard (DMBMO). Mobile phase: 60 % methanol/40 % water pH = 3.00, 1 % citric acid at 1.5 ml/min. Stationary phase: Supelcosil C-18 25 cm x 0.46 cm, 5 μ particle size.

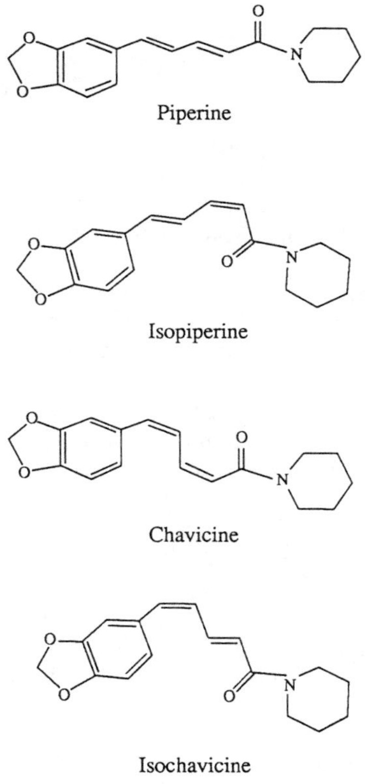

Piperine

Isopiperine

Chavicine

Isochavicine

Figure 4.   Piperinoids: Phenols found in Black Pepper.

Piperine has also been found to improve and enhance the insectidal effects of pyrethrin, allethrin, and eucalyptus oil (*20,21*). Although quite stable to heat, in the presence of light, piperine isomerizes one or more of the trans double bonds to the cis isomers (mainly isochavicine and isopiperine) which decreases the organoleptic bite. Typical commercial black pepper contains 2 % to 6 % piperine. Typical commercial extracts (oleoresin black pepper) contain 30 % to 40 % piperine.

Ginger. A different pungent spice, the ginger root is used both fresh and processed to deliver a delightful flavor to both foods & drink.

Two families of phenols with similar structures are the gingerols and shogaols which are responsible for the pungency of ginger (*22*) (Figure 5). In fresh ginger, gingerols are the prominent species, while in commercial oleoresins, the shogaols are the major pungent phenols (*23*). This is the reason for one of the major flavor differences between fresh ginger and processed ginger. During drying and processing, the thermally unstable gingerols either dehydrate to the corresponding shogaols or degrade by a retro-aldol condensation to zingerone and the corresponding aldehyde. The shogaols are also somewhat unstable, and can undergo pyrolytic breakdown to zingerone and the corresponding aldehydes (*24*) (Figure 6). Because of this instability, HPLC is the method of choice for the analysis of these compounds (*25,26*).

Apart from shogaols and zingerone, found in small amounts in fresh ginger and larger amounts in the derived stored products, other related components have been detected in ginger. The paradols, which are also pungent (found in larger quantity in a related Zingiberaceae, <u>Ammonomum maleguета</u>, <u>Roscoe</u>) (*27*) lack the beta-hydroxy ketone group found in gingerol and are relatively stable. Hexahydrocurcumin and dihydrogingerol have been isolated from fractions more polar than gingerol (*28*). Also found are the methyl gingerols, methyl shogaols, demethoxy-hexahydrocurcumin and their shogaol analogues (*29*).

The threshold pungencies of the compounds [6]-gingerol and [6]-shogaol are comparable in strength to piperine, which is about 100 times less pungent than the capsaicinoids (*24,30*). Their estimated threshold values are shown in Table I. It was also found that the higher homologs, [8]- and [10]- gingerols and [8]- and [10]- shogaols, have a lower pungency response which follows the same pattern as found among the homologs of capsaicin (*31*) and paradol (*32*).

These pungent compounds can be converted to more bioactive substances in vivo by metabolic reaction which proceeds chiefly in the presence of cytochrome P-450's. The P-450 enzymes oxidatively demethylate o-methyl-phenols to generate bioactive catechols. For example, [6]-gingerol is oxidatively demethylated to [6]-norgingerol, which has a high inhibitory potency for 5-lipoxygenase because it binds to the non-heme (*33*).

Typical dried ginger contains 1 % to 4 % pungent constituents. Typical commercial extracts (oleoresin ginger) contain 10% to 30% pungent compounds.

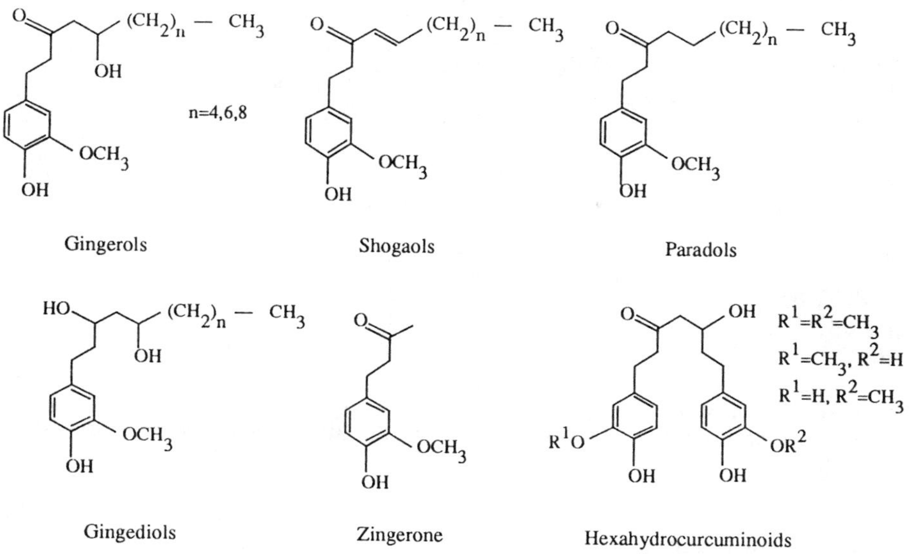

Figure 5.   Phenols found in Ginger.

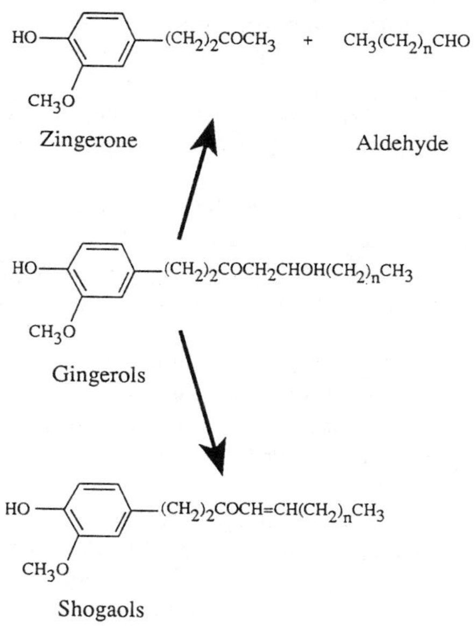

Figure 6.   Degradation reactions of gingerols.

TABLE I

PUNGENT PRINCIPLES OF CAPSICUM, BLACK PEPPER & GINGER

Threshold Pungency Values

Compound	Pungency (SHU[a])	Threshold (ppm)
CAPSAICIN	16,000,000[6]	0.06
PIPERINE	100,000[22]	10
[6]-GINGEROL	60,000[22]	17[30]
[6]-SHOGAOL	160,000[22]	7[30]

[a]Scoville Heat Units (SHU) are defined as the reciprocal of the highest dilution (threshold) at which a panel would definitely recognize the pungency sensation.

Non-Pungent Principles

Cloves, Cinnamon, Nutmeg, Mace & Vanilla. Different from the very pungent spices above, cinnamon comes from the bark of a tree, cloves from the bud of a flower, nutmeg from the kernel of a seed, mace from the aril surrounding the nutmeg seed & vanilla from a fermented fruit, the vanilla bean. They are used in most of our sweet foods to accent their flavor.

Unlike the pungent phenols of low volatility above, these phenolic flavorants (Figure 7) are volatile and found in the essential oils. They help give the characteristic aromas and flavors to cinnamon, cloves, nutmeg, mace and vanilla.

The major flavor compound in cinnamon is cinnamaldehyde, which is not a phenol. However, a minor flavor compound, coumarin, contains a masked phenol and is necessary to give natural cinnamon it's typical flavor. Eugenol, a major flavor component of cloves is also found in cinnamon in minor quantities.

Coumarin is also found in cloves and nutmeg. The major phenolic flavor compounds in cloves are eugenol and the demethoxy analogue, chavicol. Eugenol is a well known topical analgesic usually used to alleviate the pain of toothaches. It is the active ingredient and listed on the label of many over-the-counter toothache drops.

The major flavor compounds in nutmeg and mace are myristicin and safrole. These are masked phenols; they have the same methylene-dioxy group as piperine discussed above. Safrole is also a minor component of black pepper. Myristicin is one of the compounds in nutmeg responsible for its narcotic and psychomimetic effect (similar to alcoholic intoxication) (34). However, the dosage ordinarily involved in narcotic use is 50-100 times that

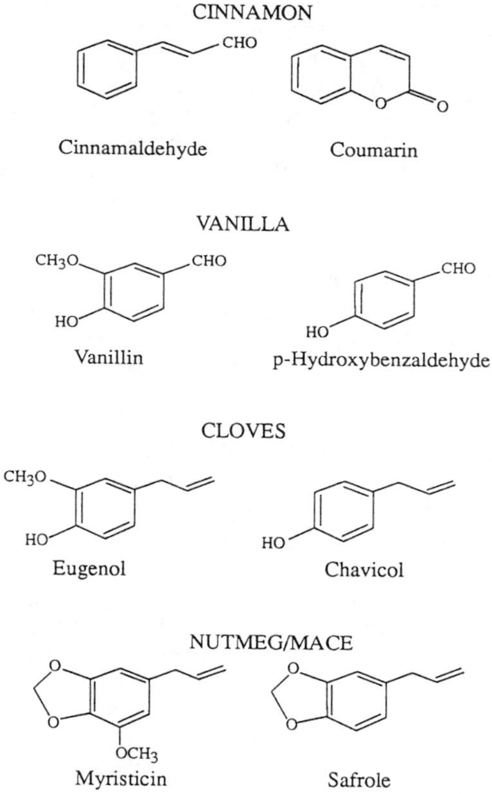

Figure 7.   Volatile phenols found in Cinnamon, Cloves, Nutmeg, Mace, & Vanilla.

encountered in any conceivable flavor use and the narcotic usage is not repeated because of unpleasant side-effects: headache, cramps & nausea (35).

There is concern about coumarin (36,37) and safrole (38) as they cause liver damage in rats, and the concentration in foods in Europe will be limited to 2 ppm and 1 ppm, respectively, in finished foods (39). Metabolism studies (40) in rat and humans show coumarin to be degraded to different metabolites, 7-hydroxycoumarin (man) and o-hydroxyphenylacetic acid (rat). Only o-hydroxyphenylacetic acid is a potent inhibitor of glucose-6-phosphatase in the liver microsomes. It appears that rats are not an appropriate model for toxicity in man. Recently, coumarin and its analogues have been tested as an anticarcinogen to the more potent carcinogenic chemicals (41).

Also found in cloves is vanillin, the major flavorant of vanilla extracts. Vanilla also contains the demethoxy analogue, para-hydroxybenzaldehyde.

Eugenol is found at 0.04% to 0.2% in ground cinnamon, at 1% to 20% in ground cloves, at 2% to 6% in oleoresin cinnamon and at 60 % to 90 % in oleoresin cloves and 70-90% in cinnamon leaf oil. Myristicin is found at 1% to 3% in ground nutmeg, at 2% to 6% in oleoresin nutmeg.

Turmeric. Turmeric is a yellow root similar to the ginger root discussed earlier. Ground turmeric is well know in curry dishes and prepared mustard. After extraction and purification to remove the bitter oils, the yellow pigments are also used to color such diverse items as yellow cake mixes, cheese, pickles and even fruit drinks.

The yellow curcuminoids (for structures, see Tonnesen's chapter) are a unique group of pigments, being phenolic as well as a good metal chelator. An industrial use of curcumin is as a metal chelator in the spectrophotometric determination of boron ions (42). They are fairly stable as powders and in acidic media; but under alkaline conditions (43,44) or in the presence of light (45), the curcuminoids degrade easily.

The separation of curcumin, demethoxycurcumin and bisdemethoxy-curcumin by HPLC is plagued by tailing on C-18 (46) and poor reproducibility on amino columns (47). We have obtained better peak shape on C-18 columns when citric acid is used to buffer the mobile phase (T. Cooper, *to be published*).

Typically ground turmeric rhizomes contain 3 % to 8 % curcuminoids. Typical commercial extracts (oleoresin turmeric) contain 30 % to 40 % curcuminoids, as crude extract. Many forms of extract are available with carriers to ease usage.

The curcuminoids are currently being tested for anticarcinogenic activity in mouse skin (48). They have also been found to be antimutagenic in bacteria to 7,12-dimethylbenz(a)anthracene (DMBA); of the homologs, bisdemethoxycurcurmin has the greatest effect (49).

Rosemary & Sage. The parts of the rosemary & sage plants used as a spice are the dried leaves. These dried leaves are used in many poultry, lamb, veal and shellfish dishes. The delicate flavors are appreciated in sausages and salads as well as soups and breadings. The phenolic compounds in rosemary and sage are utilized for their antioxidant activity in food products. Carnosic acid, carnosol, rosmanol, and rosmaridiphenol have the same structural backbone, whereas rosmarinic acid is quite different (for structures, see Nakatani's chapter). They are all vicinal diphenols, except for rosmariquinone, which is simply the oxidized ortho-diphenol.

We have obtained a crystal structure for carnosol (Figure 8). However, the experiment was indeterminate with respect to the absolute configuration of this optically active crystal. (M. J. Heeg, *Acta Cryst.*, submitted).

Phenolic antioxidants from rosemary are nonvolatile compared with the commercial antioxidants of BHA & BHT and provide better carry-through in frying operations such as in the production of French Fries (*50*).

The amount of phenolic antioxidants is extremely varied in most commercial extracts (such as oleoresin rosemary) because only the flavor strength is standardized. However, recently new products (such as Herbalox Seasoning) have been marketed that have standardized antioxidant activity and flavoring levels.

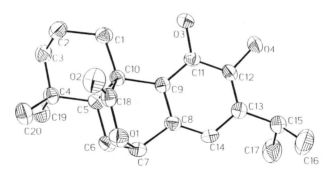

Figure 8.   Crystal structure of Carnosol.

Conclusion

In conclusion, the phenolics in a wide variety of spices have many similarities. One can see that the vicinal methoxyphenolic group is found in capsicum, ginger, cloves, vanilla, and turmeric. The demethoxy analogues of this group form the monophenols in spices. The methylene-dioxy group is found in black pepper, nutmeg and mace. Vicinal diphenols are found in the labiatae family (rosemary, sage, etc.) and are being commercially exploited as natural antioxidants.

Literature Cited

1.   Suzuki, T.; Iwai, K. In *The Alkaloids*; Brossi, A., Ed.; Academic Press: New York, NY, 1984, Vol. *23*, pp 227.

2.   Szolcsznyi, J. *Exp. Pharmacol.* 1982, *60*, 437.
3.   Carter, R.B. *Drug Develop. Res.* 1991, *22*, 109.
4.   Scoville, N.L. *J. Am. Pharm. Assoc.* 1912, *1*, 453.
5.   Govidarajan, V.S.; Narasimhan, S.; Dhanaraj, S. *J. Food Sci. Technol.* 1977, *14*, 28.
6.   Todd, P.H. Jr.; Bensinger, M.G.; Biftu, T. *J. Food Sci.* 1977, *42*, 660.
7.   Krajewska, A.M.; Powers, J.J. *J. Assoc. Off. Anal. Chem.* 1977, *70*, 926.
8.   Iwai, K.; Suzuki, T.; Jujiwake, H. *J. Chromatogr.* 1979, *172*, 303.
9.   Hoffman, P.G; Lego, M.C.; Galetto, W.G. *J. Agric. Food Chem.* 1983, *31*, 1326.
10. Law, M.W. *J. Assoc. Off. Anal. Chem.* 1983, *66*, 1304.
11. Weaver, K.M.; Luker, R.G.; Neale, M.E. *J. Chromatogr.* 1984, *301*, 288.
12. Chiang, G.H. *J. Food Sci.* 1986, *51*, 499.
13. Weaver, K.M.; Awde, D.B. *J. Chromatogr.* 1986, *367*, 438.
14. *AOAC Official Methods of Analysis*; Helrich, K., Ed.; Association of Official Analytical Chemists: Arlington, VA, 1990, 15th Edition.
15. Jurenitsch, J.; Kubelka, W.; Jentsch, K. *Planta Medica* 1960, *35*, 174.
16. Sumathikutty, M.A. *Lebensm. Wiss. u. Technol.* 1981, *14*, 225.
17. Verzele, M.; Mussche, P.; Qureshi, S.A. *J. Chromatogr.* 1979, *172*, 493.
18. Zhang, X; Li, R.; Cai, M. *J. Beijing Med. College* 1980, *12*, 83; CA 93:142676u.
19. Pei, Y.Q. *Epilepsia* 1983, *24*, 177.
20. Harville, E. *U.S. Patent 2,425,530*; 1947.
21. Inchcape, Chemco. Ltd. *Ger. Offen. 2,413,756*; 1974.
22. Govindarajan, V.S. *CRC Crit. Rev. Food Sci. Nutri.* 1982, *17*, 1.
23. Connell, D.W.; Sutherland M.D. *Aust. J. Chem.* 1968, *22*, 1033.
24. Govidarajan, V.S. *CRC Crit. Rev. Food Sci. Nutri.* 1982, *17*, 189.
25. Chen, C.C.; Kuo, M.C.; Ho C.T. *J. Food Sci.* 1986, *51*, 1364.
26. Che, C.C.; Rosen, R.T.; Ho, C.T. *J. Chromatogr.* 1986, *360*, 175.
27. Connell, D.W. *Aust. J. Chem.* 1970, *24*, 369.
28. Murata, T.; Shinohara, M.; Miyamoto, M. *Chem. Bharm. Bull Japan* 1972, *20*, 2291.
29. Harvey, D.J. *J. Chromatogr.* 1981, *212*, 75.
30. Narasimham, S.; Govindarajan, V.S. *J. Food Technol.* 1978, *13*, 31.
31. Nelson, E.K. *J. Amer. Chem. Soc.* 1919, *41*, 2121.
32. Locksley, H.D.; Rainey, D.K.; Rohan, T.A. *J. Chem. Soc. Perkin Trans I* 1972, *23*, 3001.
33. Aihara, K.; Higuchi, T; Hirobe, M. *Chem. Pharm. Bull.* 1990, *38(3)*, 842.
34. Truitt, E.B.; Callaway, E.; Braude, M.C.; Krantz, J.C. *J. Neuropsychiatr.* 1961, *2(4)*, 205.
35. Hall, R.L.; In *Toxicants Occurring Naturally in Foods*; Committee on Food Protection, Food & Nutrition Board, NRC, Ed.; National Academy of Sciences: Washington, D.C., 1973, 2nd Edition; pp 448-463.
36. Hazelton, L.W.; Tusing, T.W.; Zeitlin, B.R.; Thiessen, R.; Murer, H.K. *J. Pharmacol. Exp. Ther.* 1956, *118*, 348.
37. Sporn, A., *Igiena* 1960, *9*, 121.
38. Long, E.L.; Nelson, A.A.; Fitzhugh, O.G.; Hansen, W.H. *Arch. Pathol.* 1963, *75*, 595.

39. *Eurofood Monitor; European Community Legislation on Foodstuffs;* Amaducci, S., Ed.; Agra Europe: London, **1990**, Vol. 1, pp E-418.
40. Shilling, W.H. *Nature (London),* **1969**, *221,* 664.
41. Harvey, R.G.; Cortez, C.; Ananthanarayan, T.P.; Schmolka, S. *J. Org. Chem.* **1988**, *53,* 3936.
42. *AOAC Official Methods of Analysis;* Williams, S., Ed.; Association of Official Analytical Chemists: Arlington, VA, **1984**, 14th Edition, 20.057.
43. Tonnesen, H.H.; Karlsen, J. *Z. Lebensm. Unters. Forsch.* **1985**, *180,* 132.
44. Tonnesen, H.H.; Karlsen, J. *Z. Lebensm. Unters. Forsch* **1985**, *180,* 402.
45. Govidarajan, V.S. *CRC Critical Reviews in Food Sci. & Nutri.,* **1980**, *12,* 3.
46. Smith, R.; Witowska, B. *Analyst* **1984**, *109,* 259.
47. Tonnesen, H.H.; Karlsen, J. *Z. Lebensm. Unters. Forsch* **1986**, *182,* 215.
48. Huang, M-T.; Smart, R.C.; Wong, C-Q.; Conney, A.H. *Cancer Res.,* **1988**, *48,* 5941.
49. Mulky, N.; Amonhar, A.J.; Bhide S.V. *Indian Drugs,* **1987**, *25,* 3, 91.
50. Gray, J.I.; Crackel, R.L.; Cook, R.J.; Gastel, A.L.; Evans, R.J.; Buckley, D.S. Mich. State U., Presentation at *The Third International Conference on Ingredients & Additives,* November **1988**.

RECEIVED November 7, 1991

Chapter 10

Phenolic Compounds of *Brassica* Oilseeds

Fereidoon Shahidi

Department of Biochemistry, Memorial University of Newfoundland,
St. John's, Newfoundland A1B 3X9, Canada

Phenolic compounds of *Brassica* oilseeds occur as phenolic acids and polyphenolic tannins. Furthermore, phenolic acids may be present in the free and bound forms as well as esters and glycosides. Sinapic acid was the major phenolic acid present in rapeseed, canola and mustard seeds and was present mainly in the esterified form. Phenolic acids and tannins of *Brassica* oilseeds are responsible for the dark color and astringency of commercially-processed seed meals. Novel processing methods such as alcoholic ammoniation produced meals with a low content of phenolics which were also light in color and bland in taste.

Amongst *Brassica* oilseeds, canola/rapeseed crops are most important and rank as the third source of vegetable oilseed crops in the world (*1*). Rapeseed is used for the production of a high quality oil and a feed-grade meal. The meal has a high protein content and a reasonably well-balanced amino acid composition, thus possessing a favorable protein efficiency ratio (*2*). The content of phenolic compounds in rapeseed flour is much higher than those in other oilseeds by at least an order of magnitude (Table I).

Presence of phenolic compounds in oilseed proteins may result in the production of a dark colored meal upon commercial processing and is also

Table I.   Content of phenolics in some oilseed flours [a]

Oilseed Flour	Total Phenolics, mg/100g
Soybean	23.4
Cotton	56.7
Peanut	63.6
Canola/Rapeseed	639.9

[a]Adapted from Ref. 14.

0097–6156/92/0506–0130$06.00/0
© 1992 American Chemical Society

responsible for the bitter taste and astringency of some products. In addition, phenolic compounds, as such or in their oxidized forms, may interact with essential amino acids and other nutrients in the meal and render them unavailable for utilization (3). Nonetheless, phenolic compounds of oilseeds often possess strong antioxidant properties and when used, together with lipid-containing ingredients of formulated foods and feeds, may exert a positive effect in retarding lipid oxidation and quality deterioration (4).

Phenolic compounds of rapeseed/canola occur as phenolic acids, tannins, flavonoids and lignins. Phenolic acids of rapeseed/canola are hydroxylated derivatives of benzoic, cinnamic and coumaric acids (Figure 1) and may occur in the free, esterified and insoluble-bound forms. Sinapine, the choline ester of sinapic acid is the most important phenolic found in canola and rapeseed. Tzagoloff (5) provided evidence that sinapine in germinating mustard seedlings is hydrolyzed to choline and sinapic acid.   While choline is an important metabolite in the methylation cycle (6), sinapic acid is required for the biosynthesis of lignin and flavonoids (7). Tannins in *Brassica* oilseeds generally occur as polymers of flavan-3-ols or flavan-3,4-diols (8) (Figure 2). These compounds are the most widely distributed group of secondary plant materials ingested with food and are known to exhibit antioxidant properties.

Phenolic Acids of Canola/Rapeseed

The total content of phenolic acids in various cultivars of canola/rapeseed ranges from 1325 to 1807 mg/100 g defatted meal, on a dry basis, while corresponding values for canola/rapeseed flour range from 624 to 1281 mg/100 g (9-13). Esterified phenolic acids account for 80-95% and the free phenolic acids constitute about 3-16% of the total content of phenolic acids in canola/rapeseed and mustard products. Insoluble-bound phenolics constitute a minor portion of the total phenolics of cruciferae oilseeds.

Free Phenolic Acids

Free phenolic acids in rapeseed flour represent a fraction (9-11%) of the total phenolic acids (9) and about 15% of those present in canola meal (13). In white mustard, free phenolic acids account for only 3% of the total phenolics (11). However, according to Krygier *et al.* (9,10) and Kozlowska *et al.* (11) only 5% of free phenolic acids was present in the Yellow Sarson rapeseed variety. Free phenolic acids make a major contribution to the taste of *Brassica* meals and flours (14).

Sinapic acid was the predominant free phenolic acid present in rapeseed (9-11,15-17). It comprised up to 91% of the phenolic acids of this fraction. In white mustard, p-hydroxybenzoic acid was more predominant (41%) than sinapic acid (20%) in the free phenolic acids fraction (11). Minor phenolic acids found in this fraction included vanillic, gentisic, protocatechuic, syringic, p-coumaric, cis- and trans-ferulic, p-hydroxybenzoic and caffeic acids. Cholorogenic acid was found only in trace quantities in rapeseed flour (10). The content of total phenolics in the free phenolic acids fraction and their contribution to the overall amount in flour and meals of selected *Brassica* oilseeds are given in Table II.

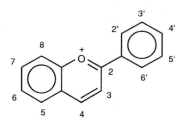

Benzoic acid

p-Hydroxybenzoic acid : 4-OH

Vanillic acid : 4-OH, 3-OCH$_3$

Syringic acid : 4-OH, 3,5-diOCH$_3$

Gallic acid : 3,4,5-triOH

Cinnamic acid

Coumaric acid : 4-OH

Caffeic acid : 3,4-diOH

Ferulic acid : 4-OH, 3-OCH$_3$

Sinapic acid : 4-OH, 3,5-diOCH$_3$

Chlorogenic acid : 3,4-diOH, ester with quinic acid

Figure 1. Structures of basic phenolic acids of *Brassica* oilseeds.

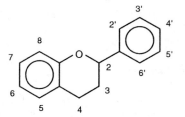

Pelargonidin : 3,5,7,4'-tetraOH

Cyanidin : 3,5,7,3',4'-pentaOH

Catechin : 3,4,7,3',4'-pentaOH

Leucocyanidin : 3,4,5,7,3',4'-hexaOH

Figure 2. Structures of basic units of condensed tannins of *Brassica* seeds.

Table II.   The content and relative proportion of free, esterified and insoluble-bound phenolic acids in some *Brassica* oilseed products [a]

Product	Phenolic Acid, mg/100 g			
	Free	*Esterified*	*Insoluble-Bound*	*Total*
Altex meal[b]	248	1458	101	1807
	(13.7)	(80.7)	(5.6)	
Tower meal[b]	244	1202	96	1542
	(15.8)	(78.0)	(6.2)	
Midas meal[c]	144.5	1524	68.7	1736.2
	(8.3)	(87.8)	(4.0)	
Mustard meal[c]	108.1	1538	22.4	1668.5
	(6.5)	(92.0)	(1.3)	
Candle flour[d]	84.5	1196.4	—	1280.9
	(6.1)	(93.4)	—	
Tower flour[d]	98.2	982	—	1080.2
	(9.1)	(90.9)	—	

[a]Values in parenthesis are as a percentage of the total content.
[b]From Ref. 13.
[c]From Ref. 17.
[d]From Ref. 10.

Esterified Phenolic Acids

Esterified phenolic acids constitute up to 80% of the total phenolic acids present in rapeseed meal and up to 80% of that in rapeseed flour (*9-11,16,18*). In mustard flour, up to 95% of the total phenolics was composed of esterified phenolic acids (*11*).

The most abundant phenolic ester in rapeseed was sinapine (Table III). At least seven other compounds that yielded sinapic acid upon hydrolysis were also identified by Fenton *et al.* (*15*). In mustard flour, p-hydroxybenzoic acid was again dominant in the soluble ester fraction, as noticed in the free phenolic acid fraction (*11*). Small amounts of p-hydroxybenzoic, vanillic, protocatechuic, syringic, p-coumaric, cis- and trans-ferulic and caffeic acids were also found in the soluble ester fraction of phenolics of rapeseed and mustard flours. Table II compares the total content of esterified phenolic acids of *Brassica* oilseeds with those in the free and insoluble-bound fractions.

Insoluble-Bound Phenolic Acids

Contribution of insoluble-bound phenolics to the total content of phenolic acids ranged from 0.6 to 1.3% in rapeseed flour (*11*). In mustard flour, this fraction accounted for 0.5% of the total. However, insoluble-bound phenolic acids in meals

Table III.   Individual phenolic acids in the esterified (major) fraction of rapeseed flour and hulls phenolics (mg/g) [a]

Phenolic Acid	Tower Hulls	Tower Flour	Sarson Flour	Candle Flour
trans-Sinapic	85.0	894.9	712.3	1196.4
cis-Sinapic	18.7	73.6	49.1	72.1
trans-Ferulic	1.9	8.7	5.9	5.5
para-Hydroxy-benzoic	0.1	1.1	0.3	1.5
Protocatechuic	0.7	0.2	——	0.5
Vanillic	0.5	trace	trace	0.6
para-Coumaric	0.6	2.2	——	trace
Caffeic	1.1	1.3	——	——
Syringic	1.4	trace	——	trace
cis-Ferulic	trace	trace	trace	——

[a]Adapted from Ref. 11.

comprised 1.3-6.2% of the total (Table II). Sinapic acid was predominant in rapeseed varieties, while p-hydroxybenzoic acid was dominant in mustard (11,16). Minor phenolic acids released from the insoluble-bound esters were p-hydroxybenzoic, vanillic, protocatechuic, and caffeic acids. Trace amounts of cis- and trans-ferulic acids were also detected. According to Zadernowski (16), the contribution of insoluble-bound phenolic acids in rapeseed meal was 1.4-2.0% of the total phenolics while their content in flours was 1.5-2.1%.

Contribution of Sinapic Acid to Phenolic Acids of *Brassica* Oilseeds

As it has already been mentioned in the previous sections, sinapine is a major phenolic compound found in the free, esterified and insoluble-bound phenolic acids of *Brassica* seeds (11,17). Table IV summarizes the content of sinapic acid in the total phenolic acids in each of the above fractions for typical seed meals from representative members of canola/rapeseed and mustard families. Sinapic acid was the dominant phenolic acid in the free and the esterified fractions. The contribution of sinapic acid to the insoluble-bound fraction was much smaller than that in other fractions. However, the proportion of the insoluble-bound fraction in the total phenolic acids of rapeseed is rather small and thus experimental errors in determination of such low quantities suffer from relatively large uncertainties.

Tannins in *Brassica* Seeds

Tannins occur either in the hydrolyzable or condensed forms in plant materials (19). Hydrolyzable tannins belong to gallotannin or ellagiotannin, depending on their hydrolytic products. While gallotannins yield gallic acid as the only phenolic moiety, ellagiotannins may produce gallic acid or its dimer ellagic acid upon hydrolysis (19). Condensed tannins are formed by polymerization of flavan-3-ols and flavan-3,4-diols (8).

Bate-Smith and Ribereau-Gayon (20) were the first to report the presence of condensed tannins in rapeseed hulls. Durkee (21) showed that rapeseed hulls contained anthocyanidins and Leung et al. (22) reported that isolated flavanols of rapeseed yield cyanidin, its n-butyl derivative, and pelargonidin upon digestion. Kaempferol-7-glucoside-3-sophoroside, a flavonol glucoside, was also detected in *Brassica* seed extracts (23). Leung et al. (22) reported that rapeseed hulls contained 0.1% condensed tannins that could be extracted with acetone and that leucocyanidin was the basic unit of this fraction. Mitaru et al. (24) reported that canola and rapeseed hulls contained 0.02-0.22% extractable tannins. Blair and Reichert (25) reported the presence of 0.09-0.39% and 0.23-0.54% of tannins in defatted rapeseed and canola colytedons, respectively. Recently, Shahidi and Naczk (18) reported that the content of condensed tannins in canola varieties was 0.68-0.72% of defatted meals (Table V).

Importance of Phenolic Acids in *Brassica* Seeds

Organoleptic Properties. The bitter taste component in *Brassica* species and *Crambe* was reported to be due to sinapine (26). Thus, glucosinolate-free canola flours or their products also have a bitter and astringent taste. The astringency of tannin-containing ingredients may be due to precipitation of proteins in the mouth. Durkee (21) suggested that the dark color of *Brassica* seed meals was due to flavonoid tannins. Oxidation of these phenolics during heat-processing gave rise to products which were dark colored themselves, or reacted with other components of the meal to produce substances which had a dark color.

Interaction with Food Components and Nutritional Implications. While some phenolic compounds may possess beneficial therapeutic effects in high doses upon injection, phenolics and tannins are generally regarded as antinutritional factors

Table IV.   The content of sinapic acid and its contribution to the total phenolic acids in the free, esterified and insoluble-bound phenolic fraction of selected *Brassica* seeds [a]

| Meal | Sinapic Acid, mg/100g | | | |
	Free	Esterified	Insoluble-Bound	Total
Triton	43.2	1172.0	8.0	1223.2
	(70.2)	(96.7)	(15.6)	(92.3)
Midas	183.5	1081.0	5.1	1189.6
	(71.6)	(70.9)	(7.4)	(68.5)
Hu You 9	92.3	1139.0	6.7	1238.0
	(77.5)	(96.4)	(17.2)	(92.4)
Mustard	92.3	1116.0	7.2	1215.5
	(85.4)	(72.6)	(32.1)	(72.9)

[a]Values in parentheses refer to the percentage of sinapic acid in the total phenolic acids of each fraction. Adapted, in parts, from Ref. 17.

Table V.   Content of condensed tannins in some *Brassica* oilseeds [a]

Seed variety	Tannins, mg/100g meal (as Catechin equivalents)
Altex	767 ± 37
Regent	763 ± 13
Tower	772 ± 13
Triton	696 ± 36
Westar	682 ± 9
Midas	556 ± 13
Hu You 9	426 ± 21

[a]Adapted from Ref. 18.

which reduce the utilization of other food components. The anti-nutrient nature of plant phenolics has been thoroughly studied by Griffiths (27,28), Deshpande and Salunkhe (29), Björck and Nyman (30), and Nyman and Björck (31).

Free phenolics and their oxidation products are known to interact with food proteins and inhibit the activity of enzymes such as oxidases, trypsin, arginase and lipases (32).   Interaction of phenolics with proteins may be reversible or irreversible.   The amount of phenolic compounds bound to proteins by hydrogen bonding may account for up to a third of the dry weight of proteins (33).   Phenolic groups in tannins may either interact with the keto-imide moities of proteins via hydrogen bonding or by hydrophobic bonding resulting from the co-alignment of the phenolic groups of tannins and the aromatic side-chains of proteins (34).

In canola and rapeseed, cinnamic acid and its esters are the preferred substrates for phenol oxidase.   Other dihydroxy phenolics such as caffeic and chlorogenic acids may be oxidized to ortho-quinones by copper-containing enzymes of plant tissues.   Once formed, ortho-quinones may react non-enzymatically to polymerize or bond to aminothiol or methylene groups.   The amino group of lysine and thioether group of methionine are commonly attacked to render the amino acids nutritionally unavailable to monogastric digestive systems (19).   Tannins bind to enzymes and other proteins by hydrogen bonding to carbonyl groups to form insoluble complexes.   Generally, hydrolysable tannins are more reactive than condensed tannins.   The oxidized derivatives of the former group form covalent bonds with proteins and these are resistant to enzymatic and microbial degradation (35,36).

Binding of tannins to proteins is highly specific for both substrates (37,38). Generally, proline-rich proteins with a conformationally loose structure strongly bind tannins.   Animal collagen and cereal prolamines are therefore most susceptible to complexation with tannins.   Thus, harmful nutritional effects of tannins due to possible non-specific interactions could be minimized (39).

In rapeseed, tannins have been reported to cause tainting of eggs by perhaps blocking the metabolism of trimethylamine (TMA) by inhibiting the activity of TMA oxidase (39-41).   Furthermore, addition of rapeseed tannins to soybean-containing diets of chickens resulted in a decrease in both protein absorption and metabolizable energy in the chicks (42).   Although rapeseed tannins have been shown to inhibit

α-amylase activity (24), some *in vitro* studies have indicated a stimulatory rather than an inhibitory effect on protein digestion by tannins. Partial denaturation of proteins by tannins might be responsible for the phenomenon (43,44).

Studies on the nutritional effects of polyphenols in monogastrics have indicated that inclusion of condensed tannins, from faba beans, leads to growth depression in chicks and ducklings (27,45). Nyman and Björck (31) studied protein utilization as well as lipid digestibility by rats in the presence of tannic acid and catechin. The apparent digestibility was affected only by tannic acid whereas catechin reduced the true digestibility. Furthermore, *in vitro* lipase activity was reduced by tannic acid (27). However, the activity of proteolytic enzymes was decreased by both tannic acid and catechin. As a result, high tannin content decreased the weight gain of rats. *In vitro* studies also demonstrated a reduction in protein digestibility (46).

Interaction of amylose and amylopectin as well as starches from different sources indicated that both tannic acid and catechin formed association complexes with starch molecules, thus reducing their digestibility (29). However, a study by Nyman and Björck (31) showed no effect by tannic acid or catechin on starch digestion or fiber fermentation in rats. Phenolic acids have also been reported to have flatulence-inhibiting activity, perhaps due to their interaction with soluble sugars (47).

Polyphenols are also known to precipitate a wide range of essential minerals, thus lowering their bioavailability (48). In particular, inhibition of iron absorption by tannin-rich materials (49-52) has been reported.

Caffeic acid and flavonoids in tea have been reported to complex with thiamin rendering it unavailable, however, addition of ascorbic acid prevented this problem (53,54). Precipitation of vitamin B_{12} by tannic acid has also been reported and has been implicated as a possible cause of anemia (55).

Antioxidant Properties. While a large body of literature on the antioxidant activity of soybean and other oilseed flours and/or their extracts has been reported (56) no such study has been carried out for rapeseed and canola products. Recently however, Zadernowski *et al.* (57) reported that different fractions of ethanolic solutions of phenolic acids, separated in the process of concentrate preparation from rapeseed of Polish varieties, had considerable antioxidant activity in a linoleate/β-carotene system. Our own work (unpublished) has indicated a pronounced antioxidant effect for rapeseed meals added to comminuted meat samples at 0-5% addition levels. However, aqueous and methanolic extracts of the meals had a much lower activity. Inhibition of meat lipid oxidation in samples of comminuted pork due to the addition of rapeseed meal or its extracts as compared with α-tocopherol and BHT is given in Figure 3. Detailed studies in this area are in progress in our laboratories.

Health Effects. The beneficial health effects of flavonoids and flavonoid-related compounds may, in part, be due to their free radical scavenging and chelating ability. However, some phenolic compounds may have anti-mutagenic and anti-carcinogenic properties. Ellagic acid, a dimer of gallic acid, has been shown to fit in the latter category. In addition, many phenolic acids have been shown to

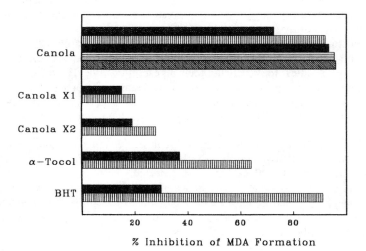

% Inhibition of MDA Formation

Figure 3. Inhibition (%) of malondialdehyde (MDA) formation by canola meal at 0.5%, ▣; 1.0%, ▥; 1.5%, ■; 2%, ▤; and 5.0%, ▨ and by canola extracts of 0.5%, ▣ and 1.0%, ▥ of canola meal by water (Canola X1) or 85% methanol (Canola X2). Values are compared with those exerted by α-tocopherol (α-Tocol) and butylated hydroxytoluene (BHT) at 30 ppm, ▣, and 200 ppm, ▥, levels.

inhibit N-nitrosamine formation. Kikugawa *et al.* (*58*) reported that the potency
of inhibition of N-nitrosamine formation by plant phenolics and ascorbic acid was:

Caffeic acid > *Ferulic acid* > *p-Coumaric acid*

> *Ascorbic acid* > *Cinnamic acid*

Thus, phenolic compounds of rapeseed/canola belonging to both the phenolic acids
and flavonoids may be considered as possessing beneficial health effects. However,
their deleterious effects on sensory quality of processed meals and flours can not
be ignored.

Effect of Ammoniation on the Phenolics of *Brassica* Seeds

The content of phenolic compounds in *Brassica* seeds may be reduced by treatment
with gaseous, aqueous or alcoholic ammonia. McGregor *et al* (*59*) reported that
gaseous ammoniation of *Brassica juncea* mustard meal removed up to 74% of its
sinapine content. Treatment of *Brassica napus* with ammonia was shown to have
an even greater effect, removing some 90% of sinapine from the seeds (*40,60*).
Meanwhile, 0.2 M ammonia in ethanol removed 39-82% of phenolics from canola
meals (*61*).

The methanol-ammonia-water/hexane extraction of canola and rapeseed
resulted in the removal of some 72.4% of their total phenolic acids. Removal of
esterified phenolic acids was achieved somewhat better than that for the free
phenolic acid fraction while the content of insoluble-bound phenolic acids remained
unchanged (*13*). The results of these findings are summarized in Table VI.
Reduction in the total phenolic acids in esterified and free phenolic acid fractions
of canola and rapeseed corresponded with the removal of sinapic acid from these
seeds.

Methanolic ammoniation of canola and rapeseed also resulted in effective
reduction of their tannin content. Thus, the treated canola and rapeseed meals
contained from 4 to 33% of condensed tannins originally present in them (*18*).
Recently, Gandhi *et al.* (*62*) have shown that ammonia depolymerized the tannins

Table VI.   **Removal of phenolic acids and tannins by methanol-ammonia-water/hexane extraction** [a]

Meal	Removal, %	
	Total Phenolic Acids	Tannins
Altex	72.0	84.9
Regent	74.6	77.5
Tower	75.2	78.0
Midas	71.1	79.3
Mustard	82.1	—

[a]Adapted from Ref. 13 and 18.

present in salseed meal and that the processed meal so prepared was nontoxic and palatable. However, Fenwick *et al.* (*60*) reported that the treatment of *Brassica napus* meals with ammonia or lime did not affect their tannin content to any great extent.

Ammoniated meals generally had a light beige color and were bland in taste. Removal of phenolics such as sinapine and flavonoid tannins is at least partially responsible for quality improvement of prepared meals from this process.

Reserch Needs

In spite of the numerous publications on the subject for soybean and other oilseeds, the available information on the effects of phenolics in *Brassica* seeds is still fragmentary. Information on the interaction of rapeseed phenolics, both simple and condensed, with food components and their sensory and nutritional effects are required. Furthermore, detailed studies on the antioxidant role of canola and rapeseed phenolics in lipid-containing foods and primarily in emulsified and restructured meat products are deemed beneficial.

Some of the research needs pertaining to the methodologies for tannin determination have been described by Martin-Tanguy *et al.* (*63*) and Deshpande and Cheryan (*64*). However, no systematic study regarding the suitability of these methods for quantification of rapeseed tannins has been carried out. Furthermore, studies on the mechanism and specificity of interaction of canola and rapeseed tannins in the foods and feeds are deemed necessary.

Literature Cited

1.   Shahidi, F. In: *Canola and Rapeseed: Production, Chemistry, Nutrition and Processing Technology.* Shahidi, F., Ed. Van Nostrand Reinhold, New York. **1990.** pp. 3-13.
2.   Shahidi, F.; Naczk, M.; Hall, D.; Synowiecki, J. *Food Chem.* **1991.** In Press.
3.   Ohlson, R.; Anjou, K. *J. Amer. Oil Chem. Soc.* **1979.** 56:431-437.
4.   Shahidi, F. *Bull. Liason Groupe Polyphenols* **1988.** 14:357-358.
5.   Tzagoloff, A. *Plant Physiol.* **1963.** 38:202-206.
6.   Byerrum, R.U.; Wing, R.E. *J. Biol. Chem.* **1953.** 205:637-642.
7.   Neish, A.C. *Ann. Rev. Plant Physiol.* **1960.** 11:55-80.
8.   Haslam, E. **1966.** *Chemistry of Vegetable Tannins.* Academic Press: London.
9.   Krygier, K.; Sosulski, F.; Hogge, L. *J. Agric. Food Chem.* **1982a.** 30: 330-334.
10.   Krygier, K.; Sosulski, F.; Hogge, L. *J. Agric. Food Chem.* **1982b.** 30:334-336.
11.   Kozlowska, H.; Rotkiewicz, D.A.; Zadernowski, R.; Sosulski, F.W. *J. Amer. Oil Chem. Soc.* **1983.** 60:1119-1123.
12.   Dabrowski, K.J.; Sosulski, F.W. *J. Agric. Food Chem.* **1984.** 32:128-130.
13.   Naczk, M.; Shahidi, F. *Food Chem.* **1989.** 231:159-164.

14. Kozlowska, H.; Naczk, M.; Shahidi, F.; Zadernowski, R. In: *Canola and Rapeseed: Production, Chemistry, Nutrition and Processing Technology.* Shahidi, F., Ed. Van Nostrand Reinhold, New York. **1990.** pp. 193-210.
15. Fenton, T.W.; Leung, J.; Clandinin, D.R. *J. Food Sci.* **1980.** 45:1702-1705.
16. Zadernowski, R. *Technologia Alimentorum* **1987.** 21, Supplement F:3.
17. Shahidi, F.; Naczk, M. *Bull. Liason Groupe Polyphenols* **1990.** 15:236-239.
18. Shahidi, F.; Naczk, M. *J. Food Sci.* **1989.** 54:1082-1083.
19. Ribéreau-Gayon, P. **1972.** *Plant Phenolics.* Oliver and Boyd Co., Edinburgh.
20. Bate-Smith, E.C.; Ribéreau-Gayon, P. *Qualitas Plant. et Materiae Vegetabiles* **1959.** 5:189-198.
21. Durkee, A.B. *Phytochemistry* **1971.** 10:1583-1585.
22. Leung, J.; Fenton, T.W.; Mueller, M.M.; Clandinin, D.R. *J. Food Sci.* **1979.** 44:1313-1316.
23. Durkee, A.B.; Harbone, J.B. *Phytochemistry* **1973.** 12:1085-1089.
24. Mitaru, B.N.; Blair, R.; Bell, J.M.; Reichert, R.D. *Can. J. Animal Sci.* **1982.** 62:661-663.
25. Blair, R.; Reichert, R.D. *J. Sci. Food Agric.* **1984.** 35:29-35.
26. Sosulski, F.W.; Hambert, E.S.; Lin, M.J.Y.; Card, J.W. *Can. Inst. Food Sci. Technol. J.* **1977.** 10:9-13.
27. Griffiths, D.W. *J. Sci. Food Agric.* **1979.** 30:458-462.
28. Griffiths, D.W. *J. Sci. Food Agric.* **1982.** 33:847-851.
29. Deshpande, S.S.; Salunkhe, D.K. *J. Food Sci.* **1982.** 47:2080-2081,2083.
30. Björck, I.M.; Nyman, M.E. *J. Food Sci.* **1987.** 52:1588-1594.
31. Nyman, M.E.; Björck, I.M. *J. Food Sci.* **1989.** 54:1332-1335, 1363.
32. Sosulski, F.W. *J. Amer. Oil Chem. Sci.* **1979.** 56:711-715.
33. Loomis, W.D.; Battaile, J. *Phytochemistry* **1966.** 5:423-428.
34. Butler, L.G. In: *Food Proteins.* Kinsella, J.E. and Soucie, W.G., Eds. Published by the Amer. Oil Chem. Soc.: Champaign, IL. **1984.** pp. 402-409.
35. Van Buren, J.P.; Robinson, W. *J. Agric. Food Chem.* **1969.** 17:772-777.
36. Van Sumere, C.F.; Albercht, J.; Dedonder, A.; De Pooter, H.; Pé, I. In: *Chemistry and Biochemistry of Plant Proteins.* Harbone, J.E.; Van Sumere, C.F. Academic Press: **1975.** pp. 211-264.
37. Hagerman, A.E.; Butler, L.G. *J. Biol. Chem.* **1981.** 256:4494-4497.
38. Asquith, T.; Mehansho, H.; Rogler, J.; Butler, L.; Carlson, D.M. **1985.** *Federation Proceedings.* Abstract No. 4016. 44:1097.
39. Butler, E.J.; Pearson, A.W.; Fenwick, G.R. *J. Sci. Food Agric.* **1982.** 33:866-875.
40. Fenwick, G.R.; Pearson, A.W.; Greenwood, N.M.; Butler, E.J. *Anim. Feed Sci. Technol.* **1981.** 6:421-431.
41. Fenwick, G.R.; Curl, C.L.; Butler, E.J.; Greenwood, N.M.; Pearson, A.M. *J. Sci. Food Agric.* **1984.** 35:749-756.
42. Yapar, Z.; Clandinin, D.R. *Poultry Sci.* **1972.** 51:222-228.
43. Mole, S.; Waterman, P.G. *J. Chem. Ecol.* **1985.** 11:1323-1332.

44.   Oh, H.I.; Hoff, J.E. *J. Food Sci.* **1986.** 51:577-580.

45.   Marquardt, R.R. **1989.** In: *Report on Workshops on Antinutritional Factors in Legume Seeds.* Wageningen, The Netherlands.

46.   Ikeda, K.; Oku, M.; Kusano, T.; Yasumoto, K. *J. Food Sci.* **1986.** 51:1527-1530.

47.   Rackis, J.J.; Honig, D.H.; Sessa, D.J.; Steggerda, F.R. *J. Agric. Food Chem.* **1970.** 18:977-982.

48.   Faithfull, N.T. *J. Sci. Food Agric.* **1984.** 35:819-826.

49.   Hallberg, L.; Rossander, L. *Hum. Nutr: Appl. Nutr.* **1982.** 36A:116-123.

50.   Gillooly, M.; Bothwell, T.H.; Torrance, J.D.; MacPhail, A.P.; Derman, D.P.; Bezwoda, W.R.; Mills, W.; Charlton, R.W.; Mayet, F. *Br. J. Nutr.* **1983.** 49:331-342.

51.   Brune, M.; Hallberg, L.; Skanberg, A.-B. *J. Food Sci.* **1991.** 56:128-131,167.

52.   Brune, M.; Rossander, L.; Hallberg, L. *Eur. J. Clin. Nutr.* **1989.** 43:547-557.

53.   Rungruangsak, K.; Tosukhowong, P.; Panijpan, B.; Vimokesant, S.L. *Amer. J. Clin. Nutr.* **1977.** 30:1680-1685.

54.   Somogyi, J.C. *Wld. Rev. Nutr. Diet.* **1978.** 1129:42-59.

55.   Carrera, G.; Mitjavila, S.; Derache, R. **1973.** Cited in: Singleton, V. *Adv. Food Res.* 27:149-242 (1981).

56.   Pratt, D.E. In: *Flavor Chemistry of Fats and Oils.* Min, D.B. and Snouse, T.H. Eds. American Oil Chemists' Society, **1985.** pp. 145-153.

57.   Zadernowski, R.; Nowak, H.; Kozlowska, H. **1991.** Abstract No. C-40, p. 125. *Rapeseed in a Changing World.* 8th International Rapeseed Congress. July 9-11, Saskatoon, SK. Canada.

58.   Kikugawa, K.; Hakamada, T.; Hasunuma, M.; Kurechi, T. *J. Agric. Food Chem.* **1983.** 31:780-785.

59.   McGregor, D.I.; Blake, J.A.; Pickard, M.D. In: *Proceedings of the 6th International Rapeseed Conference.* Vol. II. Paris. **1983.** pp. 1426-1432.

60.   Fenwick, G.R.; Spinks, E.A.; Wilkinson, A.P.; Heaney, R.K.; Legoy, M.A. *J. Sci. Food Agric.* **1986.** 37:735-741.

61.   Goh, Y.K.; Shires, A.R.; Robblee, A.R.; Clandinin, D.R. *Br. Poultry Sci.* **1982.** 23:121-128.

62.   Gandhi, V.M. Cheriyan, K.K.; Mulky, M.J.; Menon, K.K.G. 1975. J. Oil Technol. Assoc. India 7:44-47.

63.   Martin-Tanguy, J.; Guillaume, J.; Kossa, A. *J. Sci. Food Agric.* **1977.** 28:757-765.

64.   Deshpande, S.S.; Cheryan, M. *J. Food Sci.* **1987.** 52:332-334.

RECEIVED February 11, 1992

Chapter 11

Chemistry of Curcumin and Curcuminoids

Hanne Hjorth Tønnesen

Department of Pharmaceutics, University of Oslo, P.O. Box 1068,
Blindern, 0316 Oslo 3, Norway

The interest in curcumin as a natural coloring agent, as a pharma–
ceutical excipient and as a drug/drug model is increasing. The
physicochemical properties of curcumin strongly influence its
interaction with other molecules, affecting the analytical behaviour
and possibly also the in vivo reaction mechanisms. Physicochemical
properties, analytical aspects and stability of curcumin and
curcuminoids with respect to the utility of these compounds
are discussed

The natural dye curcumin (C.I.75300) extracted from the rhizomes of various
Curcuma species (Zingiberaceae) is widely used as a coloring agent in food, drugs
and cosmetics. Curcumin is also reported to have various pharmacological effects
(1). Curcuma extracts contain three different diarylheptanoids, curcumin,
demethoxycurcumin and bisdemethoxycurcumin (Figure 1). Curcumin is usually
the main constituent. In rare cases demethoxy– and bisdemethoxycurcumin amount
to 70% of the total curcuminoids in a sample (2). Commercially available "pure"
curcumin consists of a mixture of the three naturally occurring curcuminoids, with
curcumin as the main constituent (Tønnesen,H.H.,Z.Lebensm.Unters.Forsch.,in
press).
    The interest in curcumin as a natural colorant, as a pharmaceutical excipient and
as a drug/drug model seems to be increasing. Knowledge of the physicochemical
and analytical properties of curcumin is important to ensure optimal utility of the
compound as a coloring agent, as a pharmaceutical additive and in quality
assurance using biological experiments.

Structural Studies of Curcumin and Curcuminoids

The physicochemical properties and pharmacological effects of a compound are
consequences of its structure. As crystallographic studies provide information about
the three–dimensional structure of a compound in the solid state, it should be kept
in mind that the biological effects of a substance are related to its physicochemical
properties in solution. For biologically active compounds such as curcumin it is

0097–6156/92/0506–0143$06.00/0
© 1992 American Chemical Society

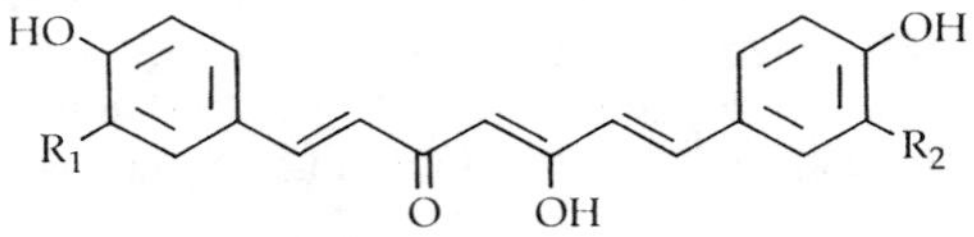

$R_1 = R_2 = OCH_3$ Curcumin
$R_1 = OCH_3$, $R_2 = H$ Demethoxycurcumin
$R_1 = R_2 = H$ Bisdemethoxycurcumin

Figure 1. The main colored compounds isolated from Curcuma longa L.

therefore of importance to obtain knowledge about their structural properties in solution (e.g.tautomerism, solvation, complexing abilities).

The crystal structures of curcumin and various other curcuminoids (both naturally occurring and synthetic) have been determined (*3–8*). In the crystals studied there were marked differences in electron delocalization and intramolecular hydrogen bonding in the fragment –CO–HC=COH–, as well as in intermolecular hydrogen bonding. In the curcumin molecule there were no significant differences in the lengths of the C–C bonds or in the C–O bonds in the enol ring (Figure 2). The hydrogen atom seemed to be statistically evenly distributed <u>between</u> the two oxygen atoms instead of being bonded to one unique oxygen atom. There is a degree of conjugation between the enol ring and one of the aromatic rings in the molecule, the two rings being essentially coplanar. The second aromatic ring has less conjugation with the enol ring system, and makes a dihedral angle of about 45° (Figure 3) (*3*). In the curcumin crystal both oxygen atoms of the enol ring are engaged in intermolecular hydrogen bonding as shown in Figure 2. In bis-demethoxycurcumin (hydrate, methanol complex) the hydrogen atom in the keto-enol structure is bonded to one unique oxygen atom, the other oxygen atom being engaged in intermolecular hydrogen bonding (*4,5*). This introduces an element of asymmetry to the molecule. The deviation from coplanarity between the terminal phenyl groups is only 16°. The crystal structure of demethoxycurcumin has not been obtained so far. When recrystallized from organic solvents demethoxy-curcumin precipitates in an amorphous form.

Of the conformeric tautomeric forms possible for β–di-ketones it appears that curcumin, demethoxy– and bisdemethoxycurcumin adopt the <u>cis</u>–enol form in solution (*9–13*). Not only the keto–enol unit, but also the phenolic hydroxyl groups can interact with the solvent. Intramolecular hydrogen bonding can be demonstrated by changes in the UV–vis spectrum of the curcuminoids in various organic solvents (Tønnesen,H.H.,to be published). The curcuminoids have low solubility in water.

The curcuminoids are powerful complexing agents, the keto–enol units being the reactive units of the molecules. The red coloration obtained when curcumin reacts with boric acid is now the basis for the most sensitive tests for boron (*13–31*). Various other metals (Zn,Sn,K,Al,Cu,Ni) are also known to form chelates with the curcuminoids (*32,33*). Chelating agents are important <u>in vivo</u>, both as catalysts and as inhibitors of biochemical processes (*34*). Curcumin, demethoxy– and bis-demethoxycurcumin strongly influence the rate of the iron–catalyzed Haber–Weiss reaction <u>in vitro</u> (*35–37,40*). <u>In vivo</u> this reaction is closely related to the inflammation process (*34*). The observed anti inflammatory effect of the curcuminoids might partly be related to the chelating abilities of these compounds.

Results from <u>in vitro</u> experiments carried out in our laboratory indicate that curcumin also interacts strongly with certain macromolecules of biological origin (e.g. serum proteins, albumin, hyaluronic acid). The importance of such interactions under <u>in vivo</u> conditions is not clear.

It is further known that curcumin, demethoxy– and bisdemethoxycurcumin strongly interact with silicic acid, which is important for chromatographic assays. This will be discussed in detail in the following section.

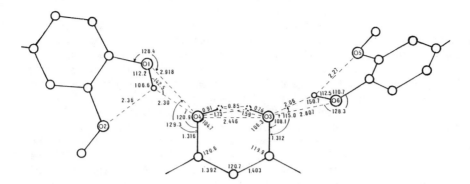

Figure 2. Geometry of the enol ring and the hydrogen bond system of curcumin in the crystal. (Reproduced with permission from ref. 3. Copyright 1982.)

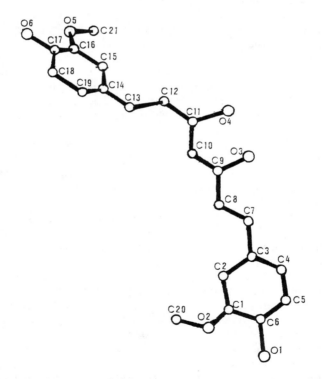

Figure 3. Three–dimensional structure of curcumin in the crystal

Analysis of Curcumin and Curcuminoids

A variety of methods for quantitation of the curcuminoids have been reported (*1*). Most of these are spectrophotometric methods, expressing the total color content of the sample. Commercially obtained curcuma products contain mixtures of curcumin, demethoxy– and bisdemethoxycurcumin. For an exact determination of the curcumin content a preseparation of the three curcuminoids is required.

The curcuminoids isolated from Curcuma species (mainly Curcuma longa L.) exhibit strong absorption from 420nm to 430nm in organic solvents. The official methods for assaying curcumin or curcuma products as food colour additives are based upon direct spectrophotometric absorption measurements (*38,39*). The evaluation of the total amount of curcuminoids in a sample by use of direct absorption measurements is only valid if the calculations are based on reference values obtained from pure standards (Tønnesen,H.H.,*Z.Lebensm.Unters. Forsch.*,in press). It should however be kept in mind that the presence of other compounds absorbing in the region of 420nm–430nm will strongly influence the results.

A direct fluorimetric method for the assay of curcumin in food products has been reported (*41*). The difficulties in obtaining reproducible results can be ascribed to the difference in fluorescence intensity in organic solvents of curucumin and the two demethoxycompounds. At fixed excitation and emission wavelengths (420nm/470nm) the relative fluorescence intensities of curcumin, demethoxy– and bisdemethoxycurcumin in ethanol are 1:2.2:10.4 at equimolar concentrations (*42*). Unless these differences are taken into account small changes in sample composition will lead to large variations in the "curcumin" content calculated.

To increase the molar absorptivity of curcumin, intensly colored dissociation forms or colored complexes can be developed by reaction with alkalis, strong mineral acids or boric acid (*41,43–48*). The colors formed are, however, found to be very unstable, and severe fading is reported after 5–10 minutes with the exception of the boric acid complexes (*31,48*, Tønnesen,H.H., *Z.Lebensm. Unters.Forsch.*,in press). Interference from co–extractives or the two other curcuminoids should be accounted for.

A preseparation of the curcuminoids can be accomplished by thin–layer chromatography (TLC) or high–pressure liquid chromatography (HPLC)(*42,46,48–58*, Tønnesen,H.H., *Z.Lebensm. Unters.Forsch.*,in press). The separation of the curcuminoids is strongly dependant on the chromatographic conditions (*58*). Curcumin and the related 1,3–diketones are shown to adsorb strongly onto the silicic acid used as the solid support in TLC and HPLC. By removing one of the keto groups from a diketone the adsorption to silica gel can be eliminated (*58*). The adsorption is therefore ascribed to intermolecular hydrogen bonding between the keto–enol unit of the β–diketones and the silicic acid. Quantitative analysis of curcumin and related compounds by TLC or HPLC will be difficult to carry out unless the chromatographic support is properly deactivated, e.g. the number of free silanol groups are kept at a minimum. HPLC systems based on C–18 stationary phases do not completely resolve the three curcuminoids (*49–52*). A reproducible separation of the colored compounds can be achieved by the use of an amino-bonded stationary phase (HPLC,TLC, Figures 4–5), provided that the water content of the system is kept below 10% (*42,56–58*,Tønnesen,H.H., *Z.Lebensm.Unters.*

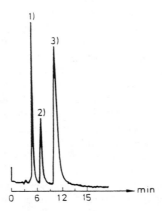

Figure 4. HPLC chromatogram of the three curcuminoids separated on a Nucleosil NH_2 column with ethanol/water (96:4) as mobile phase. 1)Bisdemethoxycurcumin 2)Demethoxycurcumin 3)Curcumin

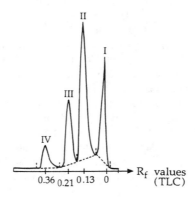

Figure 5. TLC chromatogram of the three curcuminoids separated on an aminobonded stationary phase (Merck) with ethanol/water (96:4) as mobile phase. I)Curcuminoids adsorbed to the stationary phase at the application spot II)Curcumin III)Demethoxycurcumin IV)Bisdemethoxycurcumin

Forsch.,in press). A TLC system based on an amino–bonded stationary phase, however, seems to have a catalytic effect upon curcumin degradation. To obtain reproducible results the experimental conditions must be carefully controlled (Tønnesen,H.H.,*Z.Lebensm.Unters.Forsch.*, in press).

HPLC in combination with fluorescence detection is at present the most sensitive method for the determination of curcumin, the detection limit lying in the picogram range (*42*). With respect to in vivo studies a more sensitive method for the detection of trace amounts of curcumin will be needed. This can be obtained by use of tracer techniques (*44,59*). The introduction of radioactivity to the curcumin molecule will in many countries limit the biological experiments to animal studies. Alternative chromatograpic methods for the separation and assay of the curcuminoids have therefore been evaluated recently. These include methods based on gas–liquid chromatograpy (GC) and supercritical fluid chromatograpy (SFC). GC–methods provide no alternative to HPLC due to the low volatility and thermal lability of the curcuminoids. As demonstrated by others (*60*) and also in our laboratory (Tønnesen,H.H., unpublished data), the curcuminoids form a "sticky", glassy state without defined melting points upon heating.

The SFC–systems examined do not resolve the three curcuminoids (*61*). The detection limit of curcumin was 1ng (FID). The curcuminoids were irreversibly adsorbed by columns with incomplete modification of the solid support. In conclusion it seems difficult to develop an alternative chromatographic method to HPLC /fluorescence detection for quantitative determination of trace amounts of curcumin in biological samples.

Spectroscopic methods (IR,NMR,MS) are widely used for identification and characterization of the curcuminoids (*1,10–12,27,62–64*). NMR has also been tried out for quantitative determinations (*10*). Mass spectrometry (MS) is often the method of choice when trace amounts of organic compounds are to be detected. The detection limit for curcumin in biological samples by MS needs to be determined, then the possibility of using quantitative MS (direct inlet) in curcumin analysis could be evaluated.

It should be kept in mind that the strong interactions observed between curcumin and silanol groups also occur in a glass container. Unless precautions are taken, curcumin in solution will adsorb strongly to the container wall, leading to inaccurate results (*10*).

Stability of Curcumin and Curcuminoids

The potential use of curcumin as a coloring agent, as a pharmaceutical excipient or as a drug/drug model is closely related to the stability of the compound. Curcumin is sparingly soluble in water (*36*). The compound is, however, soluble in alkali or in glacial acetic acid. Since most food and drug formulations contain significant amounts of water, curcumin will not color such products without the aid of a chemical emulsifier or an increase in pH. The color of curcumin in aqueous media or in organic solvents is not constant due to its degradation or changes in its dissociation forms. In solution at pH 1–7 the color is yellow. In the pH range 7 to 9 the color is brownish–red or deep red. The indicator property of curcumin has been commercialized in "curcuma paper", a common pH indicator (*65,66*).

The pK$_a$ values for the dissociation of the three acid protons in curcumin were found to be 7.80, 8.5 and 9.0 (67). Further experiments on the 1,7–diarylheptanoids should be carried out to determine the dissociation sequence of the three acid protons (13,30,31). Curcumin in aqueous solutions is exposed to hydrolytic degradative reactions (68). Below pH 7 the degradation rate is low compared to that at a higher pH (Figure 6). Ferulic acid, feruloylmethane, vanillin and acetone have been identified as the main degradation products. The brownish–yellow hue of a decomposed curcumin solution is ascribed to the formation of condensation products from the feruloylmethane part of the curcumin molecule (68). A kinetic study of curcumin in ethanol/water as a function of temperature has also been reported (69).

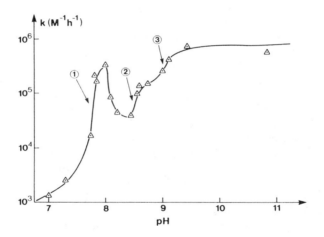

Figure 6. pH-rate profile for the hydrolytic degradation of curcumin. The inflection points (1,2,3) indicate the pK$_a$ values for the three acid protons. (Reproduced with permission from ref. 67. Copyright 1985.)

From the above results it is appearant that curcumin is not suitable for use in many products due to decomposition upon storage under alkaline conditions. This also excludes the possibility of raising the pH in certain formulations to improve the solubility of curcumin. Many food products, however, still require the use of alkaline components for proper preparation (70). Various complexes have been prepared to improve the color stability of curcumin (33,70–73), but it is mainly used as a coloring agent in dry food mixes.

Curcumin is demonstrated to have a stabilizing effect on certain photolabile drugs in formulations or in serum samples (74,75). This has been ascribed to a pure "filter" effect by curcumin in the UV–A part of the spectrum. Curcumin itself exhibits poor stability to light (76–78). Curcumin and the two demethoxy–compounds decompose when exposed to UV/vis light both in solution and in the solid state (*56,78*). The main degradation product formed after exposing curcumin to visible light is a yellow cyclization material (*78*). The apparent color of a curcuma product will not change during exposure to light above 400nm although this first degradation step occurs. When exposed to UV–light, curcumin is rapidly decolourized.

Curcumin acts as a photosensitizer in the formation of oxygen radicals and undergoes a self–sensitized decomposition, but the fading also involves other mechanisms independent of oxygen (*78*). This affects the utility of curcumin in several ways. It is not possible to stabilize curcumin photochemically by excluding oxygen or by the addition of radical quenchers. Curcumin is not a suitable additive for products stored in transparent, uncolored packing materials, but is photo–chemically stable in brown glass bottles, provided that other oxygen radical sources are excluded.

Curcumin has been shown to act as a sensitizer in the photooxidation of fatty acids (*79*) and to be phototoxic for certain bacteria (*80*). The potential use of curcumin and other curcuminoids as excipients in photon–activated drug delivery systems or as sensitizers in photodynamic therapy systems is now under investigation in our laboratory.

Conclusion

A natural product whether used in food preparations or in drug formulations should be well characterized in respect to its physicochemical and analytical properties. The keto–enol unit in the curcuminoids strongly influences their interactions with other molecules, affecting the analytical behavior and possibly also the in vivo reaction mechanisms. Due to degradative processes, precautions should be taken to avoid high pH and light exposure of the colored compounds. The curcuminoids take part in the formation and quenching of radicals, increasing their potential as possible drugs/drug models.

Literature Cited

1. Tønnesen, H. H. Ph.D. Thesis, University of Oslo, Oslo, 1986
2. Tønnesen, H. H.; Karlsen, J.; Adhikary, S. R.; Pandey. R. *Z. Lebensm. Unters. Forsch.*, **1989**, *189*, 116–118
3. Tønnesen, H. H.; Karlsen, J.; Mostad, A. *Acta Chem. Scand.*, **1982**, *B36*, 475–479
4. Tønnesen, H. H.; Karlsen, J.; Mostad, A.; Andersen, U.; Rasmussen, P. B.; Lawesson, S. O. *Acta Chem. Scand.*, **1983**, *B37*, 179–185
5. Karlsen, J.; Mostad, A.; Tønnesen, H. H. *Acta Chem. Scand.*, **1988**, *B42*, 23–27
6. Görbitz, H.; Mostad, A.; Pedersen, U.; Rasmussen, P. B.; Lawesson, S. O. *Acta Chem. Scand.*, **1986**, *B40*, 420–429
7. Mostad, A.; Pedersen, U.; Rasmussen, P. B.; Lawesson, S. O. *Acta Chem. Scand.*, **1984**, *B38*, 479–484
8. Mostad, A.; Pedersen, U.; Rasmussen, P. B.; Lawesson, S. O. *Acta Chem. Scand.*, **1983**, *B37*, 901–905
9. Pedersen, U.; Rasmussen, P. B.; Lawesson, S. O. *Liebigs Ann. Chem.*, **1985**, 1557–1569

10. Unterhalt, B. *Z. Lebensm. Unters. Forsch.*, **1980**, *170*, 425–428
11. Kuroyanagi, M.; Natori, S. *Yakugaku Zasshi*, **1970**, *90*, 1467–1470
12. Roughley, P. J.; Whiting, D. A. *J. Chem. Soc., Perkin Trans. I*, **1973**, 2379–2388
13. Quint,P. Ph.D. Thesis, Universität zu Münster, Münster, 1974
14. Rao, A. S.; Divakar, S.; Seshadri, R. *Indian J. Chem.*, **1988**, *27B*, 926–928
15. Drew, R. E. *J. Fish. Res. Board. Can.*, **1975**, *32*, 813–816
16. Aznarez, J.; Bonilla, A.; Vidal, J. C. *Analyst*, **1983**, *108*, 368–373
17. Fukui, S.; Hirayama, T.; Nohara, M.; Kobayashi, K.; Kawamura, T.; Iwaida, M.; Ito, Y.; Ogawa, S.; Kakiuchi, Y.; Yamazaki, H.; Ono, N. *Eisei Kagaku*, **1983**, *29*, 323–328
18. Mair, J. W.; Day, H. G. *Anal. Chem.*, **1972**, *44*, 2015–2017
19. Thierig, D.; Umland, F. *Frezenius Z. Anal. Chem.*, **1965**, *211*, 161–169
20. Umland, F.; Thierig, D.; Müller, G. *Frezenius Z. Anal. Chem.*, **1965**, *215*, 401–407
21. Sclumberger, M. E. *Bul. Soc. Chim. France*, **1866**, *5*, 194–202
22. Quint, P.; Umland, F. *Frezenius Z. Anal. Chem.*, **1979**, *295*, 260–270
23. Umland, F.; Pottkamp, F. *Frezenius Z. Anal. Chem.*, **1968**, *241*, 223–234
24. Quint, P.; Umland, F.; Sommer, H. D. *Frezenius Z. Anal. Chem.*, **1977**, *285*, 356–358
25. Roth, H. J.; Miller, B. *Arch. Pharm.*, **1964**, *297*, 660–673
26. Uppström, L, R.; Östling, G. *Anal. Lett.*, **1976**, *9*, 311–324
27. Miyamoto, M. *Bull. Chem. Soc. Jpn.*, **1963**, *36*, 1208–1213
28. Hayes, M. R.; Metcalfe, J. *Analyst*, **1962**, *87*, 956–969
29. Spicer, G. L.; Strickland, J. D. H. *J. Chem. Soc.*, **1952**, 4644–4650
30. Spicer, G. L.; Strickland, J. D. H. *J. Chem. Soc.*, **1952**, 4650–4652
31. Dyrssen, D. W.; Novikov, Y. P.; Uppström, L. R. *Anal. Chim. Acta*, **1972**, *60*, 139–151
32. Arrieta, A.; Dietze, F.; Mann, G.; Beyer, L.; Hartung, J. *J. Prakt. Chem.*, **1988**, *330*, 111–118
33. Maing, Y. I.; Miller, I. European Patent, 1981, EP 0025637 A1
34. Halliwell, B.; Gutteridge, J. M. C. Eds. *Free radicals in biology and medicine*; Oxford University Press, Oxford, 1985, pp.119–125, 150, 283
35. Tønnesen, H. H. *Int. J. Pharm.*, **1989**, *50*, 91–95
36. Tønnesen, H. H. *Int. J. Pharm.*, **1989**, *51*, 259–261
37. Kunchandy, E.; Rao, M. N. A. *Int. J. Pharm.*, **1989**, *57*, 173–176
38. British Standard Methods of test for Spices and Condiments, 1983, Part 13
39. WHO Food Additive Series, Rome, 1976, no.7, pp.75–78
40. Kunchandy, E.; Rao, M. N. A. *Int. J. Pharm.*, **1990**, *58*, 237–240
41. Karasz, A. B.; DeCocco, F.; Bokus, L. *J. Assoc. Off. Anal. Chem.*, **1973**, *56*, 626–628
42. Tønnesen, H. H.; Karlsen, J. *J. Chromatog.*, **1983**, *259*, 367–371
43. Krishnamurthy, N.; Mathew, A. G.; Nambudiri, E. S.; Shivashankar, S.; Lewis, Y. S.; Natarajan, C. P. *Trop. Sci.*, **1976**, *18*, 37–45
44. Holder, G. M.; Plummer, J. L.; Ryan, A. J. *Xenobiotica*, **1978**, *8*, 761–768
45. Ravindranath, V.; Chandrasekhara, N. *Toxicol.*, **1980**, *16*, 259–265
46. Luckner, M.; Bessler, O.; Luckner, R. *Die Pharmazie*, **1967**, *22*, 371–375

47. Satyanarayana, M. N.; Chandrasekhara, N.; Rao, D. S. *Res. Ind.*, **1969**, *14*, 82–83
48. Janssen, A.; Gole, T. *Chromatographia*, **1984**, *18*, 546–549
49. Asakawa, N.; Tsuno, M.; Hattori, T.; Ueyama, M.; Shinoda, A.; Miyake, Y.; Kagei, K. *Yakugaki Zasshi*, **1981**, *101*, 374–377
50. Singh, S.; Khanna, M.; Sarin, J. P. S. *Indian Drugs*, **1981**, *18*, 207–209
51. Amakawa, E.; Hirata, K.; Ogiwara, T.; Ohnishi, K. *Bunseki Kagaku*, **1984**, *33*, 586–590
52. Smith, R. M.; Witowska, B. A. *Analyst*, **1984**, *109*, 259–261
53. Jentzsch, K.; Spiegl, P.; Kamitz, R. *Sci. Pharm.*, **1970**, *38*, 50–58
54. Lehmann, G.; Gerhardt, V.; Collett, P. *Z. Lebensm. Unters. Forsch.*, **1970**, *194*, 345–348
55. Jentzsch, K.; Spiegl, P.; Kamitz, R. *Sci. Pharm.*, **1968**, *36*, 251–256
56. Khurana, A.; Chi–Tang, H. *J. Liquid. Chromatogr.*, **1988**, *11*, 2295–2304
57. Tønnesen, H. H.; Karlsen, J. *Z. Lebensm. Unters. Forsch.*, **1983**, *177*, 348–349
58. Tønnesen, H. H.; Karlsen, J. *Z. Lebensm. Unters. Forsch.*, **1986**, *182*, 215–218
59. Plummer, J. L. Ph.D. Thesis, The University of Sydney, Sydney, 1977
60. Marmion, D. M. *J. Assoc. Off. Anal. Chem.*, **1970**, *53*, 244–249
61. Baastø, M. B. Thesis, University of Oslo, Oslo, 1991
62. Govindarajan, V. S. *CRC Critical Reviews in Food Science and Nutrition*, **1980**, *12*, 199–301
63. Khalique, A.; Amin, M. N. *Sci. Res.*, **1967**, *4*, 193–197
64. Bellamy, L. J.; Spicer, G. S.; Strickland, J. D. H. *J.Chem. Soc.*, **1952**, 4653–4656
65. Conn, H. J., Ed. *Biological Stains*, 9th ed., The Williams and Wilkins Company, Baltimore, 1977, pp.474–475
66. Thoms, H., Ed. *Handbuch der praktischen und wissenschaftlichen Pharmazie*, Urban und Schwarzenberg, Berlin, 1925, Vol. 2, p.300
67. Tønnesen, H. H.; Karlsen, J. *Z. Lebensm. Unters. Forsch.*, **1985**, *180*, 402–404
68. Tønnesen, H. H.; Karlsen, J. *Z. Lebensm. Unters. Forsch.*, **1985**, *180*, 132–134
69. Ràcz, I.; Spiegl, P. *Sci. Pharm.*, **1972**, *40*, 251–259
70. Leshik, R. R. Eur. Pat. Appl., EP 0037204, 1981
71. Stransky, C. E. U.S. Patent 4138212, 1977
72. Schranz, J. L. U.K. Patent Application GB 2132205A, 1984
73. Wattenwyl, R.von Patent DE 2925364, 1978
74. Tønnesen, H. H.; Karlsen, J. *Int. J. Pharm.*, **1987**, *38*, 247–249
75. Tønnesen, H. H.; Karlsen, J. *Int. J. Pharm.*, **1988**, *41*, 75–81
76. Coulson, J. *Int. Flavours Food Add.*, **1978**, *9*, 207–216
77. Auslander, D. E.; Goldberg, M.; Hill, J. A.; Weiss, A. L. *Drug Cosmet. Ind.*, **1977**, *121*, 55–60, 138, 140
78. Tønnesen, H. H.; Karlsen, J.; van Henegouwen, G.B. *Z. Lebensm. Unters. Forsch.*, **1986**, *183*, 116–122
79. Scieberle, P.; Haslbeck, F.; Laskawy, G.; Grosch, W. *Z. Lebensm. Unters. Forsch.*, **1984**, *179*, 93–98
80. Tønnesen, H. H.; de Vries, H.; Karlsen, J.; van Henegouwen, G. B. *J. Pharm. Sci.*, **1987**, *76*, 371–373

RECEIVED June 25, 1992

Chapter 12

Phenolic Compounds in Botanical Extracts Used in Foods, Flavors, Cosmetics, and Pharmaceuticals

Mostafa M. Omar

Dr. Madis Laboratories, Inc., South Hackensack, NJ 07606

Phenolic compounds constitute by far the largest and most widespread group of secondary plant products. They display a great variety of structures, ranging from simple compounds containing a single aromatic ring to highly complex polymeric substances such as tannins and lignin. Many drug plants are still used today just as much as in the dim past. Although we have learned only in recent decades to master the art of extraction, separation, isolation and structure elucidation thereby arriving at the true active principles of many plants, the modern use of isolated, standardized and precisely dosed isolates or secondary synthetic products is often similar to the use of the whole plant extract. This text illustrates phenolic plants which are indicative of the great potential provided by the use of the plants. Many are known only from the medical aspect, but many others are used to a much greater extent for beverages, spices and through perfumery cosmetics.

Ginger and Hops are used pharmaceutically, but much larger amounts are used in the beverage industry. Thyme is a useful gastric agent, but is indispensable for flavoring and oral hygiene.

However, phenolic plants of one type or another are used daily by each of us in the form of medicines, cosmetics or spices.

Outlined below are certain characteristics of the chemistry, pharmacology and sources of several plant extractives and constituents of current interest to the food, drug and cosmetic industries. Most contain phenolic structure types which play a role in their uses.

ARTICHOKE: (Cynara scolymus L., family Compositae)

Artichoke is a perennial, herbaceous plant of Mediterranean origin. The cauline leaves are the part of the plant used contain chlorogenic acid, 1,3-di-0-caffeylquinic acid, (cynarine), cryptochlorogenic, neochlorogenic, 3,5-di-0-caffeylquinic acids, caffeic acid and three flavonoids (luteolin 4'-0-glucoside, luteolin 7-0-glucoside and luteolin 7-0-rhamno-glucoside. (Fig.1)

The artichoke was popular in Roman time, both for its therapeutic properties,

0097–6156/92/0506–0154$06.00/0
© 1992 American Chemical Society

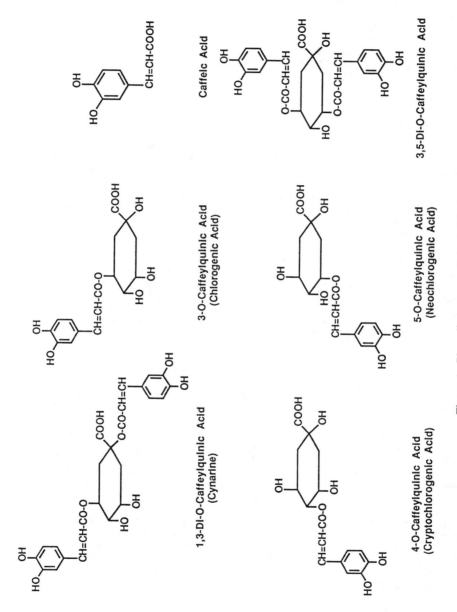

Figure 1. Phenolic compounds in Artichoke

and as a food. The leaves with their bitter taste contain the active principles used pharmacologically. The Artichoke possesses cholagogic, liver protective, nutritive, tonic, stomachic, astringent, diuretic and hypoglycaemicizing properties. Preparations of the drug, in form of fluid extracts and tinctures are used for the excretion of bile and anti-toxic functions of the liver, and as a diuretic for malfunctions of the urinary system, and in arteriosclerosis. Some are also used to reduce forms of puritis, urticaria and eczyma. The dried leaves are used in preparation of special liqueurs. Artichoke has been commercially used in Natural Balance capsules.(1-3)

BERBERIS: (*Berberis vulgaris* L., *family Berberidaceae*)

Berberis is a spiny bush reaching about 2.5 m, and bearing yellow flowers, followed by red berries among the leaves. Both have an orange-yellow inner surface. The stem bark, root bark, the root and the berries are the parts used. The plant contains alkaloids of the isoquinoline type, mainly berberine, berbamine and derivatives, berberrubine, bervulcine, columbamine, isotetrandrine, jatorrhizine, magno-florine; oxyxanthine and vulvracine, may also be present are chelidonic acid, resin and tannins. (Fig.2)
Berberis plant has been used as an antipyretic, anti-haemorrhagic, antiinflammatory and as an antiseptic. Berberine alkaloid has highly bactericidal, amoebicidal and trypanocidal actions. The plant extract is active in vitro and in animals against cholera. Berberine has some anticonvulsant and uterine stimulant activity; it stimulates bile secretion and has sedative and hypotensive effect in animals. Berbamine is also strongly antibacterial against some strains of Staphylococcus aureus, E. coli, S. viridans, Pseudomonase aeruginosa and Salmonella typhi. It has been shown to increase white blood cells and platelet counts in animals with leukocytopaenia and has been used in China to treat patients with leukopaenia due to the chemotherapy. Berbamine is also used to treat essential hypotension. The berberis extract is found in Elixir of Chelidonium (Indigestion Mixture). (4)

BOLDO: (*Peumus boldus* Molina, *family Monimiaceae*)

Boldo is the dried leaf of Peumus boldus Mol., an evergreen shrub, frequenting the meadows of the Andes in Chile, where its yellowish-green fruit is eaten and its bark is used in tanning. Its wood is employed in charcoal making. It has a bitter aromatic taste and lemony camphoraceous odor. The leaves and the bark of the plant are used. Boldo contains isoquinoline type alkaloids (0.7%), including boldine, isocorydine, N-methyl-laurotetanine, norisocorydine, isoboldine, laurolistine and reticuline.(Fig.3) The volatile oil, containing mainly p-cymene, 1,8-cineole, ascaridole and linalool.
The leaves of Boldo are used as a diuretic by stimulating secretion and excretion of urine and also as a liver stimulant. They are used in chronic liver or hepatic torpor (inactivity of the liver) problems. They are used as an antiseptic. This herb has also been found valuable in cases of inflammation of the bladder and for destruction of gonorrhea germs in the urinary tract. The flavor industry employs boldo derivatives in several formulations for liqueurs, bitters and in special compounded aromas for flavoring drugs. Boldo is commercially used in slimming tablets (Boldo Aid to Slimming Tablets).(5-8)

BUCKTHORN: (*Rhamnus frangula* L., *family Rhamnaceae*)

Buckthorn is a shrub or small tree, often thorny with elliptical finely toothed

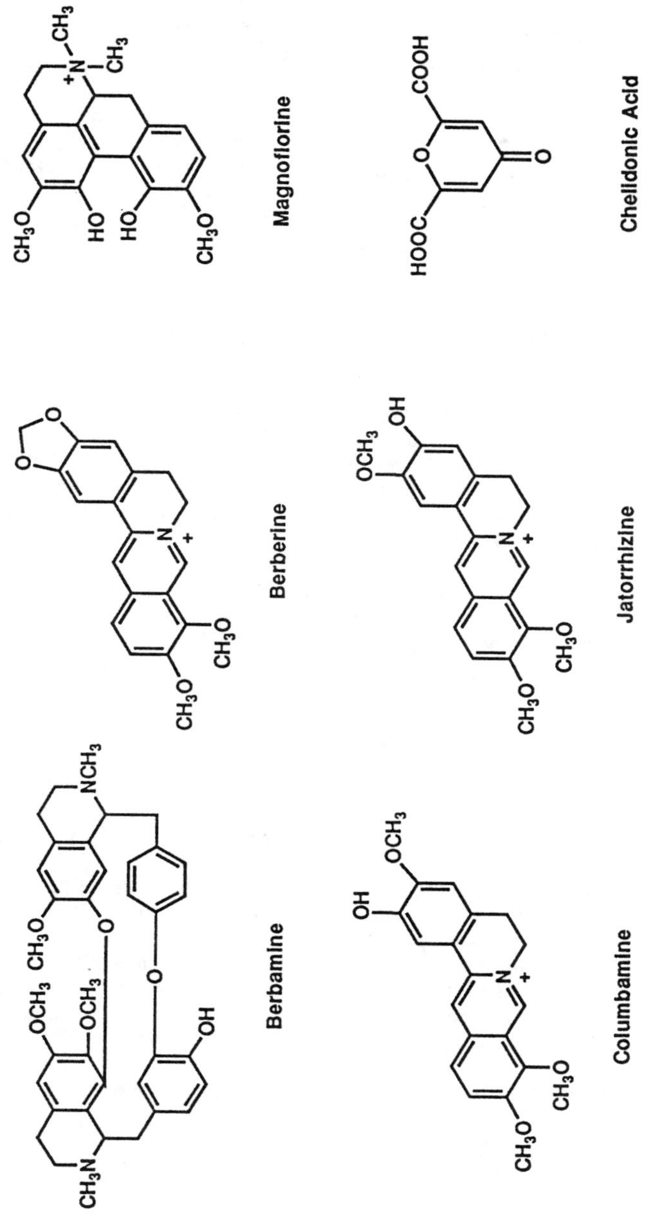

Figure 2. Phenolic compounds in Berberis

leaves. The berries are globular 8-10 cm diameter and black in color. The bark has glossy reddish or greenish brown cork. Berries and the bark are the parts used. They contain an anthraquinone glycosides including emodin, aloe-emodin, chrysophanol and rhein. Frangula-emodin (Frangulin A), rhamnicoside alaterin and physcion have also been reported as constituents. Naphlolide glycosides of the sorigenin type and flavonoid glycosides are present in the bark.(Fig.4)

The berries are used to make syrup of blackthorn which is a laxative. It is used in veterinary practice. The bark may be found as an adulterant of other Rhamnus species. An extract of the berries has been shown to produce tumor necrosis in mice.(9-10)

CHICORY: (*Chichorium intybus* L., *family Compositae*)

Chicory root is brownish with tough, loose, reticulated, white layers surrounding a radiate woody column often crowned with remains of the stem. The root is the part used. It contains up to 58% inulin. Sesquiterpene lactones including lactucin and lactupicrin (= intybin). The root also contains coumarins; chicoriin, esculetin, esculin, umbelliferone and scopoletin.(Fig.5) The roasted root contains a large number of flavor ingredients including acetophenone and the B-carboline alkaloids harman and norharman. (11-15)

Chicory is used as diuretic, tonic and laxative. A decoction has been used for liver complaints, gout and rhumatism. An alcoholic extract has been shown to depress heart rate and amplititude in vitro and has anti-inflammatory effects in vitro. The roasted root is used in coffee mixtures and substitutes. Flavors in which the chicory extract is used are butter, caramel, chocolate, coffee, maple, nut, root beer, sarsaparilla, vanilla, winter green and birch beer.

GENTIAN: (*Gentiana lutea* L., *family Gentianaceae*)

Gentian occurs in commerce as cylinderical pieces of roots and rhizomes. The taste is initially sweet and then bitter. Gentian contains iridoids including amarogenin, gentiopicroside (=gentiopicrin) and swertiamarin. Xanthones such as gentisein, gentisin, isogentisin and 1,3,7-trimethoxyzanthone. Also gentian contains gentianine and gentialutine alkaloids, phenolic acids including acids including gentisic, caffeic, protocatechuic, syringic and simple acids.(Fig.6) Sugars such as gentianose and gentiobiose, and traces of volatile oil are also present.(16-22)

Gentian is used as a bitter tonic. It is the most popular of all gastric stimulantsand very widely used to improve digestion, stimulate the appetite and treat all types of gastrointestinal disorders including dyspepsia, gastritis, heartburn, nausea and diarrhoea. Gentiopicroside has been shown to stimulate gastric secretion in animals and gentian extract is reported to be choleretic. Gentian has been used in Neurelax Tablets, Valerian & Scullcap Compound, Tonic and Nervine Essence, Stomach and Liver Medicinal Tea Bags.

GINGER: (*Zingiber officinale* Roscoe, *family Zingiberaceae*)

Ginger is well-known. The rhizome is freed from rootlets and may be peeled before drying. Some types are sun-bleached to improve appearance. The odor and taste are very characteristic, aromatic and pungent. The rhizome is the part used. Ginger contains approximately 1-2% volatile oil containing mainly zingiberene and bisabolene with zingiberol, curcumene, camphene, citral, cineol, borneol, linalool, methylheptenone and many other minor components. The

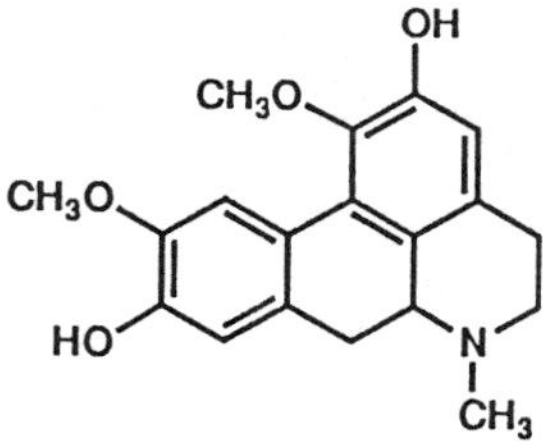

Boldine

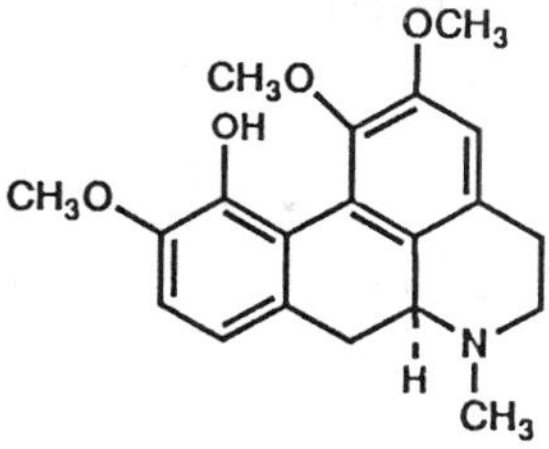

Isocorydine

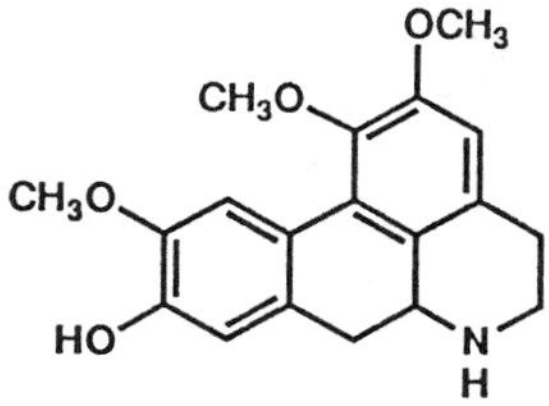

Laurotetanine

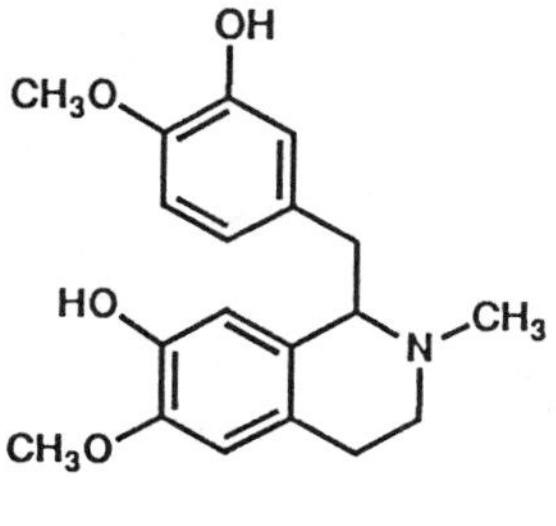

Reticuline

Figure 3. Phenolic compounds in Boldo

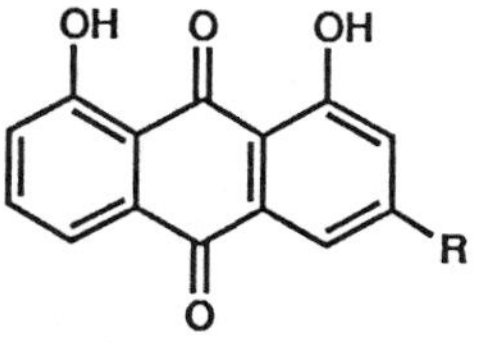

Chrysophanol	R = CH$_3$
Aloe-Emodin	R = CH$_2$OH
Rhein	R = COOH

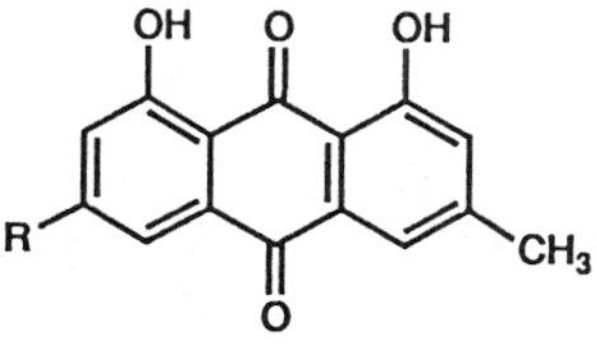

Frangula-(Rheum-)	
Emodin	R = OH
Physcion	R = OCH$_3$

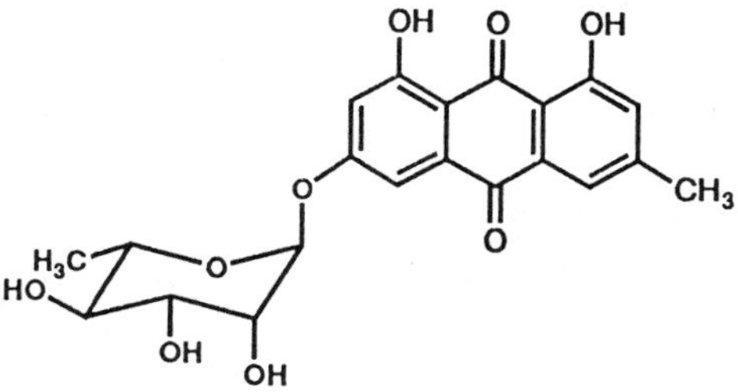

Frangulin A

Figure 4. Phenolic compounds in Buckthorn

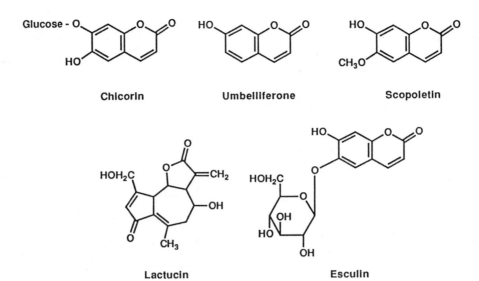

Chicorin   Umbelliferone   Scopoletin

Lactucin   Esculin

Figure 5. Phenolic compounds in Chicory

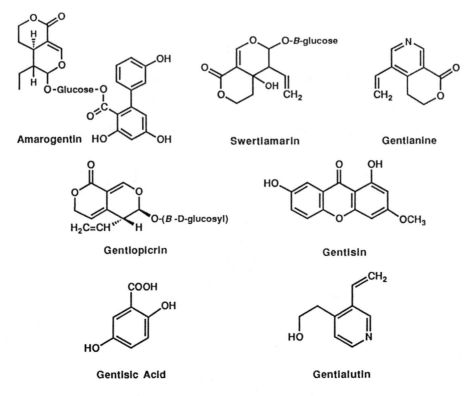

Amarogentin   Swertiamarin   Gentianine

Gentiopicrin   Gentisin

Gentisic Acid   Gentialutin

Figure 6. Phenolic compounds in Gentian

pungent principles in the ginger are a mixture of phenolic compounds with carbon side chains consisting of seven or more carbon atoms, refered to as gingerol, gingerdiols, gingerdiones, dihydrogingerdiones and shogaols.(Fig.7) The shogaols are produced by dehydration and degradation of the gingerols and are formed during drying and extraction. The shogaols are twice as pungent as the gingerols, which accounts for the fact that dried ginger is more pungent than fresh ginger.(23-23)

Ginger is used as a carminative, anti-emetic, spasmolytic, anti-flatulent, antitussive and antipyretic. Gingerol and shogaol have been shown to supress the gastric concentrations and in a recent study capsules containing the dried rhizome were found to be superior to an antihistamine (dimenhydrinate) in preventing the gastrointestinal symptos of motion sickness. Both fresh and dried rhizome supress gastric secretion and reduce vomitting. Gingerol and shogaols also have sedative, antipyretic, analgesic and transient hypotensive actions.

Commercially ginger is used in Elder Flowers and Peppermint with composition Essence, Stomach and Liver Medicinal Tea Bags, Hydrastis Compound Digestive Tablets and as Ginger Root Capsules.

HOPS: *(Humulus lupulus L., family Cannabinaceae)*

Hops is a climbing plant with perennial root. It is widely found in Europe and England growing in hedges and thickets. Hops contain volatile oil, tannins, sugars, fatty acids and resin. The volatile oil (0.3 to 1.0%) consists mainly of the terpene humulene (-caryophyllene), with -caryophyllene, myrecene, farnesene, 2-methylbut-3-ene-2-o1, 3-methyl-but-2-ene-1-al, 2,3,5-trithiahexane with traces of acids such as 2-methyl-propanoic and 3-methylbutanoic which increases significantly in concentration in stored extracts. Hops also contain flavonols mainly glycosides of kaempferol and quercetin. About 3-12% resin composed of -bitter acids such as humulone, cohumulone, adhumulone (Fig.8) and - bitter acids such as lupulene, colupulone, xanthohumol and adlupulone. Hops is also found to contain estrogenic substances of undertermined structure. Two of these have molecular weights from 66-80,000.(33-41)

The hops products are used as sedative, tranquilizers, hypnotics, tonics, diuretics, anodyne and aromatic bitters. The sedative and tranquillizing activity is well established in a variety of animal tests. It is due at least in part to the 2-methylbut-3-ene-2-ol. Hops pillows are popularly used for sleeplessness. Other pharmacological actions of hops include spasmolytic and anti-microbial activity. The bitter acids are anti-microbial. Hops are used in commercial products such as Anased, Neurelax, Blood pressure Tablets and in Nervine Medicinal Tea Bags.

THYME: *(Thymus vulgaris L. & Thymus serpyllum L., family labiatae)*

Thyme is indigenous to the Mediterranean region but found in gardens elsewhere. Both Thymes are small bushy herbs. Thyme contains a volatile oil (1-2%), resin, tannins and gums. Thyme oil yields not less than 40% by volume of phenols. The major constituent is thymol with lesser amount of carvacrol in both species; also 1,8-cineol, borneol, geraniol, linalool, bornyl and linalyl acetate, thymol methyl ether and pinene are present. Thyme contains also flavonoids; apigenin, luteolin, thymonin, naringenin and others in addition to labiatic acid, caffeic acid and tannins.(Fig.9)

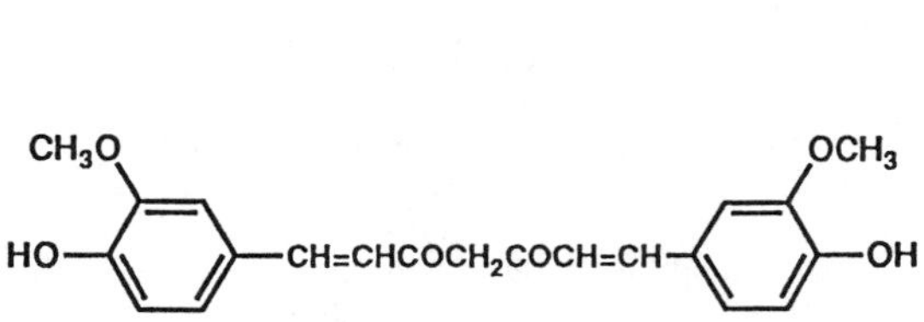

Curcumin

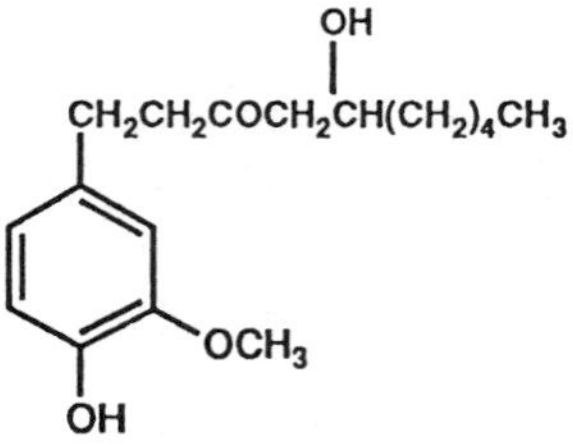

Gingerol

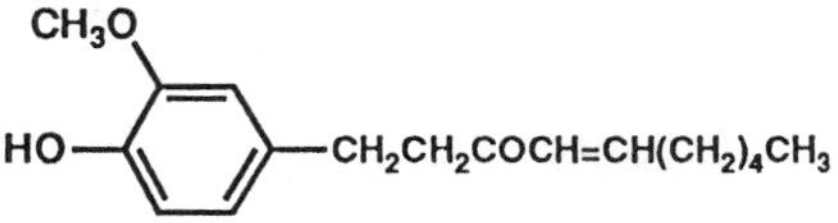

Shogaol

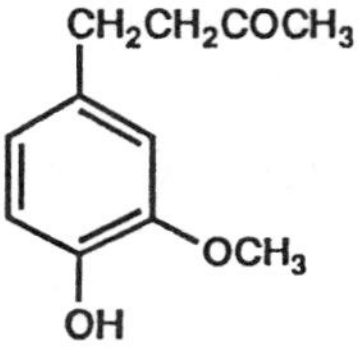

Zingerone

Figure 7. Phenolic compounds in Ginger

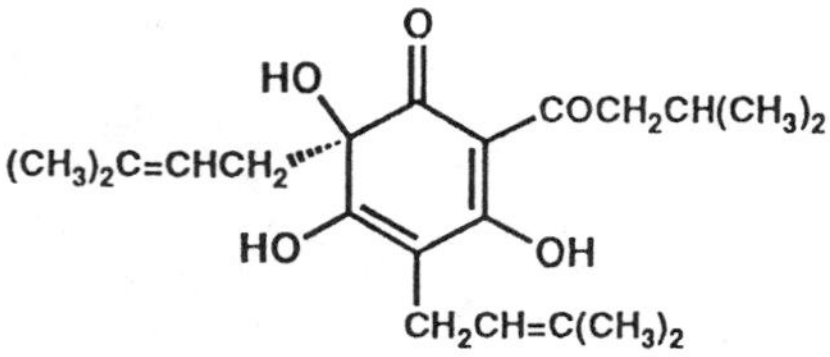

Humulon

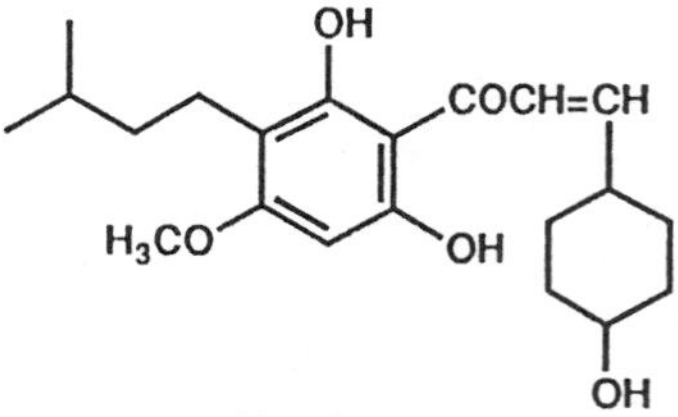

Xanthohumol

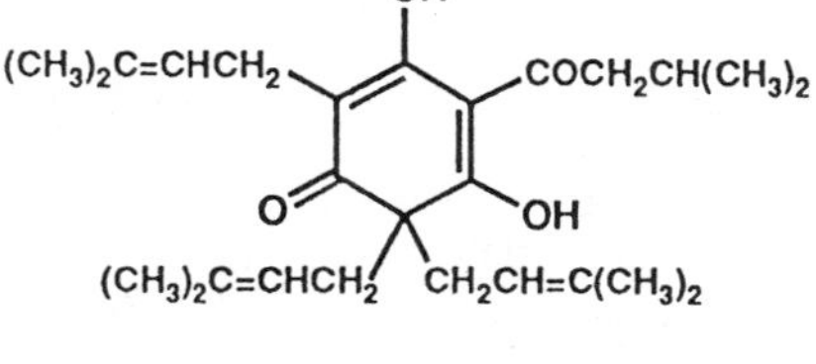

Lupulon

Figure 8. Phenolic compounds in Hops

Thyme is a carminative, antiseptic, antitussive, expectorant, spasmolyhtic and antifungal. Thymol is a urinary tract antiseptic, anthelmintic, a counter irritant for use in topical antirheumatic preparations. It has many other useful actions including as a larvicide. The herb is widely used as flavor and it is a popular ingredient of mouth washes and toothpastes. Thyme is used in commercial products such as Bronc-Ease Capsules.(42-45)

UVA-URSI: (*Arctostaphylos uva-ursi L. Sprent., family Eriaceae*)

Uva-Ursi is a small ever-green shrub found throughout central and northern Europe and North America. The leaves have no marked odor, but are strongly astringent and somewhat bitter. Uva-Ursi leaves contain both tannins and gallic acid. They also contain arbutin, methyl-arbutin, ursone, quercetin, myricetin, uvaol and ursolic acid.(Fig.10)
Uva-Ursi leaves are used in disease of the urino-genital tract as stimulant, diuretic, and antiseptic. They resemble buchu in their action but more astringent. Commercially they are used in Antitis Tablets, Diuretab, Kasbah Remedy, Blood Pressure Medicinal Tea Bags, Liver and Bile Medicinal Tea Bags.(46-49)

WHITE PINE: (*Pinus strobus L., family Pinaceae*)

White Pine is the dried inner bark of <u>Pinus</u> <u>strobus</u> L. (Fam.Pinaceae). The White pine is the principle timber pine of the northern USA and Canada. The outer corky layer of the bark is removed before the inner bark is dried. The White Pine taste is mucilagenous and astringent.
White Pine bark contains coniferin, coniferyl alcohol, pinostrobin, pinomyricetin, pinoquercetin, pinoresinol, cryptostrobin, dihydropinosylvin and pinosylvin.(Fig.11)
Uses of white pine bark includes as an expectorant, demulcent and diuretic. The bark is used mainly in the form of a syrup for coughs and colds. It is contained in the product Prunicodeine tablets with the usual dose of 2 g. and in Bronc-Ease Capsules.(50-52)

WITCH-HAZEL: (*Hamamelis virginiana L., family Hamamelidaceae*)

Witch Hazel is the dried leaf or bark of Hamamelis virginiana L. (Fam. Hamamelidaceae). The plant is a shrub or small tree attaining a height of 8 meters found particularly in low damp woods. The flowers appear in the fall at the same time as the ripening of the fruits of the previous year. The taste is very astringent. Leaves, twigs and the bark are the parts used. Witch-hazel contains 8-10% tannins, composed mainly of gallotannins with some condensed catechins and proanthocyanins. The presence of hamamelitannin is disputed. It also contains flavonoids quercetin, kaempferol, astragalin, myricitrin; volatile oil, containing hexanol, n-hexen-2-al, -and -ionones.(Fig.12)
The bark contains 1-7% tannins, mainly the hamamelitannins with some condensed tannins such as d-gallocatechin, 1-epigallotannin and 1-epicatechin. It also contains saponins, volatile oil and resin. Witch-hazel is used as astringent and haemostatic. Witch-hazel extract is very highly regarded for the treatment of haemorrage, piles and varicose veins. The distilled extract, known as witch-hazel water, is used as a treatment for sprains and bruises, spots and blemishes. It is also used in eye drops, after shave lotions and in cosmetic preparations. When mixed with equal parts of rose water it is used as a skin tonic. Distilled witch-hazel is reputed not to contain tannins.(53-55)

Thymol

Carvacrol

Naringenin

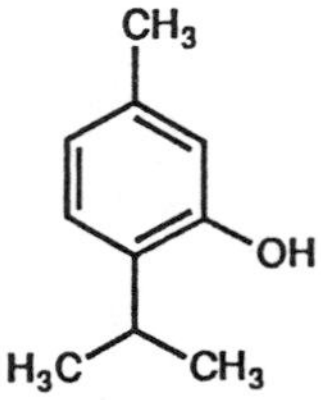

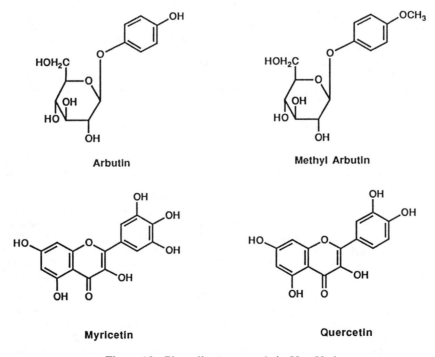

Figure 9. Phenolic compounds in Thyme

Arbutin

Methyl Arbutin

Myricetin

Quercetin

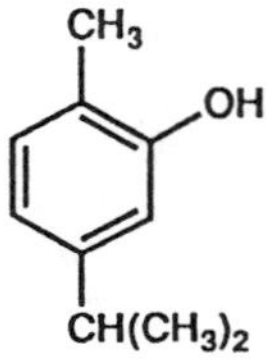

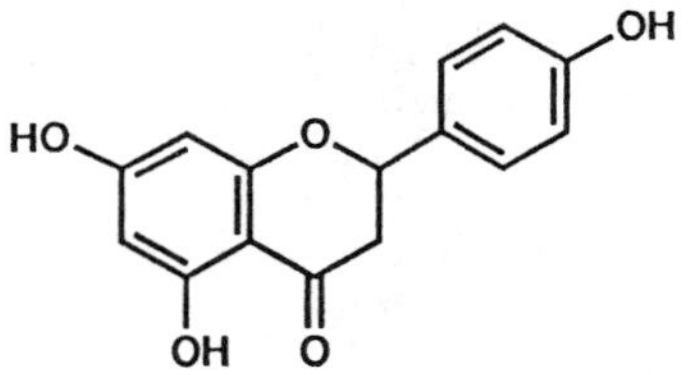

Figure 10. Phenolic compounds in Uva-Ursi

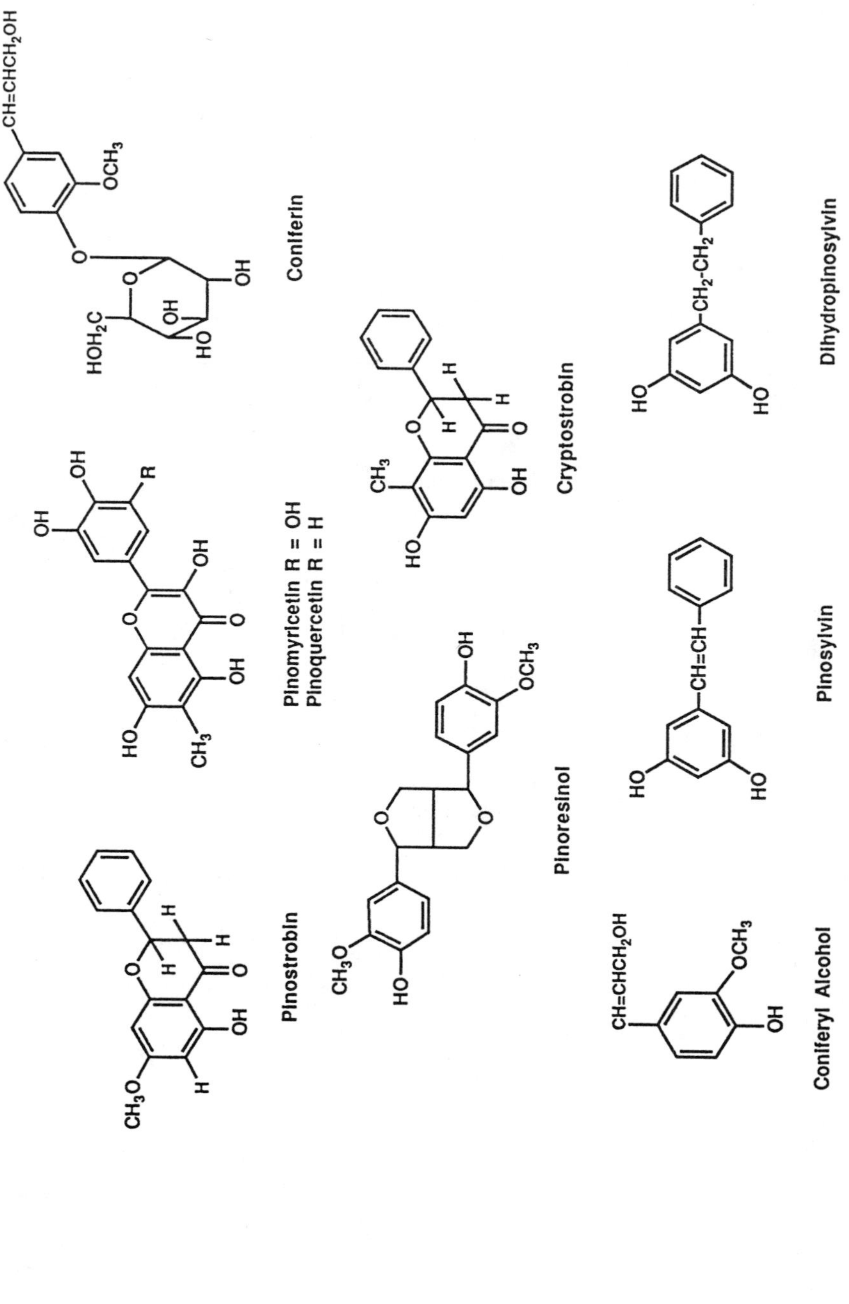

Figure 11. Phenolic compounds in White Pine

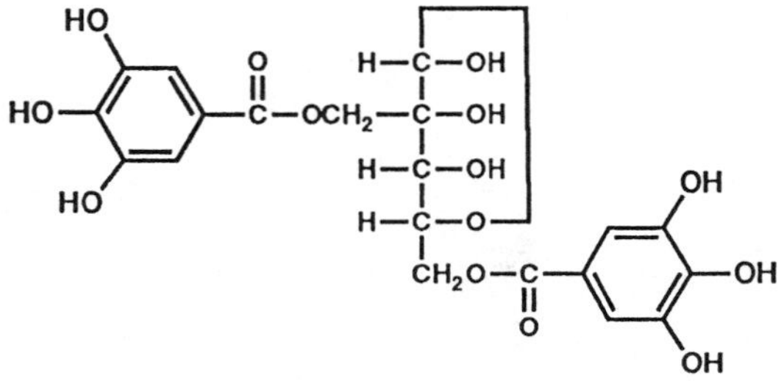

Hamamelitannin

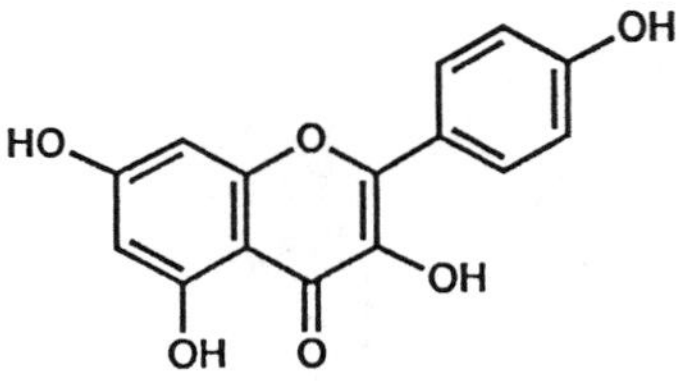

Kaempferol

Figure 12. Phenolic compounds in Witch-Hazel

Literature Cited:

1. Bombardelli, E.; Gabetta, B.; Martinelli, E.M. Fitoterapia **1977**, 134.
2. Litchfield, J.T.; Wilcoxon, F.J. Pharmacol. Exp. Ther. **1949**, 96.
3. Chabrol, E.; Charonnat, R.; Maximin, M.; Waitz, R.; Poirin, J. C.R. Soc. Biol. Franc. **1931**, 1020-1022, 1100-1102.
4. Heinz, A. Drogenkunde; 8th ed., Hoppe Pub. de Gruyter **1973**.
5. Charles, C. Thomas An Atlas of Medicinal Plants of Middle America; J.F. Morton Pub.; USA **1981**, 140-152.
6. Urzua, A.; Acuna P. Fitoterapia **1983**, 4, 175.
7. Bombardelli, E.; Martinelli, E.M. Fitoterapia **1976**, 47, 3.
8. Kreitmar, H. Pharmazie **1952**, 7, 205.
9. Thomson, R.H. Naturally Occuring Quinones; Academic Press, **1971**, 2nd.ed.
10. Rauwald, H.W.; Just, H.D. Planta Med. **1981**, 42, 244.
11. Wagner, H. The Biology and Chemistry of the Compositae, V.N. Heywood ed., Academic Press, **1977**.
12. Camara, B.; Moneger, R Phytochem. **1978**, 17, 91.
13. Proliac, A.; Blanc, M. Helv. Chem. Acta **1976**, 58, 2503.
14. Balbaa, S.; Hilal, S. Planta Med. **1973**, 24, 133.
15. Benoit, P.S. Lloydia **1976**, 39, 160.
16. Inouye, H.; Blanc, M. Tet. Letter **1968**, 4429.
17. Inouye, H. Chem. Pharm. Bull. **1970** 18, 1856.
18. Bricout, J.; Lefort, D. Phytochem. **1974**, 13, 2819.
19. Lewis, J.R.; Gupta, P.J. Chem. Soc. Chem. Comm. **1971**, 4, 629.
20. Atkinson, J.E. Tetrahedron **1969**, 25, 1507.
21. Marco, J.L.; Bertelli, D.J. Phytochem. **1979**, 24, 2567.
22. Denford, K.E. Experientia **1973**, 29, 939.
23. Narasimhan, S.; Gouinarajan, V.S. J. Food Tech. **1978**, 13, 31.
24. Hikino H. Economic and Medicinal Plant Research, Academic Press, UK, **1985**, Vol. 1.
25. Suckawa, M.J. Pharmacobio. Dyn. **1984**, 7, 11, 836.
26. Mowrey, D.B.; Clayson, D.E. Lacet **1982**, 655.
27. Girardon, P.; Toth, L. Planta Med. **1985**, 51, 6, 533.
28. Abdo, M.S.; Al-Khafawi, A.A. Planta Med. **1969**, 17, 14.
29. Kikuchi, F. Chem. Pharm. Bull. **1982**, 30, 754.
30. Al-Meshal, I.A.; Saad, B. Fitoterapia **1985**, 56, 4, 232.
31. Farnsworth, N.R. J. Pharm. Sci. **1975**, 64, 4, 535.
32. Hartley, R.D. Phytochem **1968**, 7, 1641.
33. Hartley, R.D.; Fawcett, C. Phytochem **1968**, 7, 1395.
34. Mior, M. Phytochem **1980**, 19, 2201.
35. Kumai, A.; Okamoto, R. Toxicol. Lett. **1984**, 21(2), 203.
36. Karawya, M.S.; Kawi, M. Lloydia **1969**, 32, 76.
37. Hansel, R.; Nicholas, H. Planta Med. **1982**, 45(4), 224.
38. Dupaigne, P. Planta Med. Phytother. **1974**, 8, 104.
39. Schmalreck, A.F.; Evans, A. Can. J. Microbiol. **1975**, 21, 205.
40. Bhardwaj, D.K.; Singh, R. Curr. Sci. **1977**, 46, 753.
41. Miguel, J.D. J. Agric. Food Chem. **1976**, 24, 833.
42. Svendsen, A.B.; Karlsen, J. Planta Med. **1966**, 14, 376.
43. Montes, G.M. An Real Acad. Farm. **1981**, 47(3), 238.
44. Van Den Broucke, C.O. Pharm. Weekbl. **1983**, 5(1), 9.
45. Jahodar, L.; Yagura, T. Pharmazie **1981**, 36(2), 294.
46. Pelter, A.; Hansel, R. Tet. Lett. **1968**, 19, 2911.

47. Greger, H. Phytochem. **1978**, 17, 806.
48. Frohne, D. Planta Med. **1970**, 18, 1.
49. Ibeder, R.M. J. Food Sci. **1962**, 27, 455.
50. Ammon, H.P.; Handel, M. Planta Med. **1981**, 43, 105, 209 and 313.
51. Zinkel, D.F. Phytochem. **1972**, 11, 425.
52. Bernard, P.; Witchle, M. J. Pharm. Belg. **1982**, 26, 661.
53. Ohta, Y.; Busses, W. Tet. Lett. **1966**, 52, 6365.
54. Sharma, B.R.; Sharma, P. Planta Med. **1981**, 43, 102.
55. Freidrich, H.; Krugen, N. Planta Med. **1974**, 25, 138.

RECEIVED November 20, 1991

Flavor Of Phenolic Compounds

Chapter 13

Contribution of Phenolic Compounds to Smoke Flavor

Joseph A. Maga

Department of Food Science and Human Nutrition, Colorado State University, Fort Collins, CO 80523

The roles of hemicellulose, cellulose and lignin thermal degradation on the formation of phenolic-based wood smoke volatiles are discussed. Compositional differences between hardwoods/softwoods and heartwood/sapwood as related to smoke volatile formation are also described. Phenols are the major contributors to wood smoke aroma, but other compound classes are also important. Various phenols (guaiacol, 4-methylphenol, 2,6-dimethoxyphenol) have been described as possessing smoky aromas.

The smoking of certain foods probably represents one of the earliest known forms of food processing and preservation. However, in modern society, with many forms of preservation available, foods are now primarily smoked to provide unique flavors in the finished product. As a result, various smoke source media, such as hickory, maple and mesquite, are now commonly available to provide diverse smoked flavors.

Through the years, various researchers have attempted to characterize the compounds that are responsible for the unique sensory properties associated with wood smoke. One would predict that this would be a somewhat simple task based on the relatively few number of components found in wood. However, when one considers factors such as wood moisture content, temperature of smoke generation, instability of intermediates, and the low odor thresholds for some of the compounds produced, a very complex and confusing picture can result. One must also consider the fact that when wood smoke is exposed to food, food components can also serve as intermediates in the formation of smoke flavor compounds detected at the time of food consumption.

As a result, to date, over 400 compounds have been identified as volatile components of wood smoke. As can be seen in Table I, most of the identified compounds are carbonyls and phenols. However, as with all flavor chemistry, the presence of a compound does not necessarily mean that it makes any significant contribution to flavor. Therefore, if one were to evaluate the actual flavor contribution of each of the

0097–6156/92/0506–0170$06.00/0
© 1992 American Chemical Society

410 compounds identified, it becomes quickly apparent that phenolic compounds are major contributors to wood smoke flavor.

Table I.   Compound Classes Identified in Wood Smoke

Class	Number Identified
Carbonyls	131
Phenols	75
Acids	48
Furans	46
Alcohols	22
Esters	22
Lactones	16
Miscellaneous	50
Total	410

Therefore, the major objectives of this presentation are to discuss the role of smoke generation variables, including wood composition, on the formation of phenolic compounds in the resulting smoke and to present data on their sensory properties.

Wood Composition

The major component of most woods is cellulose, which is a long chain glucose polymer. Hemicellulose is another major component and is usually composed of a combination of five-carbon sugars such as arabinose and xylose, and six-carbon sugars including glucose, mannose and galactose. The third major component is lignin, which can be classified as a phenolic-based compound containing various possible combinations of differently bound hydroxy- and methoxy-substituted phenylpropane units. Other minor components in most woods include volatile oils, terpenes, fatty acids, simple carbohydrates, polyhydric alcohols, nitrogen and phenolic compounds, and inorganic constituents (1).

With some wood sources, bark composition should also be considered. Bark is primarily composed of suberin, which is an ester of various aliphatic hydroxy acids and numerous phenolic acids bound in various configurations.

Hardwood Versus Softwood

From a structural standpoint, hardwoods are more complex than softwoods. Also, most hardwoods have a higher proportion of pentosans than hexosans in their hemicellulose fraction than softwoods. The lignin fraction also differs between hardwoods and softwoods with approximately three more syringlypropane units than guaiacylpropane units in hardwood as in softwood, and in general, softwoods have more lignin than hardwoods. The mannan content of most softwoods is normally 10-15%, while in hardwoods, it rarely exceeds 3%. Also, the

xylan content of hardwoods (10-20%) is usually twice that of softwoods. Softwoods also contain more resin extractives than hardwoods.
    Because of these compositional differences, wood smoke generated
from hardwood and softwood differ in their relative amounts of
phenolics. As seen in Table II, hardwood smoke is higher in syringol
derivatives while softwood smoke has a greater proportion of guaiacol
derivatives.

Table II.   Phenolic Differences in Hardwood
Versus Softwood Wood Smoke

Compound	Relative % in Hardwood Smoke	Softwood Smoke
Guaiacol	8	10
4-Methylguaiacol	7	10
4-Ethylguaiacol	3	5
4-Vinylguaiacol	3	5
4-Allylguaiacol	2	4
Syringol	12	3
4-Methylsyringol	9	3
4-Vinylsyringol	3	1
4-Allylsyringol	3	1
4-Propenylsyringol	10	2

SOURCE: Adapted from ref. 2.

Heartwood Versus Sapwood

The amount of heartwood (center of tree) to sapwood (outer portion of
tree) can significantly vary with different woods and also within the
same species. The faster a tree grows, the less heartwood it
contains. Also, the proportion of heartwood decreases with tree
height. From a compositional standpoint, heartwood is lower in
moisture than sapwood; and with softwoods, heartwood has less lignin
and cellulose but more extractives than sapwood, while no consistent
trend is apparent in hardwoods.

Thermal Reactions

Among the major wood components, hemicellulose is the first to undergo
thermal destruction, followed by cellulose and finally lignin. The
temperatures required for these reactions are summarized in Table III.
    From a theoretical standpoint, smoke can be generated at a
relatively low temperature representing primarily hemicellulose
degradation; but under these conditions, lignin may not be completely
degraded and thus the resulting smoke will have a different composition than if a higher temperature were utilized.
    The presence or absence of air can also significantly influence
the end products formed during the thermal degradation of wood. In

the absence of air, a combination of water, organic liquids and gases are released, leaving a mixture of tar, pitch and charcoal. In the presence of air, combustion results, degrading both volatile and nonvolatile components as well as producing additional volatiles.

The influence of wood smoking temperature on resulting smoke phenol content can perhaps best be appreciated by viewing the data shown in Table IV. As temperature is increased, phenol content also increases. However, at higher temperatures the phenol content begins to gradually decline, probably due to thermal instability.

Table III. Thermal Degradation of Wood

°C	Reaction
90-170	Free and bound water losses
200-260	Hemicellulose degradation
260-310	Cellulose degradation
310-500	Lignin degradation
>500	Secondary reactions

SOURCE: Adapted from ref. 2.

Table IV. Wood Smoking Temperature Versus Phenol Content

Temperature (°C)	Content (mg/100g)
400	350
450-500	800
500-600	2960
700-800	2810
900-1050	1945

SOURCE: Adapted from ref. 3.

Hemicellulose Degradation

Hemicellulose, upon thermal degradation, primarily yields furan derivatives and aliphatic carboxylic acids. However, both furans and acids are not very thermally stable, especially at higher temperatures, and thus each undergo extensive oxidation. As seen in Figure 1, both the acetyl and O-methyl groups of hemicellulose can react to produce volatile degradation products.

Cellulose Degradation

The thermal degradation of cellulose occurs by two distinct temperature dependent pathways. At lower temperatures, bond scission, elimination of water, formation of free radical, carbonyl, carboxyl and hydroperoxide groups, and the evolution of carbon monoxide and dioxide gasses occur. At higher temperatures, a series of reactions including transglycosylation and fission occur producing a series of anhydro sugars and lower molecular weight compounds. As seen in Figure 2, numerous intermediates can form, thereby producing a variety of catalysts that can be involved in further classical carbohydrate thermal degradations.

Lignin Degradation

Phenols and derivatives are most directly related to lignin degradation. Initial degradation by fission occurs with the heterocyclic furan, pyran rings and ether linkages of lignin. This eventually results in the production of guaiacol, which in turn can further degrade to form phenol and cresols. Since hardwoods have higher amounts of methoxy substitutions than softwoods, the dimethoxy rings can result in the formation of syringol and an entire series of para-substituted compounds in addition to guaiacol.

The formation of ferulic acid is also a key step in lignin degradation since it can undergo further decarboxylation to produce the series of compounds shown in Figure 3. In the presence of air, oxygenated compounds such as vanillin are formed.

As can be seen from above, wood composition and smoke generation temperature can significantly influence the resulting volatile composition of wood smoke.

Phenols and Wood Smoke

Various researchers (4-15) have reported that phenols, as a compound class, are the major contributors to wood smoke aroma. However, other researchers tend to dispute the above claim since they reported on other compound classes that possess typical smoke aroma properties. As examples, the data summarized in Tables V and VI show that phenols are not totally responsible for smoke aroma.

There is no question that phenols can be generated at relatively high concentrations in wood smoke, and when their relatively low odor thresholds are considered, there is no question that they do make a significant contribution to wood smoke aroma. As seen in Table VII, guaiacol, 4-methylguaiacol, syringol and phenol are the major phenols found in wood smoke. In the case of commercial liquid smokes, up to 1.06 mg/ml of phenols have been reported (17).

Table V. Contribution of Smoke Subfractions
To Wood Smoke Aroma

Smoke Subfraction	Aroma Quality
Phenolic only	Good
Phenolic plus carbonyl	Better
Phenolic plus neutral plus basic	Best

SOURCE: Adapted from ref. 13

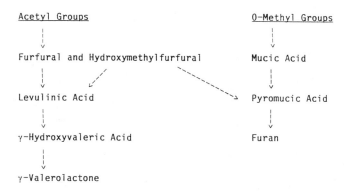

Figure 1. Thermal degradation of hemicellulose.

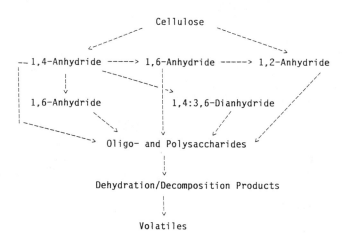

Figure 2. Thermal degradation of cellulose.

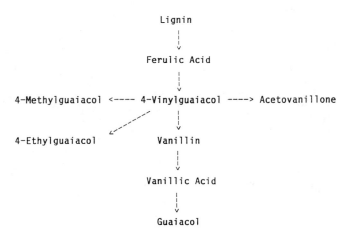

Figure 3. Ferulic acid thermal degradation.

Table VI.   Aroma Intensities of Various
Wood Smoke Subfractions

Wood Smoke Source	Phenolic	Basic	Carbonyl	Neutral
Apple	3.40	5.32	4.02	2.28
Chestnut	3.21	3.44	3.86	1.82
Cherry	3.59	4.61	5.40	4.13
Red Oak	3.17	4.07	6.03	2.92
White Oak	3.44	3.52	5.46	2.63

SOURCE: Adapted from ref. 16.

Table VII.   Major Phenols Identified in Wood Smoke

Compound	Whole Smoke (mg/l)	Vapor Phase (mg/l)
Guaiacol	417	32
4-Methylguaiacol	333	14
Syringol	392	7
Phenol	59	6
Total	1201	59

SOURCE: Adapted from ref. 5.

It should be noted that wood smoke is not the only source of phenols in certain smoked foods.   Spices such as cinnamon, thyme, pepper, nutmeg, marjoram and cloves have also been shown to produce phenols when heated (15).

Sensory Properties

Various researchers have attempted to evaluate the sensory properties of individual phenols relative to wood smoke.   For example, in Table VIII it can be seen that among the three compounds evaluated, by far the most potent was 4-methylguaiacol in both odor and taste sensitivity.   A panel was also asked to characterize the taste and odor of the same three compounds and 58% of the panel reported that guaiacol had a smoky taste while 72% thought that 2,6-dimethoxyphenol had a smoky odor (18).   From these data, it would appear that phenols have different odor and taste properties thus complicating their overall degree of influence relative to smoke flavor.   Flavor is commonly defined as a combination of aroma and taste properties.   The odor properties of various phenols have been characterized; and, as seen in Table IX, various descriptors related to wood smoke have been reported.

Phenol compound volatility also is a factor relative to smoke flavor contribution (20).   The low boiling phenol portion (60-90°C) obtained from the distillation of smoke condensate was found to

primarily contain phenol, cresols and quaiacols and was described as
having a hot and bitter taste; whereas the fraction distilled at 92-
132°C and containing isoeugenols and syringols had a pure and
characteristic smoke flavor. In turn, the high boiling phenol
fraction (132-200°C) had an acid, chemical sensory property.

Table VIII. Reported Flavor Thresholds (PPM) and
Corresponding Flavor Indices* of Phenols

Compound	Taste Threshold	Taste Index	Odor Threshold	Odor Index
4-Methylguaiacol	0.065	6,400	0.09	4,600
Guaiacol	0.013	90,000	0.021	58,800
2,6-Dimethoxyphenol	1.65	1,400	1.85	1,200

* Concentration in smoke condensate ÷ mean threshold concentration

SOURCE: Adapted from ref. 18.

Table IX. Odor Descriptions of Various Phenols

Description	Compounds
Pungent	Phenol, O-, M- and P-cresol; 2,3- and 2,4-Xylenol
Cresolic	2,6-, 3,4- and 3,5-Xylenol; 2-Ethyl-, 3-ethyl- and 5-methylphenol; 2,3,5-Trimethylphenol
Sweet, smoky	Guaiacol; 3-Methyl- and 4-ethyl-guaiacol
Weak, phenolic	3-Methylguaiacol
Woody	4-Allylguaiacol
Smoky	2,6-Dimethoxyphenol
Heavy, burnt	2,6-Dimethoxy-4-methyl and ethyl, propyl, propenylphenol
Heavy, sweet, burnt	Pyrocatechol; 3-Methyl-, 4-methyl- and 4-ethylpyrocatechol

SOURCE: Adapted from ref. 19.

Conclusions

Phenolic compounds do make major contributions to smoke flavor, and the amounts and types of these compounds produced during some generation are primarily dependent upon the wood source and smoke generation temperature.

Literature Cited

1.   Maga, J. A. *Smoke in Food Processing*; CRC Press: Boca Raton, FL 1988; 1-27.
2.   Baltes, W.; Wittkowski, R.; Sochtig, I.; Block, H.; Toth, L. *In The Quality of Foods and Beverages*; Charalambous, G.; Inglett, G., Eds.; Academic Press: New York, NY, 1981; 1-14.
3.   Toth, L. *Fleischwirtschaft*. 1980, *60*, 728.
4.   Knowles, M. E.; Gilbert, J.; McWeeny, D. J. *J. Sci. Food Agric.* 1975, *26*, 189.
5.   Kornreich, M. R.; Issenberg, P. *J. Agric. Food Chem.* 1972, *20*, 1109.
6.   Lustre, A. B.; Issenberg, P. *J. Agric. Food Chem.* 1970, *18*, 1056.
7.   Radecki, A.; Lamparczyk, H.; Grzybowski, J.; Halkiewicz, J.; Bednarek, P.; Boltrukiewicz, J.; Rapicki, W. *Bromatol. Chem. Toksykol.* 1976, *9*, 327.
8.   Radecki, A.; Bednarek, P.; Boltrukiewicz, J.; Cacha, R.; Rapicki, W.; Grzybowski, J.; Halkiewicz, J.; Lamparczyk, H. *Bromatol. Chem. Toksykol.* 1975, *8*, 179.
9.   Radecki, A.; Grzybowski, J. *Bromatol. Chem. Toksykol.* 1981, *14*, 129.
10. Radecki, A.; Grzybowski, J.; Halkiewicz, J.; Lamparczyk, H. *Acta Aliment. Pol.* 1977, *3*, 203.
11. Luten, J. B.; Ritskes, J. M.; Weseman, J. M. *Z. Lebensm. Unters. Forsch.* 1974, *168*, 289.
12. Fiddler, W.; Doerr, R. C.; Wasserman, A. E.; Salay, J. M. *J. Agric. Food Chem.* 1966, *14*, 659.
13. Fujimaki, M.; Kim, K.; Kurata, T. *Agric. Biol. Chem.* 1974, *38*, 45.
14. Kasahara, K.; Nishibori, K. *Bull. Jpn. Soc. Sci. Fish.* 1979, *45*, 1543.
15. Baltes, W.; Sochtig, I. *Z. Lebensm. Unters. Forsch.* 1979, *169*, 17.
16. Maga, J. A.; Japojuwo, O. O. *J. Sensory Stud.* 1986, *1*, 9.
17. Ishiwata, H.; Watanabe, H.; Watanabe, M.; Hayashi, T.; Hara, Y.; Kato, S.; Tanimura, A. *Bull. Nat. Inst. Hygienic Sci.* 1976, *94*, 112.
18. Wasserman, A. E. *J. Food Sci.* 1966, *31*, 1005.
19. Kim, K.; Kurata, T.; Fujiman, M. *Agric. Biol. Chem.* 1974, *38*, 53.
20. Toth, L.; Potthast, K. *Adv. Food Res.* 1984, *29*, 87.

RECEIVED November 7, 1991

Chapter 14

Hydroxycinnamic Acids as Off-Flavor Precursors in Citrus Fruits and Their Products

Michael Naim[1], Uri Zehavi[1], Steven Nagy[2], and Russell L. Rouseff[3]

[1]Department of Biochemistry and Human Nutrition, Faculty of Agriculture, Hebrew University of Jerusalem, Israel
[2]Florida Department of Citrus, Citrus Research and Education Center, Lake Alfred, FL 33850
[3]Citrus Research and Education Center, University of Florida, Lake Alfred, FL 33850

During maturation, processing and storage of fruits and vegetables, hydroxycinnamic acids may become potential precursors for a variety of vinyl phenols which contribute desirable or objectional aroma to important food products. In citrus juice, 4-vinyl guaiacol (PVG) is a major detrimental off-flavor with a taste threshold of 0.075 ppm. Research conducted at our laboratories in recent years has revealed that in grapefruit and oranges, hydroxycinnamic acids occur mainly in bound forms. The peel contained the major portion of these acids compared to the endocarp, and the flavedo contained higher concentrations than the albedo. In most cases, hydroxycinnamic acids content was in the following order: ferulic acid > sinapic acid > coumaric acid > caffeic acid. PVG was produced in citrus juice from degraded free ferulic acid which was released from bound forms during processing and storage. Aroma quality of stored orange juice was related to changes in free ferulic acid and PVG content. Bound or conjugated forms of ferulic acid differ significantly in their ability to serve as precursors for ferulic acid, and this ability depends on the type of linkage with ferulic acid. Although the content of free ferulic acid in citrus fruits may decrease during the season, that present at the end of the season still exceeds the threshold level of PVG.

Phenolic acids have been implicated as possibly influencing the toxicological, nutritional, color, sensory and antioxidant properties of foods with which they are associated (1-3). They have been shown to be directly involved in the

0097–6156/92/0506–0180$06.00/0
© 1992 American Chemical Society

biosynthesis of coumarins, flavonoids and lignin (4). Hydroxycinnamic acids (HCA), a major class of phenolic acids, occur naturally as either free or in a wide range of bound or conjugated forms containing organic acids, sugars, amino compounds, lipids, terpenoids and other phenolics (4, 5). Ferulic and coumaric acids were reported (6, 7), although apparently not yet found in citrus, to be covalently attached to polysaccharides. The release of free ferulic acids from these polymers could be similar to the release of other bound hydroxycinnamic acids. Questions concerning the mode of linkage between phenolic acids and carbohydrates in plant cell walls have attracted considerable attention in recent years. These acids are considered to be present as phenolic - carbohydrate esters because they are released following alkaline hydrolysis (8, 9). Modern chemical and physical (particularly NMR and MS) methods (5, 8, 9) have been extensively employed in structure elucidation of phenolic - carbohydrate esters.

Vinyl phenols are produced from free HCA's in fruits, vegetables and their products and contribute to either desirable or objectionable aroma of important food products, such as fried soy products, cauliflower, cooked asparagus, tomato, roasted peanuts, cooked corn, and dried mushrooms (see reference 1 for detailed references). HCA and their derivatives, the potential precursors of vinyl phenols, are present in all plant-derived food systems and in most diets. Their levels may vary dramatically and are affected by factors such as germination, ripening, storage and processing (1).

The occurrence of vinyl phenols in citrus products during processing and storage may be particularly significant with regard to their possible role as off-flavors. Some objectionable volatiles of citrus products can significantly affect acceptance since they reach their flavor threshold levels under typical processing and storage conditions, whereas other volatiles may be irrelevant as off-flavors (10). In stored orange and grapefruit juices, PVG is a major detrimental compound with a flavor threshold of 0.075 ppm (11).

Distribution of Phenolic Acids in Citrus Fruit and Juice

HCA's and their bound forms are found in most citrus fruit parts, and their extraction and isolation are mainly based on polarity, acidity and hydrogen-bonding capacity of the hydroxyl group(s) attached to the aromatic ring. A variety of solvents are used for extraction, but with the exception of HCA bound to insoluble carbohydrate and protein, the vast majority of HCA may be extracted with methanol and/or ethanol mixtures. Following extraction, the preparation is filtered, reduced in volume and, generally, purified by extraction with ethyl acetate and passed through a silica gel column by the procedures of Naim et al. (12). The column eluate is concentrated to a small volume and subjected to HPLC analysis. Optimum HPLC conditions have been developed by Rouseff et al. (13) for the separation of the five cinnamic acids: caffeic, coumaric, sinapic, ferulic and cinnamic. Figure 1 demonstrates a separation using an Alltech Adsorbosphere HS, 5μ C-18 column, 25 cm x 4.6 mm i.d. and eluting with CH_3CN buffered at pH 2 under gradient conditions (20% to 90% CH_3CN) over 30 minutes (13).

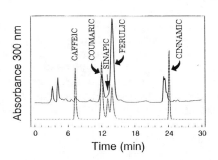

FIGURE 1.   Chromatogram of hydrolyzed orange juice (——) and standards of equal concentration (- - - -). Reproduced with permission from ref. 13.   Copyright 1990 John Wiley and Sons.

It is evident that most HCA's are found in citrus fruit parts in bound forms *(12, 14)*. In a recent study by Peleg et al. *(15)*, the distribution of bound and free phenolic acids in grapefruit and oranges was quantitated as noted in Tables 1 and 2. In all fruit parts, only small quantities occurred as free acids while most were present in bound forms that could be liberated by hydrolysis. In most cases, the bound hydroxycinnamic acids content was in the following order: ferulic acid > sinapic acid > coumaric acid > caffeic acid. Ferulic acid content was high compared with other acids. Values obtained for caffeic acid were apparently low due to decompositon during hydrolysis *(16)*. Peels contained the major quantity of cinnamic acids when compared to the endocarp, and the flavedo was richer in acids than the albedo. These results were expected in view of the fact that in general, citrus peels contain higher levels of many constituents than the corresponding juice or other edible portions *(17)*. Orange peels contained more cinnamic acids than grapefruit peels but the contents in juice sacs and endocarp of both fruits were similar. A concentration gradient of HCA was observed from the flavedo inward. Therefore, the technological processes to which fruits are exposed and, particularly, during juice extraction may affect the amount of HCA transferred into the juice; thus, affecting the amount of off-flavor precursors in citrus products.

The observed concentration of bound forms of cinnamic acids in juice was not affected by the harvest date. However, a trend towards an increase in the content of bound sinapic and coumaric acids (grapefruit and orange) and caffeic (grapefruit) was suggested *(15)*. This may be related to biochemical changes that occur when fruits reach their maximum size and growth is terminated *(17)*. Conversely, the content of free cinnamic acids was reduced as a function of harvest time, though statistical significance was not evident for all cases. It is pertinent to note that, although the content of free ferulic acid decreased during the season, that present at the end of the season still exceeded the flavor threshold level of PVG. This indicates a continual source of potential precursors for the formation of objectional aroma components during processing and storage of juice products.

Significance of Ferulic Acid as a Precursor for PVG

Tatum et al. *(11)* identified α-terpineol, 2,5-dimethyl-4-hydroxy-3-(2H)-furanone and PVG in stored juices. Detection limits for these compounds in juice were 2.5, 0.1, and 0.075 ppm, respectively *(11)*. When the above three compounds were added collectively to freshly processed single-strength orange juice (SSOJ), they imparted an aged, off-flavor aroma, similar to that observed in stored juice. The most detrimental component was PVG which contributed an "old-fruit" or "rotten" flavor to the juice. PVG accumulated during short- *(12)* and long-term storage of orange juice *(18)*. The accumulation of PVG in orange juice during storage was dependent on time and temperatures. When orange juice was stored at

Table 1. Content (mg/kg) of hydroxycinnamic acids (bound and free) in grapefruit

Fruit parts	Sinapic Bound	Free	Ferulic Bound	Free	Coumaric Bound	Free	Caffeic Bound	Free
Peel	31.8	2.1	155.5	4.9	24.3	2.4	10.1	1.7
Albedo	2.0	1.2	24.9	1.5	6.6	1.0	5.8	1.4
Flavedo	29.8	0.9	130.6	3.4	17.7	1.4	4.2	0.3
Juice sacs	3.5	0.7	27.8	0.5	16.7	1.5	11.1	0.7
Endocarp	5.1	0.3	24.5	0.3	2.5	0.9	9.8	1.8

Reprinted with permission from ref. 15. Copyright 1991.

Fruits harvested randomly (mid season) from various sections of 4 trees to provide 1 kg material. Values derived from HPLC analyses (300 nm). Flavedo values were calculated.

Table 2. Content (mg/kg) of hydroxycinnamic acids (bound and free) in oranges

Fruit parts	Sinapic		Ferulic		Coumaric		Caffeic	
	Bound	Free	Bound	Free	Bound	Free	Bound	Free
Peel	95.1	5.4	178.4	3.2	76.7	0.5	7.3	0.2
Albedo	46.2	0.1	27.2	0.5	5.1	0.0	3.5	0.0
Flavedo	48.9	5.3	151.8	2.2	71.6	0.5	3.8	0.2
Juice sacs	8.6	0.1	28.0	0.1	5.3	0.0	3.1	0.0
Endocarp	10.8	0.1	21.3	0.1	4.4	0.0	1.8	0.0

Reprinted with permission from ref. 15. Copyright 1991.

Fruits harvested randomly (mid season) from various sections of 4 trees to provide 1 kg material. Values derived from HPLC analyses (300 nm). Flavedo values were calculated.

elevated temperatures (e.g., 50°C) for more than 3 months *(18)*, PVG decreased, presumably due to its transformation to dimer and trimer forms *(19)*. We have hypothesized that PVG is formed in citrus products from free ferulic acid following its release from bound forms *(12)*. A mechanism by which cinnamic acids are decarboxylated to vinyl phenols has been proposed *(20, 21)*. Laboratory pasteurization of SSOJ increased the content (p < 0.05) of free ferulic acid from 185 μg/L to 316 μg/L *(12)*. Similarly, free ferulic acid content in commercial, pasteurized samples of SSOJ following incubation (28 days) was temperature dependent: 2.9, 3.7, and 9.5 mg/L at 4°, 35°, and 50°C, respectively. These data provide evidence that ferulic acid in orange juice was released from bound forms during storage, and pasteurization of fresh juice accelerates this release. About 200 ppb PVG (more than twice the detection threshold) can be found in commercial single strength orange juice following pasteurization (Figure 2). As little as 14 days of storage of single-strength juice at 35°C resulted in PVG accumulation of more than 3-fold threshold levels. Furthermore, PVG accumulation in stored orange juice was accelerated by the addition of exogenous ferulic acid (Figure 2). Degradation of ferulic acid under similar conditions was evident in model solutions of orange juice (MOJ) incubated with added ferulic acid *(12)*. Concomitantly, the addition of ferulic acid to orange juice resulted in a reduction of aroma quality, most probably due to the accelerated accumulation of PVG (Figure 3).

Source of Free Ferulic Acid in Citrus Juice and Mechanisms For Its Release and PVG Accumulation

Potential sources of free ferulic acid in oranges and grapefruit are bound forms of the acid; five of them are currently known, namely, feruloylputrescine, feruloylglucose, feruloylglucaric, diferuloylglucaric and feruloylgalactaric acids (see Figure 4) *(8, 9, 22, 23)*. Bound or conjugated forms of ferulic acid, however, depending on the type linkage with ferulic acid, may differ significantly in their ability to serve as precursors for ferulic acid and the resulting objectional aroma of PVG *(24)*. Employing feruloylputrescine and feruloylglucose in MOJ revealed (Figure 5) that the release of ferulic acid following incubation was observed in both cases, but the rate of release was significantly higher during incubation of feruloylglucose. This was anticipated for a glucosyl ester compared to an amide under acidic conditions. It is apparent that in relation to its concentration, feruloylglucose may be particularly important in releasing free ferulic acid and, subsequently, the production of PVG in stored citrus juice. It should be noted that the conversion of the recently isolated feruloylglucaric, diferuloylglucaric or feruloylgalactaric acids *(6, 7)* to free ferulic acid and then to PVG shown in Figure 4 has yet to be demonstrated.

Evidence that PVG is formed from ferulic acid was recently provided with MOJ incubated with ferulic acid *(25)*. PVG was not found in MOJ

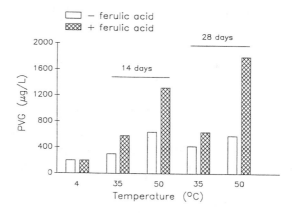

Figure 2. PVG content in single strength orange juice incubated for 14 and 29 days with and without addition of 10 mg/L ferulic acid. (Adapted from ref. 12.)

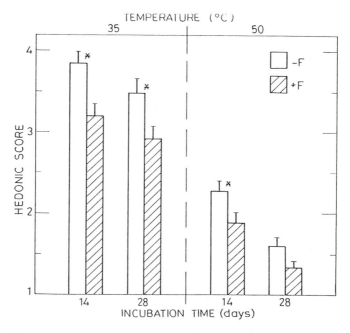

Figure 3. Hedonic scores given to the aroma of single strength orange juice incubated with and without addition of ferulic acid (F). A value of 1 was assigned for objectional aroma, 5 for the most pleasant, and numbers in-between for other preferences. Values are the mean and SEM of 44 replicates for each data point. *Indicates a significant difference between scores given to samples incubated with ferulic acid versus those incubated without addition of ferulic acid. (Reproduced with permission from ref. 12. Copyright 1988 Institute of Food Technologists.)

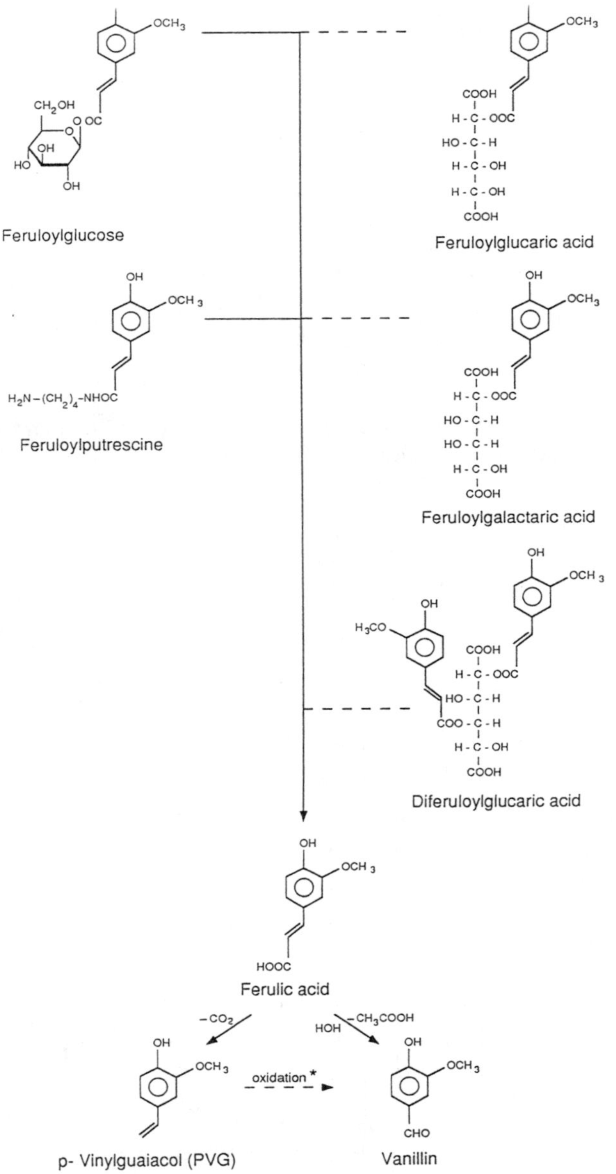

FIGURE 4.   Possible pathways for the release of free ferulic acid from its bound forms in citrus juice and the degradation to either PVG or vanillin. Dotted lines indicate hypothetical pathways.

* Could not be demonstrated in our MOJ experiments.

incubated without ferulic acid. Vanillin, another ferulic acid degradation product, was also detected. A recent study *(26)* indicated the presence of vanillin in processed orange juice. It has been suggested *(20)*, that upon the decarboxylation of ferulic acid, the obtained PVG is further oxidized to vanillin (Figure 4). However, since vanillin was not produced in MOJ incubated with added PVG, but which did not contain ferulic acid *(Peleg et al., unpublished)*, one may hypothesize that in citrus juice, ferulic acid is converted to vanillin (retro-aldol) via a direct route without intermediate PVG formation (Figure 4). More importantly, PVG can be formed from ferulic acid under nitrogen atmosphere *(20)*. Nitrogen atmosphere and the presence of butylated hydroxytoluene (BHT) did not appear to significantly inhibit PVG accumulation in stored MOJ containing ferulic acid, although the formation of browning products were significantly reduced *(25)*. Further, the presence of Cu^{++} ions did not stimulate, but rather inhibited PVG accumulation (though stimulated vanillin formation). One may, therefore, hypothesize that PVG is formed from free ferulic acid in citrus juice mainly by ionic mechanisms and factors which affect PVG formation may be different from those of sugar degradation during storage of orange juice.

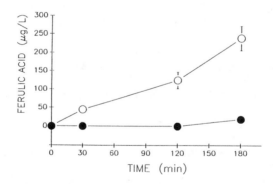

FIGURE 5.   Ferulic aid content in model system of orange juice solutions incubated with either feruloylputrescine (●) or feruloylglucose (o). Adapted from ref. 24.

Acknowledgments

This work was supported by Grant No. I-1528-88 from BARD, The Untied States - Israel Binational Agricultural Research and Development Fund. The collaboration by Mrs. Hanna Peleg, Dr. Benjamin J. Striem, Dr. Joseph Kanner and Dr. Hyoung S. Lee is acknowledged. We thank Dr. Erwin Glotter for helpful suggestions.

Literature Cited

1.   Maga, J. A. *CRC Critical Reviews in Food Science and Nutrition* **1978**, *10*, 323-372.
2.   Herrmann, K.   *Chem. Mikrobiol. Technol. Lebensm.*   **1990a**, *12*, 137-144.
3.   Herrmann, K.   *Chem. Mikrobiol. Technol. Lebensm.*   **1990b**, *12*, 161-167.
4.   Harborne, J. B.   In *Encyclopedia of Plant Physiology*; Bell, E. A.; Charlwood, B. V. Ed.;   Springer-Varlag: Berlin, Heidelberg, New York, **1980**, pp 329-402.
5.   Ishii, T.; Hiroi, T. *Carbhoydr. Res.* **1990**, *196*, 175-183.
6.   Fry, S. C. *Biochem J.* **1982**, *203*, 493-504.
7.   Fry, S. C. *Annu. Rev. Plant Physiol.* **1986**, *37*, 165-186.
8.   Risch, B.; Herrmann, K.; Wray, V.   *Phytochemistry*   **1988**, *27*, 3327-3329.
9.   Risch, B.; Herrmann, K.; Wray, V.; Grotjahn, L. *Phytochemistry* **1987**, *26*, 509-510.
10. Handwerk, R. L.; Coleman, R. L.   *J. Agric. Food Chem.*   **1988**, *36* 231-236.
11. Tatum, J. H.; Nagy, S.; Berry, R. E. *J. Food Sci.* **1975**, *40*, 707-709.
12. Naim, M.; Striem, B. J.; Kanner, J.; Peleg, H. *J. Food Sci.* **1988**, *53*, 500-503, 512.
13. Rouseff, R. L.; Putnam, T. J.; Nagy, S.; Naim, M.   In *Flavour Science and Technology*; Bessiere, Y.; Thomas, A. F., Eds. John Wiley and Sons: New York, **1990**, pp 195-198.
14. Stohr, H.; Herrmann, K. *Z. Lebensm. Unters-Forsch.* **1975a**, *159*, 305-306.
15. Peleg, H.; Naim, M.; Rouseff, R. L.; Zehavi, U. *J. Sci. Food Agric.* **1991**, (in press).
16. Cilliers, J. J. L.; Singleton, V. I. *J. Agric. Food Chem.* **1989**, *37*, 890-896.
17. Sinclair, W. B. *The Biochemsitry and Physiology of The Lemon and Other Citrus Fruits.* University of California, Division of Agriculture and Natural Resources, Oakland, **1984**.
18. Lee, H. S.; Nagy, S. *J. Food Sci.* **1990**, *55*, 162-163, 166.
19. Klaren-Dewite, M.; Frost, D. J.; Ward, J. P. *Recl. Trav. Chim. Pays-Bas Belg.* **1971** *90*, 906-911.
20. Fiddler, W.; Parker, W. E.; Wassermann, A. E.; Doerr, R. C. *J. Agric. Food Chem.* **1967**, *15*, 757.
21. Pyysalo, T.; Torkkeli, H.; Honkanen, E.   *Lebensm-Wiss u Technol.* **1977**, *10*, 145-147.
22. Wheaton, T. A.; Stewart, I. *Nature* 1965, *206*, 620-621.
23. Reschke, A.; Herrmann, K. *Z. Lebensm. Unters-Forsch.* **1981**, *173*, 458-463.

24. Peleg, H.; Striem, B. J.; Naim, M.; Zehavi, U. *Proc. Int. Citrus Congress* Goren, R.; Mendel, K., Eds; Balaban Rehovot and Margraft, Weikershim, **1988**, *4*, 1743-1748.

25. Peleg, H.; Naim, M.; Rouseff, R. L.; Nagy, S.; Zehavi, U. *15th International Congress of Biochemistry*; Jerusalem, 4-8 August, **1991**. Abstr., p 96.

26. Martin, A. B.; Acree, T. E.; Hotchkiss, J.; Nagy, S. *J. Agric. Food Chem.* **1992** (in press).

RECEIVED January 13, 1992

Chapter 15

Phenolic Compounds in Maple Syrup

Thomas L. Potter and Irving S. Fagerson

Mass Spectrometry Facility, College of Food and Natural Resources,
Chenoweth Laboratory, University of Massachusetts, Amherst, MA 01003

"Acid/neutral" dichloromethane extracts of
two grade A medium amber maple syrups were
prepared and analyzed by gas chromato-
graphy/mass spectrometry. In excess of 133
compounds were detected. Phenolics accounted
for 41 of these compounds and over 70 percent
of the total mass of compounds detected. The
composition of the syrup extracts was qualitatively similar but exhibited substantial
quantitative differences. The data indicate
that during syrup production oxidation of
phenolic lignin monomers in maple sap is
limited. Implications for product flavor and
quality are discussed.

Maple syrup is derived from evaporative concentration
of the spring sap flow of the sugar maple (Acer
saccharum March). It is unique in that it is the only
commercially available food derived solely from the sap
of deciduous trees. Under current classifications, the
U.S. Department of Agriculture (USDA) recognizes two
syrup grades, A and B, and, within grade A, light,
medium and dark amber.

The syrup is comprised primarily of sucrose with
smaller amounts of glucose and fructose (1). Key flavor
constituents include vanillin, 3-methyl-2-hydroxy-
cyclopenten-2-one (cyclotene), 2,5-dimethyl-4-hydroxy
2(2H)-furanone (furaneol) and related compounds (2,3).
These compounds are not present in appreciable quantities in sap and are presumed derived from the sugars
and other sap constituents during evaporative
concentration (3).

Investigations with model systems have served to
demonstrate that formation of flavor compounds during
maple syrup production follows two pathways. One

0097–6156/92/0506–0192$06.00/0
© 1992 American Chemical Society

involves the thermal decomposition of sugars under
alkaline conditions to form cyclotene, furaneol and
related compounds. Shaw et al. (5,6) have reported
that these compounds are formed upon heating alkaline
sucrose and fructose solutions. Given the reported
detection of trace quantities of free amino acids in
maple saps (1), non-enzymatic browning reactions may
also contribute (7).

The second flavor production pathway involves a
different set of precursor compounds, namely, lignin
monomers present in sap. Oxidation and alkaline
hydrolysis of these phenolic compounds are presumed
responsible for vanillin and syringaldehyde formation.
Fiddler et al. (8) have reported that vanillin is a
major product of the thermal decomposition of ferulic
acid in air. Vanillin is also found in coniferyl
alcohol preparations left exposed to air (Potter and
Fagerson unpublished results).

The significance of oxidation has also been
illustrated by Underwood and Filipic, (3). They showed
that alkaline nitrobenzene oxidation of sap extracts,
"lignaceous material" isolated from sap, and chloroform
extracts of maple syrup greatly increased vanillin and
to a lesser extent syringaldehyde content. The latter
observation indicates that there is a substantial
quantity of unoxidized phenolic materials in maple
syrups and led in part to our investigation.

In the current work, we report identification of
phenolic and related flavor compounds in
dichloromethane extracts of two grade A medium amber
Massachusetts syrups.
The data has confirmed hypotheses that sap lignin
monomers are present in relatively high concentration
in maple syrup.
Implications for improving syrup flavor and quality are
discussed.

Materials and Methods

Materials. Syrup A was purchased at a local market.
Label information provided on its container listed the
product as Grade A medium amber. Syrup B, also labeled
as grade A medium amber, was obtained directly from a
local producer. The syrups were produced in the 1990
syrup season.

Sample Preparation. One hundred grams of syrup were
diluted to 500 ml with distilled-deionized water and
spiked with 25 ug each of naphthalene-d8 and phenol-d6
using 15 ul of acetone to aid dissolution. Solution pH
was then adjusted to 2 with 6N sulfuric acid, followed
by serial extraction in a separatory funnel with 3 X 50
ml aliquots of "distilled in glass" dichloromethane.
The resulting "acid/neutral" extract was concentrated

by rotary evaporation and taken to its final volume
(0.3 ml) under a stream of dry nitrogen. Emulsions
formed during extraction were broken by freezing.
Solvents were obtained from Burdick and Jackson
(Muskegeon, MI) and deuterated standards from Aldrich
Chemical Co. (Milwaukee, WI). Identical extraction
conditions were applied to a reagent blank consisting
of 500 ml of distilled-deionized water. All extracts
were stored at -20°C prior to GC/MS analysis.

GC/MS Analysis. GC/MS analysis was performed with a
Hewlett-Packard 5985B system equipped with a 30 m X
0.33 mm I.D. DB-WAX (J + W Scientific, Folsom, CA)
fused silica capillary column (0.25 micron film).
Helium inlet pressure on the column was fixed at 104
kPa and injection was in the Grob splitless mode at
250°C. The oven temperature profile was as follows:
60°C(hold 1 minute) to 240°C at 2°C/min (hold 29
minutes). The column was interfaced to the mass
spectrometer through an open-split interface
(Scientific Glass and Engineering, Austin, TX). Mass
spectra were obtained at 70eV over 33 to 350 amu at a
mass filter scan rate of 300 amu/sec. The instrument
was tuned to meet standard perfluorotributylamine
criteria.

Reference Compounds. The corresponding alcohols and aldehydes of ferulic, coumaric, syringic, sinapic,
vanillic, homovanillic, and homoveratric acids were
prepared by LiAlH$_4$ reduction in ether. Products were
isolated by dichloromethane extraction after quenching
with water. Typically, the alcohol, aldehyde,
dihydroalcohol, and in some cases the allyl phenols
were obtained. All reagents and other reference
compounds were obtained from Aldrich Chemical Co.
(Milwaukee, WI).

Results and Discussion

Total ion current chromatograms obtained from the
analysis of the syrup extracts are shown in Figure 1.
In excess of 133 compounds were detected. Eighty-four
were assigned structures with 48 confirmed by analysis
of reference compounds. Phenolics accounted for 41 of
the compounds detected.

Overall, results were in qualitative agreement
with published studies of maple syrup volatiles
composition (3,4). The exception is in the much larger
number of com-pounds detected in the current work. A
minor difference noted is that Kallio et al. (10)
reported homovanillic acid to be among the major
volatile constituents of maple syrup but did not report
detection of dihydroconiferyl alcohol. We detected only
trace amounts of the acid and relatively high

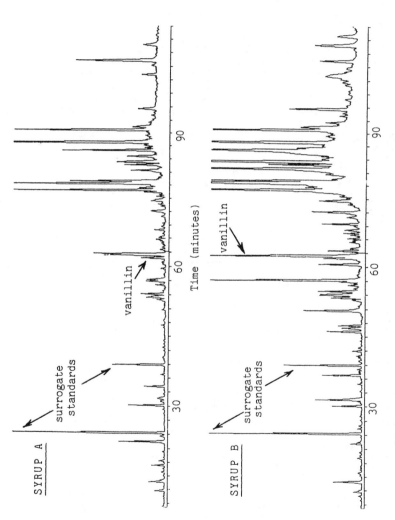

Figure 1. Total ion current chromatograms of maple syrup dichloromethane extracts

concentrations of the alcohol. We also obtained nearly identical mass spectra for these compounds under electron impact conditions and observed that their chromatographic separation could not be achieved using capillary columns coated with non-polar liquid phases. Thus, peak mis-assignment by Kallio et al. (10) may provide a possible explanation.

The Table I data show that the estimated total concentration of compounds in the syrup extracts ranged from 5.8 to 14.7 ug/g. Concentration estimates were based on the assumption of unit total ion current response to naphthalene-d8. Recovery of the surrogate standards, naphthalene-d8 and phenol-d6, ranged from 41 to 69 percent. Given this, concentration values reported are expected to represent minimum values. The Table I data also show that "phenolics" accounted for 72 to 78 percent of the total mass of compounds detected.

Table I. Classes of compounds detected in maple syrups

	Concentration (ug/g)[a]	
	A	B
sucrose pyrolysis products[b]	0.24	0.44
fatty acids[c]	0.23	0.49
phenolics[d]	4.52	10.6
unknowns	0.68	2.28
miscellaneous plasticizers[e]	0.11	0.91
total 5.79 14.7		

[a] Concentration based on assumed unit total ion current to the surrogate standard, naphthalene-d8; [b] Compounds reported as sugar degradation products include various lactones, furanones, cylcopentenones and related compounds; [c] free fatty acids; [d] compounds with a hydroxy-phenyl moiety and/or aromatic carboxylic acids, alcohols and aldehydes; [e] miscellaneous phthalate and adipate ester plasticizers.

The concentrations of major syrup phenolics are summarized in Table II. As indicated, vanillin and syringaldehyde, phenolic oxidation products of and major contributors to syrup flavor were detected at much lower concentrations than their presumed precursors, coniferyl, dihydroconiferyl and dihydrosinapyl alcohol. This observation provides quantitative support to Underwood and Filipic's

experimental results (3). In short, there is a
significant portion of the lignaceous material in maple
sap which is not oxidized during evaporation.

Table II. Principal phenolic and related flavor
compounds in maple syrups

Compound[a]	Concentration (ug/g)[b]	
	A	B
phenethyl alcohol	0.03	
eugenol		0.07
isoeugenol isomer	0.01	0.08
isoeugenol isomer		0.09
vanillin	0.09	0.90
syringaldehyde	0.34	0.93
dihydroconiferyl alcohol	1.04	3.30
coniferyl aldehyde	0.09	0.40
syringyl alcohol	0.15	0.76
coniferyl alcohol	1.27	1.31
dihydrosinapyl alcohol	0.48	0.86
vanillic acid	0.03	0.12
sinapyl aldehyde + homovanillic acid	0.04	0.21
4-allyl-3,5-dimethoxytoluene	0.31	0.12
syringic acid	0.06	0.22

[a] all compounds confirmed by analysis of reference
compound except 4-allyl-3,5-dimethoxytoluene; [b]
concentration based on assumed unit response to the
surrogate standard, naphthalene-d8.

Another notable feature of the Table II data is
the observed quantitative difference in the relative
composition of the syrups extracts. Syrup B had a ten-
fold greater concentration of vanillin and three-fold
greater concentrations of syringaldehyde and
dihydroconiferyl alcohol. These differences demonstrate
the quantitative variability among syrups within the
same grade and produced in the same season and
geographic region.
    In terms of sensory impact, it is interesting to
note that the taste detection threshold for vanillin in
water has been reported to be in the 0.5 to 1 parts per
million range (9). These observations indicate that
vanillin might contribute to the taste of syrup B but
not A. Implications for syrup aroma can be inferred,
since Kallio et al. (4) found that vanillin provided
the most significant attribute for maple syrups in
aroma panel tests.

A possible explanation of the compositional differences observed is variation in oxidation conditions during sap evaporation. For example, processing variables, such as foam control may have a significant impact on the surface area of contact with air. Thus, reduced foam control may promote phenolics oxidation and vanillin production. Analysis of a syrup produced in a jacketed steam kettle with no foam control showed that vanillin and was the principal phenolic constituent (Potter and Fagerson unpublished data).

While substantial variation was noted in the "phenolics" concentration in the syrups, concentrations of key sugar degradation products detected were similar. Cyclotene, furaneol, and isomaltol were detected at 15 to 27, 7 to 9 and 2 to 4 ng/g, respectively. This relative uniformity in composition suggests that these compounds contribute to the base syrup flavor while vanillin and other compounds express subtle differences between products.

Conclusions

A wide array of natural phenolics and other compounds were identified in the syrups studied. Among the phenolics, various lignin monomers and their partial oxidation products were prominent. The latter compounds play a key role in maple syrup flavor.

Data obtained has provided quantitative support for observations that oxidation of sap phenolic constituents does not approach completion during syrup production. This presents the possibility that, other things being equal, differences in evaporator design and syrup production practices may produce syrups of differing sensory qualities. We conclude that in depth studies of oxidation processes during syrup production may assist in the development of more uniform and higher quality products.

Another feature of the study is that it demonstrates the potential for detecting syrup adulteration by gas chromatographic analysis. Analyses focused on detecting coniferyl and dihydroconiferyl alcohols and other natural products provide a means for differentiating between natural and adulterated products. Such analyses may replace and/or supplement carbon isotope ratio measurements described by Morselli and Baggett (11).

Acknowledgments

Support for this work was provided by the Massachusetts Agricultural Experiment Station. We thank W. Coli for donation of a syrup sample. Contribution no. 3003 from the Massachusetts Agricultural Experiment Station.

Literature Cited

1. Morselli, M.F. and Feldheim, W. **1988**. *Z. Lebens. Unters.Forsch. 186*, 6-10.
2. Underwood, J.C.; Willets, C.O.; Lento, H.G. **1961**. *J. Food Sci. 26*, 288-290.
3. Underwood, J.C.; Filipic, V.J. **1964**. *J. Food Sci. 29*, 814-818.
4. Kallio, H.; Rine, S.; Pangborn, R.M.; and Jennings, W. *Food Chem.* **1987**, *24*, 287-299.
5. Shaw, P.E.; Tatum, J.H.; Berry, R.E. **1968**. *J. Food Agric. Chem. 16*, 979-982.
6. Shaw, P.E., Tatum, J.H. and Berry, R.E. **1969**. *J. Food Agric. Chem. 17*, 907-909.
7. Hodge, J.E.; Mills, F.D.; Fisher B.E. *Cereal Sci. Today.* **1972**, *17*, 34-38,40.
8. Fiddler, W.; Parker, W.E.; Wasserman, A.E.; and Doerr, R.C. *J. Agric. Food Chem.* **1967**, *15*, 757-761.
9. Anonymous. *Compilation of Odor and Taste Threshold Values Data.* W.H. Stahl, Ed. American Society for Testing and Materials, Philadelphia, PA. **1973**.
10. Kallio, H. Comparison and Characteristics of Aroma Compounds from Maple and Birch Sirups. *Frontiers of Flavor.* Proceedings of the 5th International Flavor Conference. G. Charalambous, Ed. **1988** , 241-248.
11. Morselli, M.F. and Baggett, K.L. *J. Assoc. Off. Anal. Chem.* **1984**, *67*, 22-27.

RECEIVED November 20, 1991

Chapter 16

Phenolic Compounds of *Piper betle* Flower as Flavoring and Neuronal Activity Modulating Agents

Lucy Sun Hwang[1], Chin-Kun Wang[1], Ming-Jen Sheu[2], and Lung-Sen Kao[3]

[1]Graduate Institute of Food Science and Technology and [2]Department of Horticulture, National Taiwan University, Taipei, Taiwan
[3]Institute of Biomedical Sciences, Academia Sinica, Taipei, Taiwan

Piper betle is a tropical plant from Southeast Asia. The flower of this plant is the major ingredient which gives flavor and other properties to a special chewing food, betel quid, which is quite popular in Taiwan. The essential oil of Piper betle flower was extracted by vacuum steam distillation, analyzed by GC and identified by GC co-chromatography, GC-MS and IR. Safrole and myrcene were the major compounds found in the essential oil (27.6% and 26.4%, respectively). Safrole, hydroxychavicol, eugenol, isoeugenol, and eugenol methyl ester were found to be the major phenolic compounds by HPLC and TLC analyses. Results of the catecholamine secretion experiment, a test for neuronal modulating activity, showed that eugenol, isoeugenol and hydroxychavicol are strong nerve stimulators while safrole and eugenol methyl ester are weak.

Piper betle L., belonging to the Piperaceae family, is a tropical plant that originated in Malaysia but also grows well in the central part of Taiwan. The flower of this plant is a major ingredient which gives flavor and other properties to a special chewing food, betel quid.

Betel quid is generally composed of three parts, betel nut, Piper betle flower and catechu mixed with lime. It was a traditional fun food for aboriginal Taiwanese. It has a strong and characteristic taste, provoking blood circulation resulting in a stimulating effect. Because of this, it has gained popularities among laborers and students in Taiwan.

There have been clinical reports pointing out the correlation between betel nut chewing and oral cancer (1-5). However, there were also reports on the antimutagenic and anticarcinogenic effects of Piper betle leaf (6,7). The leaves of Piper betle were also found to contain fungicidal and nematocidal components (8).

0097–6156/92/0506–0200$06.00/0
© 1992 American Chemical Society

The phenolic compounds, hydroxychavicol and eugenol, present in betle leaf were recently found to possess an antinitrosating property (*9,10*). The phenolic components in the flower of Piper betle have not been extensively studied. In this paper, the flavoring and nerve stimulating properties of the phenolic compounds in the flower of Piper betle will be presented.

Materials and Methods

Materials. Fresh flowers of Piper betle were purchased from farmers of Nantou County, Taiwan. They were either used immediately or stored at $4^{\circ}C$ before use.

Bovine adrenal medulla cells were prepared from fresh bovine adrenal glands according to the method of Kilpatrick et al. (*11*).

Extraction of Essential Oil. Essential oil of Piper betle was vacuum steam distilled from a total of 14 Kg of fresh flowers using an apparatus as described by Chang et al. (*12*). In order to facilitate the extraction, Piper betle flowers were suspended in refined soybean oil and blended to give a paste before steam distilling at $26^{\circ}C$ for 3 hr under 3 mmHg of vacuum.

Refined soybean oil was also vacuum steam distilled under the same condition as the Piper betle in order to differentiate volatiles of Piper betle from those of soybean oil.

Isolation and Identification of Volatile Constituents in the Essential Oil. Volatile constituents of the essential oil were separated by repeated Gas Chromatography on a 10% SE-30 column followed by a 20% Carbowax 20M column to collect the fractions for IR identification. Six major components were collected and identified by IR, MS and GC co-chromatography. The rest of the components were isolated and identified by capillary GC-MS on a 50 m x 0.2 mm ID Carbowax 20M WCOT glass capillary column and an HP 5983 Mass Spectrophotometer.

Extraction of Phenolic Compounds. Fresh flowers of Piper betle were extracted with 80% aqueous acetone (1:10, w/v) in a Waring blender for 3 min followed by soaking for 2 hr before filtration. The filtrate, containing the extracted phenolic compounds, is used in the following analysis.

TLC Analysis of Phenolic Compounds. The above phenolic extract was spotted on a silica gel G TLC plate and developed in n-hexane:ethyl acetate:acetic acid = 70:29:1 (v/v/v). The phenolic compounds were visualized by spraying a 20% sodium carbonate solution followed by a Folin-Ciocalteu phenol reagent.

HPLC analysis of Phenolic Compounds (*13*). Ten grams of fresh flowers were extracted four times with 60ml of 80% aqueous acetone. The combined extract was diluted to 250 ml with solvent. Ten ml of this extract was concentrated with a rotary evaporator to 1 ml followed by filtering through a $0.22\,\mu$m membrane. Ten μl of this filtrate was used for HPLC analysis.

The HPLC column was a 25 cm x 4.6 mm ID μ-Bondapak C18 column (Waters Assoc. USA). Glacial acetic acid:water (25:975, v/v; solvent A) and methanol (solvent B) were used for gradient elution. A linear gradient was applied, starting with 100% solvent A and finishing with 100% solvent B over a 50 min period. The flow rate was kept at 2 ml/min and the UV detector wavelength was set at 280 nm. Standard compounds were co-chromatographed with the acetone extracts for tentative identification of phenolic compounds.

Catecholamine Secretion Experiments (11). Catecholamine secretion was measured by the method of Kilpatrick et al. (11) with modifications described as follows. The physiological salt solution (PSS) contained 137 mM sodium chloride, 4.4 mM potassium chloride, 1.2 mM potassium biphosphate, 2.2 mM calcium chloride, 0.7 mM magnesium chloride, 10 mM glucose, 5 mM hepes (N-2-hydroxyethylpiperazine-N'-2-ethanesulfonic acid), and 3.6 mM sodium bicarbonate. This buffer has a pH of 7.4 and osmolarity of 300 mOsm. The bovine serum albumin (BSA) solution was prepared by dissolving 5 mg BSA in 1 ml PSS solution. High K^+ solution contained 98.2 mM sodium chloride, 56 mM potassium chloride, 0.7 mM magnesium chloride, 2 mM calcium chloride, 11.2 mM glucose, and 10 mM hepes, pH 7.4. The ^{3}H-norepinephrine (^{3}H-NE) loading buffer contained 1% ascorbic acid, and 1.2 x 10^{-4} M ^{3}H-NE in PSS. Triton X-100 (TX-100) solution contained 1% TX-100 and 2 mM EGTA (ethyleneglycol-bis-N,N,N',N'-tetraacetic acid). The pH of the solution was adjusted with potassium hydroxide to 7. The experiment was carried out according to the scheme shown in Figure 1.

Results and Discussion

Essential Oil Constituents. The essential oil obtained from the Piper betle flower by vacuum steam distillation was about 0.81% on a dry weight basis. A total of 50 peaks were found on the GC chromatogram (Fig.2). Among them, 28 peaks were identified either with IR, GC-MS or GC-co-chromatography with authentic compounds as shown in Table I. The identified components comprised about 96.24% of the total essential oil, among them 62.93% were monoterpene hydrocarbons and 30.41% were phenolic compounds. The six phenolic compounds were safrole, 2,6-di-t-butyl-4-methyl-phenol, methyl eugenol, 2-ally-phenol, eugenol and isothymol (carvacrol). Safrole was positively identified by IR and MS, it constituted 27.64% of the total essential oil and was the largest and most important flavor component of the Piper betle flower essential oil. 2,6-Di-t-butyl-4-methyl-phenol (BHT) was found to come from the soybean oil background (97.8% of the total volatiles). BHT might be added to the soybean oil as an antioxidant. The other four phenolic compounds only contributed 1.5% to the total volatiles.

Safrole is the major volatile compound in Sassafras albidum (85% of essential oil) and is also present in a variety of spices such as mace, nutmeg, cocoa, black pepper and California laurel. Sassafras oil was the major flavor ingredient in root beer until

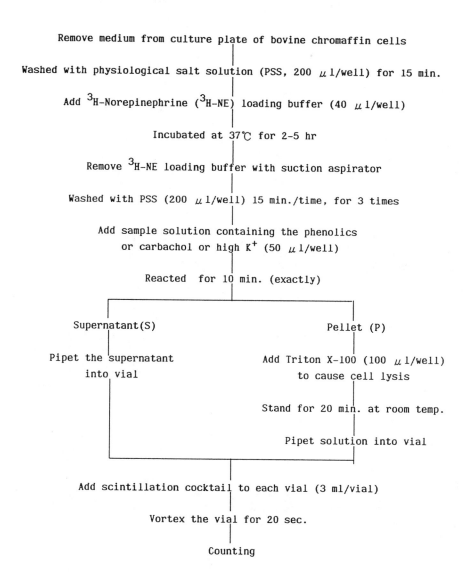

Figure 1. Scheme for chromaffin cell secretion experiment.

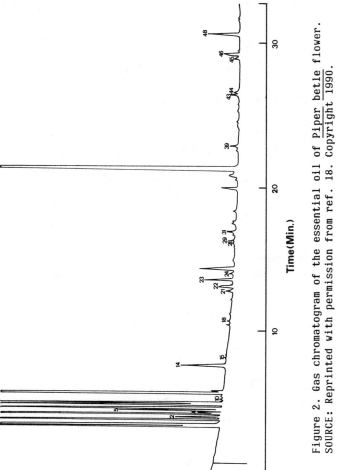

Figure 2. Gas chromatogram of the essential oil of Piper betle flower.
SOURCE: Reprinted with permission from ref. 18. Copyright 1990.

Table I. Volatile Compounds Identified in the Essential Oil of
Piper betle Flower

Park number	Compound	% Peak Area	IR	M.S.(M.W.)	Identification Co-chroma-tography
	I. Monoterpene hydrocarbon				
1	α-pinene	5.97	✓	✓(136)	✓
2	Camphene	0.49		✓(136)	✓
3	β-Pinene	8.24	✓	✓(136)	✓
4	Sabinene	0.20		✓(136)	✓
5	Δ3-carene	1.33		✓(136)	✓
6	Myrcene	26.43	✓	✓(136)	✓
7	Limonene	9.05	✓	✓(136)	✓
8	β-Phellandrene	6.63		✓(136)	
10	γ-Terpinene	0.05		✓(136)	✓
12	p-Cymene	4.47	✓	✓(134)	✓
13	Terpinolene	0.07		✓(136)	
	II. Monoterpene aldehyde				
18	Citronellal	0.04		✓(154)	✓
29	Neral	0.09		✓(152)	✓
	III. Monoterpene alcohol				
21	Linalool	0.16		✓(154)	✓
24	4-Terpineol	0.08		✓(154)	✓
31	α-Terpineol	0.11		✓(154)	✓
	IV. Monoterpene acetate				
22	Linalyl acetate	0.36		✓(196)	✓
28	Citronellyl acetate	0.06		✓(184)	✓
	V. Monoterpene ether				
9	1,8-Cineole	0.20		✓(154)	✓
	VI. Sesquiterpene hydrocarbon				
23	β-Caryophyllene	0.73		✓(204)	✓
	VII. Phenolic compounds				
37	Safrole	27.64	✓	✓(162)	✓
39	2,6-Di-t-butyl-4-methyl-phenol(BHT)*	1.27		✓(220)	
44	Methyl eugenol	0.20		✓(178)	
45	2-Allyphenol(Chavicol)	0.11		✓(134)	✓
46	Eugenol	0.46		✓(164)	✓
48	Isothymol (carvacrol)	0.73		✓(150)	✓
	VIII. Miscellaneous				
14	2,6,10-Trimethyl dodecane	0.73		✓(212)	
15	Nonyl aldehyde (C-9 aldehyde)	0.02		✓(142)	✓
43	Methyl-N-methyl anthranilate	0.32		✓(155)	✓

SOURCE: Adapted from ref. 18
* also found in soybean oil

safrole was found to be carcinogenic (14,15) and thus banned for use as a food additive in 1960 (16). The occurrence of safrole in the essential oil of Piper betle might contribute to the high incidence of oral cancer in betel quid chewers.

Phenolic Compounds in the Flower of Piper betle L. From TLC analysis of the phenolic extract, seven spots were detected (Fig.3). By comparing with authentic compounds, eugenol and hydroxychavicol were identified. To further analyze the phenolic compounds in Piper betle, HPLC with a photodiode array detector was employed. Figure 4 shows the chromatogram of the phenolic extract of Piper betle. Seven compounds were identified by co-chromatography with the authentic compounds and by comparing their UV-VIS spectra. The six phenolic compounds found in the flower of Piper betle are listed in Table II, together with their contents in a fresh sample and recoveries of the HPLC method. Safrole was found to be the major phenolic, followed by hydroxychavicol, eugenol, eugenol methyl ester, isoeugenol and quercetin.

Table II. The Contents and Recoveries of Phenolic Compounds from Fresh Piper betle Flower by HPLC Analysis

Phenolic compounds	M.W.(g)	Content (mg/g fresh wt.)	Recovery(%)
Hydroxychavicol	151	9.74	96.4
Eugenol	164	2.51	95.8
Isoeugenol	164	1.81	92.1
Quercetin	338	1.11	93.6
Eugenol methyl ester	178	1.81	93.9
Safrole	162	15.35	96.4

Catecholamine Secretion Test of Phenolic Compounds in Piper betle Flower. Under physiological conditions, chromaffin cells are stimulated by acetylcholine released from preganglionic nerve terminals. Acetylcholine binds with the nicotinic receptors on the plasma membrane which opens the receptor-associated ion channel and allows Na^+ influx. The cells are then depolarized and the voltage-sensitive calcium channels are opened to allow the influx of Ca^{+2}. This causes the increase of cytoplasmic Ca^{+2} concentration which triggers exocytosis to secrete catecholamine (17).

There are many similarities between the above process and the mechanism of neurotransmitter release (17). Since the chromaffin cells are easy to isolate in large quantities and stable in the culture medium for about two weeks, they are one of the most favorable model systems in the study of neurotransmission (11). Catecholamine secretion from chromaffin cells was therefore chosen to be the test system for examining the physiological effects, especially on the nerve system, of the phenolic compounds from Piper betle flower.

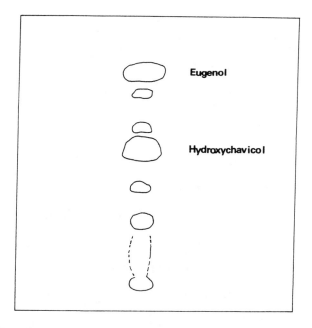

Figure 3. TLC chromatogram of the phenolic compounds from <u>Piper betle</u> flower on silica gel precoated plate. Solvent system : n-hexane : ethyl acetate : acetic acid = 70 : 29 : 1 (v/v/v). Visualizing agent : 20% Sodium carbonate solution and Folin-ciocalteu reagent.

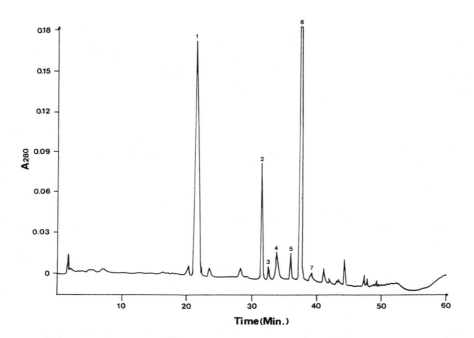

Figure 4. HPLC chromatogram of the phenolic compounds from <u>Piper betle</u> flower.   Peaks identified : 1) hydroxychavicol; 2) eugenol; 3) isoeu-genol; 4) quercetin; 5) eugenol methyl ester; 6) safrole; 7) flavone.

The standard compounds of the five major phenolics were used to stimulate the chromaffin cells, and their effects on the secretion of catecholamine were measured. In order to investigate the mechanism which causes this effect, we also studied the effects of these phenolics on carbachol- and a solution containing a high concentration of external K^+ (high K^+)- induced catecholamine secretion. Carbachol, an analog of acetylcholine, stimulates the secretion through the same mechanism as acetylcholine by binding with nicotinic receptors. High K^+, on the other hand, depolarizes the cell directly, by-passing the nicotinic receptors, and subsequently induces the secretion of catecholamine. By examining the effects of the phenolics on the secretion induced by carbachol and high K^+, we can understand whether these compounds act at the nicotinic receptor or not.

Results in Figures 5 and 6 showed that eugenol by itself caused the secretion of 93% of total cellular catecholamine at concentrations above 6 mM. By similar experiments, we obtained results for the other four phenolics. Their maximum effects on catecholamine secretion from chromaffin cells and the concentration required for inducing 50% of maximum secretion (EC_{50}) are summarized in Table III. These results clearly indicated that hydroxychavicol, eugenol and isoeugenol by themselves could induce almost total cellular catecholamine secretion, while eugenol methyl ester and safrole had no significant effect on inducing catecholamine secretion. When the chromaffin cells were viewed under a light microscope, no apparent morphological changes were observed, even at the highest phenolic concentration. This indicated that these phenolic compounds acted directly on the exocytotic machinery without causing cell lysis.

Table III. Effects of Phenolic Compounds from Piper
betle Flower on the Secretion of Catecholamine

Compounds	Max. Secretion[a] (%)	EC_{50}[b] (mM)
Hydroxychavicol	92.8 ± 0.7	0.8
Eugenol	92.6 ± 0.4	4.8
Isoeugenol	93.0 ± 0.3	4.0
Eugenol methyl ester	0.8 ± 0.4	0.7
Safrole	4.0 ± 0.5	3.0

a: Maximum catecholamine secretion from chromaffin cell induced by various compounds.
b: Concentration required for inducing 50% of maximum catecholamine secretion.

Results shown in Table IV indicate that hydroxychavicol, eugenol and isoeugenol give an inhibitory effect on the carbachol-induced secretion of catecholamine while they show a stimulatory effect on the high K^+-induced catecholamine secretion. Their effective concentration ranges are different, indicating that these phenolic compounds might have multiple sites of action on chromaffin cells.

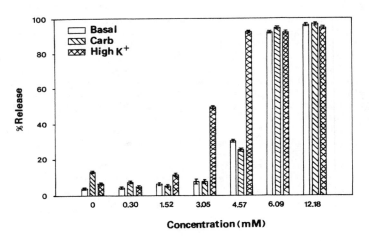

Figure 5. Catecholamine secretion test for eugenol.

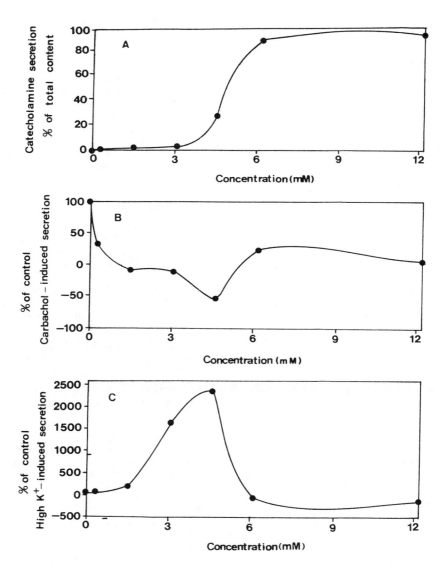

Figure 6. Effect of eugenol on catecholamine secretion from chromaffin cells. (A) basal, (B) induced by carbachol, (C) induced by high concentration of external K.

Table IV. Effects of Phenolic Compounds from Piper betle Flower
on the Carbachol and High K^+-induced Secretions

Compounds	Carbachol-induced secretion $IC_{50}{}^a$ (mM)	High K^+-induced secretion $IC_{50}{}^a$ (mM)	$EC_{50}{}^b$ (mM)
Hydroxychavicol	1.00	N.E.^{c}	7.2
Eugenol	0.25	N.E.	2.8
Isoeugenol	0.75	N.E.	2.2
Eugenol methyl ester	0.25	0.25	N.E.
Safrole	2.80	2.5	N.E.

a: Concentration required for causing 50% of inhibition relative
to control.
b: Concentration required for causing 50% of the maximum effect.
c: No effect.

Eugenol methyl ester and safrole showed significant inhibitory
effect on both carbachol- and high K^+-induced catecholamine secretion at the same effective concentration ranges (Table IV). These
results suggest that eugenol methyl ester and safrole may affect
the steps after the cells are depolarized.
From the above results, we can be sure that some of the phenolic compounds of Piper betle flower can indeed modulate the neuronal
activity. The mechanisms of their actions are not the same. The
combination of these phenolic compounds might cause mutual interaction. Further investigation is needed to elucidate these points.

Acknowledgments

We sincerely acknowledge the financial supports from the National
Science Council and the Department of Health, Republic of China.

Literature Cited

1. Suri, K.; Goldman, H. M.; Wells, H. Nature, 1971, 230, 383-384.
2. Ranadive, K. J.; Ranadive, S. N.; Shivapurkar, N. M.; Gothoskar,
   S. V. Int. J. Cancer, 1979, 24, 835-843.
3. Stich, H. F.; Stich, W. Cancer Letters, 1982, 15, 193-202.
4. Wenke, G.; Brunneman, K. D.; Hoffmann, D.; Bhide, S. V. J.
   Cancer Res. Clin. Oncol. 1984, 108, 110-113.
5. Nair, J.; Ohshima, H.; Malaveille, C.; Friesen, M.; O'Neill, I,
   K.; Hautefeuille, A.; Bartsch, H. Food Chem. Toxic. 1986, 24,
   27-31.
6. Padma, P. R.; Amonkar, A. J.; Bhide, S. V. Mutagenesis, 1989, 4,
   154-156.
7. Padma, P. R.; Lalitha, V. S.; Amonkar, A. J.; Bhide, S. V.
   Cancer Letters, 1989, 54, 195-202.
8. Evans, P. H.; Bowers, W. S.; Funk, E. J. J. Agric. Food Chem.
   1984, 32, 1254-1256.

9. Nagabhushan, M.; Amonkar, A. J.; Nair, U. J.; D'Souza, A. V.; Bhide, S. V. Mutagenesis, 1989, 4, 200-204.
10. Shenoy, N. R.; Choughuley, A. S. U. J. Agric. Food Chem. 1989, 37, 721-725.
11. Kilpatrick, D. L.; Ledbetter, F. H.; Carson, k. A.; Kirshner, A. G.; Slepetis, R.; Kirshner, N. J. Neurochem. 1980, 35, 679-692.
12. Chang, S. S.; Vallese, F. M.; Hwang, L. S.; Hsieh, O. A. L.; Min, D. B. S. J. Agric. Food Chem. 1977, 25, 450-455.
13. Mueller-Harvey, I.; Reed, J. D.; Hartley, R. D. J. Sci. Food Agric. 1987, 39, 1-14.
14. Lehman, A. J. Assoc. Food Drug Off. U. S. Q. Bull. 1961, 25, 194-195.
15. Long, E. L.; Nelson, A. A.; Fitzhugh, O. G.; Hansen, W. H. Arch. Pathol. 1963, 75, 595-604.
16. Federal Register 1960, 25, FR12412.
17. Douglas, W. W. In Handbook of Physiology; Blaschko, H.; Sayers, G.; Smith, A. D. Ed.; Sect. 7, Endocrinology, Vol. VI, The Adrenal Gland; American Physiological Society, Washington, 1975, p. 367-388.
18. Hwang, L. S.; Sheu, M. J. Food Science, 1990, 17, 298-305.

RECEIVED December 2, 1991

Chapter 17

Antioxidant Activity of Phenolic Compounds in Meat Model Systems

Fereidoon Shahidi, P. K. J. P. D. Wanasundara, and C. Hong

Department of Biochemistry, Memorial University of Newfoundland, St. John's, Newfoundland A1B 3X9, Canada

Antioxidant activity of a number of phenolic compounds of plant origin was evaluated in meat model systems. Amongst the phenolic compounds tested quercetin, morin, kaempferol, myricetin, tannic acid, eugenol, isoeugenol and ellagic acid were most effective when applied to meat at 30 or 200 ppm levels. Addition of 0.5 to 2.0% of deheated mustard flour (DMF), available commercially, or its alkanolic extracts, effectively retarded lipid oxidation and off-flavor development in meat emulsion systems. The activity of all phenolics tested and that of DMF or its extracts was superior to that of α-tocopherol at a 200 ppm addition level. Furthermore, DMF had a favorable effect on water binding characteristics of meat model systems studied.

The deleterious effects of oxidation in meat products are numerous. Oxidative reactions are responsible for changes in flavor, color, texture and nutritional value, due to the destruction of fat-soluble vitamins and essential fatty acids such as linoleic acid, in both fresh and cooked muscle foods (1). Lipids in cooked meats, particularly their phospholipids, are susceptible to autoxidation. These reactions are generally catalyzed by a number of factors such as presence of oxygen, light, heat, heavy metals, pigments, alkaline conditions and also depend on the degree of unsaturation of lipid fatty acids.

Lipid oxidation, defined as oxidative deterioration of unsaturated fatty acids, is a free radical-mediated phenomenon involving a chain reaction mechanism. The relative rate of oxidation of lipid fatty acids of meats, mainly C18, depends on their degree of unsaturation as given below:

$$C18:3\omega3\ (2500) > C18:2\omega6\ (1200) > C18:1\omega9\ (100) > C18:0\ (1)$$

Antioxidants are added to lipid-containing foods to prevent the formation of various off-flavors and development of rancidity. In meats, nitrite curing has

0097–6156/92/0506–0214$06.00/0
© 1992 American Chemical Society

shown to be an effective means of extending the shelf-life of cooked products against autoxidation.   Commonly used synthetic antioxidants have also proven effective in retarding lipid oxidation in meat products.   However, the present trend in food processing is to use natural ingredients (2).   Thus, use of naturally-occurring antioxidants for retarding meat flavor deterioration (MFD)/warmed-over flavor (WOF) development is highly desirable.   Pratt and Watts (3) reported that hot-water extracts of many vegetables were effective in retarding MFD in cooked products. These authors attributed the effectiveness of vegetable extracts to their content of flavonoids.

Flavonoids are widely distributed in plant materials and as such have considerable practical impact on the flavor retention and wholesomeness of raw and cooked meats (3).   Antioxygenic activity of herbs and spices has also been attributed to the presence and content of flavonoids and/or flavonoid-related compounds (4).   The relationship between flavone structures and their antioxidant activity has been thoroughly investigated.   Effectiveness of flavonoids in retarding lipid oxidation in fat-containing foods is apparently related to their ability to act as free radical acceptors (5-10) as well as to their chelating capacity for metal ions. Metal chelation by flavonols is due to ortho-dihydroxy (3′,4′-dihydroxy) grouping on the B ring and to the ketol structure in the C ring in their chemical structures (7,11,12).   Lack of at least one of these groups may reduce or even delete the chelating capacity of flavonoids.

In addition to flavonoids, many plant protein extenders used in meat emulsion systems contain large quantities of phenolic acids.   Phenolic compounds isolated from rapeseed processing in the preparation of protein concentrates contained a large percentage of phenolic acids.   The extracted phenolics in ethanol, after fractionation, possessed strong antioxidant effects in linoleate/β-carotene systems which were, in some cases, comparable to the effectiveness of approved food phenolics (13).

Application of some natural phenolics to meat was previously reported from our laboratories (14).   The present paper summarizes the results of studies on the efficiency of individual phenolic compounds of plant origin in meat model systems. Use of deheated mustard flour (DMF), as an example of a non-meat ingredient, a seasoning, and as a protein extender in comminuted meat products, and its effectiveness in retarding MFD was also investigated.

Materials and Methods

Materials. All flavonoids and phenolic compounds, as well as tannic acid, sesemol and other related compounds were obtained from the Sigma Chemical Company (St. Louis, Missouri) or the Aldrich Chemical Company (Milwaukee, Wisconsin). Deheated mustard flour (DMF) was a commercial product from UFL Foods Inc. (Mississauga, Ontario).   Pork loins with their subcutaneous fat removed were obtained from the Newfoundland Farm Products Corporation (St. John's, Newfoundland).

Extracts of DMF were prepared by extraction with water or 85% (v/v) methanol-water.   The content of total phenolics in DMF extracts was determined using the Folin-Denis reagent, as described by Rhee *et al.* (15,16).

Sample Preparation. In each case, 80 g comminuted (4.8 mm grind plate) pork, less the amount of additives (if $\geq 0.5\%$), was mixed with 20 g of distilled water. Selected additives were introduced, generally at a 30 or 200 ppm level of addition, as such, in aqueous, or in a 50% (v/v) ethanolic solution to meat prior to heat processing. Meat systems were then cooked, to an internal temperature of $75 \pm 1°C$, in Mason jars. Cooked meat samples were cooled to room temperature, homogenized in a Waring blender and then stored in Nasco Whirl-Pak bags at 4°C until used.

The percent inhibition of malonaldehyde (MA) production in meat samples was quantified after their steam distillation using the classical 2-thiobarbituric acid test (*17*) as given in the following equation:

$$\% \text{ inhibiion of MA production } = (1 - \frac{MA \text{ content of treated sample}}{MA \text{ content of control}}) x 100$$

Results and Discussion

Inhibition of malonaldehyde (MA) production, deduced from 2-thiobarbituric acid (TBA) values during a 3-week storage period of treated meats at 4°C, is given in Table I. The antioxidant activity of flavonoids arises (Table I) from their action as primary antioxidants by a free-radical scavenging mechanism and to a lesser extent by their chelating ability (results not shown) through the presence of free ortho-hydroxyl groups at 3' and 4' on the B ring. Due to the limited number of flavonoids tested in this study, no other structural effects were evident. In all cases, antioxidant activity of flavonoids tested was superior at a 200 ppm, as opposed to a 30 ppm level of addition to comminuted pork. Furthermore, it was evident that the antioxidant activity of flavonoids was generally governed by their chemical structural attributes.

Presence of a double bond at the C_2-C_3 position and a free hydroxyl group at the C_3 position of ring C had a marked influence on the antioxidative activity of flavonoids. Hence, naringin and naringenin, which did not possess these features, were least effective in preventing oxidation of meat lipids. In addition, presence of a free hydroxyl group at the C_7 position of ring A had no effect on the activity of flavonoids as no difference in the action of naringenin and naringen was observed. Rutin with a bound (etherified) hydroxyl group at the C_3 position was much less effective as an antioxidant than its analogue with a free hydroxyl group. Generally, presence of ortho-hydroxyl groups at the 3' and 4' positions of ring B contributed to the enhancement of flavonoids' antioxidative activity. Thus, quercetin was found to have a superior activity as compared with morin. Myricetin, with the highest number of hydroxyl groups in ring B, was found to be the most active flavonoid (at 200 ppm) in this study.

The activity of benzoic acid derivatives as antioxidants depended on their chemical structure, especially the number of free hydroxyl groups found in each compound (Table II). Of the benzoic acid derivatives tested, the activity of gallic acid was superior to that of syringic acid which in turn was better than that of

Table I   Inhibition of malonaldehyde formation by flavonoids in cooked, stored meat model systems

Ring A: 5,7-di-OH

Compound	Ring B				Ring C	Inhibition, %[a]
	2'	3'	4'	5'	3	
Kaempferol	H	H	OH	H	OH	40,96
Morin	OH	H	OH	H	OH	29,97
Myricetin	H	OH	OH	OH	OH	—,99
Naringenin[b]	H	H	OH	H	H,H	4,7
Naringin[b,c]	H	H	OH	H	H,H	4,7
Quercetin	H	OH	OH	H	OH	96,98
Rutin	H	OH	OH	H	O-rutinose	28,37

[a]At 30 and 200 ppm level of addition to meat, respectively.
[b]Saturated C_2-C_3 bond.
[c]7-O-Rhamnoglucose.

vanillic acid. This latter trend seems to be related to the number of hydroxyl groups present, and to a lesser extent on the number of oxygen-containing substituents. Thus, ellagic acid, a dimer of gallic acid, was determined to be the most effective antioxidant at a 30 ppm level of addition. In view of its desirable pharmacological and anticarcinogenic activity (*19*) application of ellagic acid to meat products may have practical significance. The activity of tannic acid, a compound consisting of glucose and three molecules of gallic acid, was moderate at a 30 ppm and excellent at a 200 ppm level of addition.

Of the cinnamic acid derivatives generally found in oilseeds, the activity of caffeic acid with two free hydroxyl group was superior to others. Activity of chlorogenic acid, an adduct of quinic acid and with two hydroxyl groups, was pronounced at the 200 ppm level (Table III). Other phenolic acids with only one free OH group were less active (Table III).

Of the other natural antioxidants examined, the activity of eugenol, isoeugenol and sesamol (Figure 1) was 95-99% at the 200 ppm level of addition. However, they were less effective at the 30 ppm level (65-79%). These activities are higher than those of ascorbic acid, α-tocopherol and butylated hydroxytoluene, BHT. The activity of BHT, a known synthetic antioxidant, was 13 and 88% at 30 and 200 ppm level of addition to meat, respectively (Figure 1).

Thus, naturally-occurring antioxidants, found in different plant materials, lend themselves to exploitation in different processing applications (*18*). Presence of flavonoid and flavonoid-related compounds in vegetables, fruits, oilseeds and other foods contributes to our intake at least 1 g/day (*20*). Thus, their possible use in meat formulations, as such or in the form of protein extenders and binders as well as spices and condiments may present an effective and attractive means of retarding lipid oxidation in muscle foods.

Deheated mustard flour (DMF) is an example of an economical meat adjunct which may be used as a spice for flavor and could also improve technological properties of wieners and bologna-type products. Its inclusion in emulsified pork model systems inhibited lipid oxidation. The percent inhibition of MA production was 34% for 0.5%, 75% for 1.0%, 95% for 1.5%, and 96% for 2.0% DMF addition after a 3-week of storage at 4°C. Based on these results, inclusion of 1.0 to 2.0% DMF in meats not only offers its intended flavor effects but it also effectively retards MFD (See Figure 2).

Extracts of DMF were also found to possess antioxidant activity. Both aqueous and 85% (v/v) methanoic solutions were used for extraction preparation. However, the antioxidant effect of 85% (v/v) methanoic extracts was superior to that of aqueous extracts (Figure 2). The antioxidant properties of extracts so obtained might in part be due to the content of their phenolic compounds (13). Thus, 85% (v/v) methanoic extracts of DMF contained 1.56% total phenolics, while the aqueous extracts contained only 1.08%. The chemical nature of the antioxidant in DMF is currently under investigation in our laboratories.

In addition to its strong antioxidant properties, DMF was also effective in reducing the cooking loss and improving the juiciness of products. The potency

Table II   **Inhibition of malonaldehyde formation by benzoic acid derivatives in cooked, stored meat model systems**

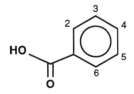

Compound	Substituent	Inhibition, %[a]
Gallic Acid	3,4,5-triOH	44,73
Vanillic Acid	4-OH, 3-OMe	21,28
Syringic Acid	4-OH, 3,5-diOMe	39,57
Ellagic Acid	Dilactone of 2 gallic acids	98,99
Tannic Acid	Ester of 3 to 5 gallic acids with glucose	56,99

[a]At 30 and 200 ppm level of addition to meat, respectively.

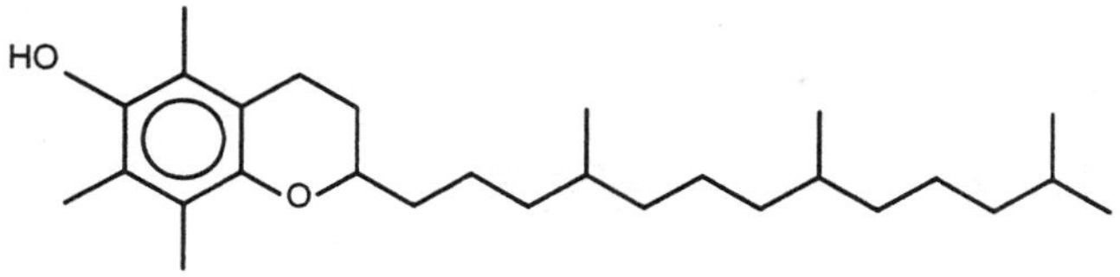

α-Tocopherol (11,56)

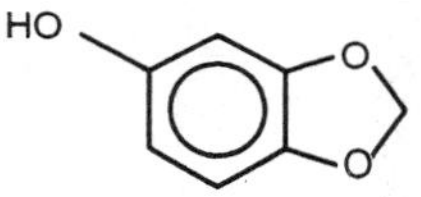

Sesamol (70,99)

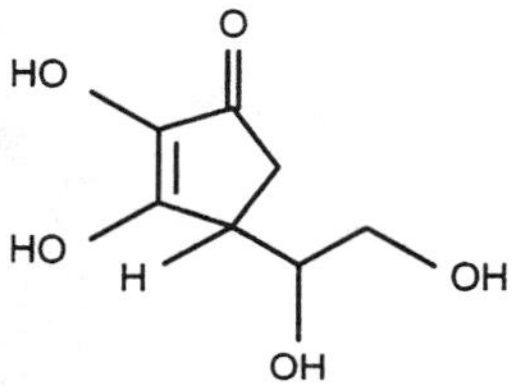

Ascorbic acid (50)

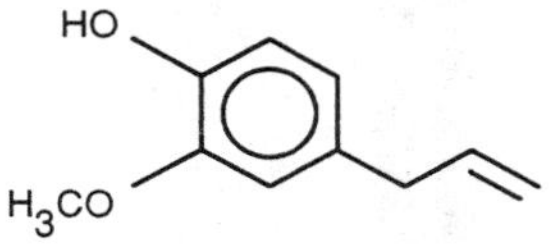

Eugenol (65,95)

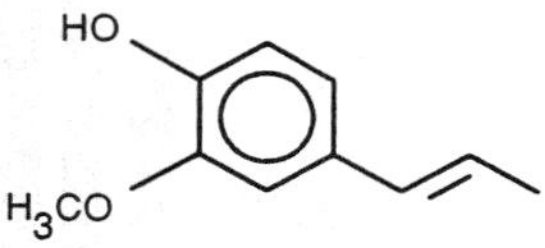

Isoeugenol (79,99)

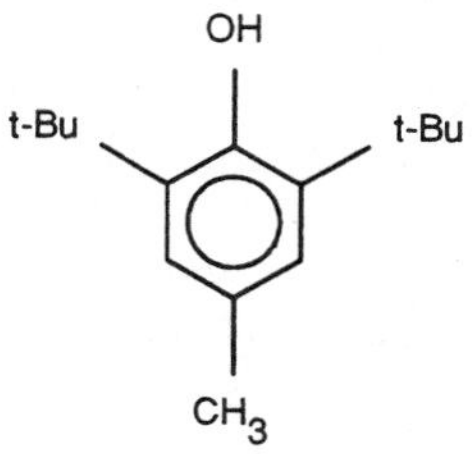

Butylated Hydroxytoluene, BHT (13,88)

Figure 1.   Some naturally occurring antioxidants and their efficiency as percent inhibition of malonaldehyde formation, as given in parentheses, at 30 and 200 ppm levels of addition, respectively; ascorbic acid was used at 500 ppm level.

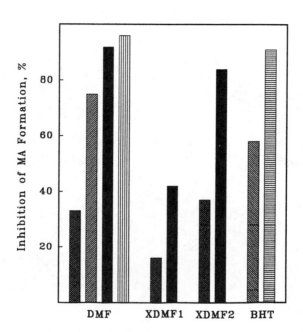

Figure 2. Inhibition of malonaldehyde (MA) formation (%) by 0-2% deheated mustard flour (DMF) at 0.5%, ▦; 1.0%, ▨; 1.5%, ■; and 2.0%, ▥ and with DMF extracts with water (XDMF1) and 85% methanol (XDMF2) from 0.6%, ▦; and 1.5%, ■ meal. Values are compared with butylated hydroxytoluene at 30 ppm, ▨ and 200 ppm, ▤, level in cooked, stored meat model systems.

Table III   Inhibition of malonaldehyde formation by cinnamic acid
derivatives in cooked, stored meat model systems

Compound	Substituent	Inhibition, %[a]
Coumaric Acid	4-OH	24,39
Caffeic Acid	3,5-diOH	57,70
Ferulic Acid	4-OH; 3-OMe	34,51
Sinapic Acid	4-OH; 3,5-diOMe	27,46
Chlorogenic Acid	3,4-diOH, ester with quinic acid	8,53

[a]At 30 and 200 ppm level of addition to meat, respectively.

of DMF at a 1-2% addition level was equivalent to that of sodium tripolyphosphate (STPP) in pork at 3000-5000 ppm, as it has been shown elsewhere (*21*).   In mechanically deboned chicken meat, its effect was close to that of 1500-3000 ppm of STPP (results not shown).

Conclusions and Future Research Needs

Phenolic compounds of natural origin and related compounds, as such or as spices or protein extenders, may offer an alternative method of meat flavor preservation. Extending of the shelf-life of products would have a definite positive effect in the operation of institutional food services, fast food chains and pre-cooked meat markets.   Work exploring natural antioxidants and the test of their efficiency is deemed necessary.   Furthermore, effectiveness of novel protein extenders, spices and their extracts as well as selected amino acids/peptide combinations may not only confer expected properties but may also function as natural antioxidant substitutes for incorporation in food systems.   Further research in this area may prove beneficial.   Furthermore, application of such natural antioxidants in muscle foods, particularly restructured and emulsified systems, may have practical significance.

Acknowledgments

Financial support from the Natural Sciences and Engineering Research Council (NSERC) of Canada is acknowledged.

Literature Cited

1.   Dziezak, J.D. *Food Technol.* **1986.** *40*, 94-97, 101-102.

2. Bailey, M.E. *Food Technol.* **1988.** *42*(6), 123-126.
3. Pratt, D.E.; Watts, B.M. *J. Food Sci.* **1964.** *29*, 27-33.
4. Pruthi, J.S. In: *Spices and Condiments: Chemistry, Microbiology and Technology.* Academic Press, New York. **1980.** pp. 17-24.
5. Heimann, W.; Reiff, F. *Fette U. Seifen* **1953.** *5*, 451-456.
6. Simpson, T.H.; Uri, N. *Chem. Ind.* **1956.** 956-957.
7. Mehta, A.C.; Seshadri, R.T. *J. Sci. Ind. Res.* **1959.** *18B*, 24-28.
8. Crawford, D.L.; Sinnhuber, R.O.; Aft, A. *J. Food Sci.* **1961.** *26*, 139-142.
9. Das, N.P.; Pereira, T.A. *J. Amer. Oil. Chem. Soc.* **1990.** *67*, 255-258.
10. Pratt, D.E.; Hudson, B.J.F. In: *Food Antioxidants.* Hudson, B.J.F., ed. Elsevier Applied Science. London and New York. **1990.** pp. 171-191.
11. Lewis, E.J.; Watts, B.M. *Food Res.* **1958.** *23*, 274-279.
12. Kelley, G.G.; Watts, B.M. *Food Res.* **1957.** *22*, 308-312.
13. Zadernowski, R.; Nowak, H.; Kozlowska, H. **1991.** Abstract No. C-40. Presented at the 8th International Rapeseed Congress. Saskatoon, Canada. July 9-11.
14. Shahidi, F. *Bull. Liason Groupe Pholphenols.* **1988.** *14*, 361-362.
15. Rhee, K.S.; Ziprin, Y.A.; Rhee, K.C. *J. Food Sci.* **1979.** *44*, 1132-1135.
16. Rhee, K.S.; Ziprin, Y.A.; Rhee, K.C. *J. Food Sci.* **1981.** *46*, 75-77.
17. Shahidi, F.; Rubin, L.J.; Wood, D.F. *J. Food Sci.* **1987.** *52*, 564-567.
18. Hayes, R.B.; Bookwalter, G.N.; Bagley, E.B. *J. Food Sci.* **1977.** *42*, 1527-1532.
19. Castonguay, A.; Lui, L.; Stoner, G.D. *Bull. Liason Groupe Polyphenols.* **1990.** *15*, 153-157.
20. Kuhnau, J. *Wld. Rev. Nutr. Diet.* **1976.** *24*, 117-191.
21. Saleemi, Z.O.; Wanasundara, P.K.J.P.D.; Shahidi, F. *J. Agric. Food Chem.* **1991.** Submitted.

RECEIVED February 11, 1992

Chapter 18

Oilseed Food Ingredients Used To Minimize Oxidative Flavor Deterioration in Meat Products

Ki Soon Rhee

Meat Science Section, Department of Animal Science, Texas A&M University, College Station, TX 22843–2471

Extensive studies have been conducted in our laboratory on utilization of oilseed food ingredients as antioxidants for meat products. Initial studies dealt with antioxidant activity, in model systems, of aqueous or alcoholic extracts of defatted flours, protein concentrates and protein isolates from glandless cottonseed, peanut and soybean. The model system studies were followed by evaluation of antioxidative effectiveness of oilseed ingredients when used in different categories of meat products. Glandless cottonseed ingredients which were higher in total phenolic concentration than other oilseed ingredients were also superior in antioxidative effectiveness in meat products. Furthermore, defatted glandless cottonseed flour incorporated into the coating of batter-breaded meat nuggets minimized oxidative flavor deterioration in the meat part of the nugget.

Oxidative Flavor Deterioration in Meat Products

Lipid oxidation is a major cause of flavor deterioration in meat and meat products. Additionally, the oxidation of lipids may adversely affect the color of fresh (raw) meat. Many studies have reported a close interrelationship between lipid oxidation and meat pigment oxidation, i.e., discoloration (1-4). The off-flavors of meat and meat products that are associated with lipid oxidation have been described as "warmed-over", "rancid", "oxidized", "stale", "cardboard", "painty", and "fish" flavors (5-7). However, "warmed-over flavor" is the most often used descriptive term for the flavor defect of cooked-stored meat (5, 7). Controlling lipid oxidation in meat and meat products is

0097–6156/92/0506–0223$06.00/0
© 1992 American Chemical Society

more important now than ever before because the use of precooked meat products by institutional foodservices, fast-food outlets, delicatessens, etc., is becoming more prevalent, along with the increased marketing of precooked frozen meat entrees in retail outlets. Lipid oxidation occurs more readily in cooked meat or thermally processed uncured meat products. In fact, warmed-over flavor can be detected in uncured, cooked meat in a few hours of refrigeration (8). Oxidative flavor deterioration in meat and meat products can be controlled or minimized by using a variety of antioxidants (9-11), including natural antioxidants (10), and by vacuum packaging. Oilseed food ingredients are one category of the natural antioxidants evaluated in meat products. This chapter is an overview of research studies conducted in our laboratory that were aimed at utilizing oilseed food ingredients in meat products as natural antioxidants.

Earlier Studies

Extensive research has been conducted since the early 1960's on the development and utilization of oilseed food products for their economic, functional or nutritional attributes. However, the antioxidant property that many oilseed food ingredients may possess has not been fully utilized in formulation of food products, including meat products. In a previous literature review on the use of oilseed ingredients in meat products (10), we found that many studies dealt with lipid oxidation in meat patties and loaves extended with oilseed ingredients, especially soy ingredients. However, in most of these studies, oilseed ingredients were added at very high levels, as high as 25-30% in the rehydrated state, for the purpose of extending the meat. The reduction of lipid oxidation in these studies, therefore, was due in large part to the meat dilution effect exerted by the nonmeat ingredients used at relatively high levels. In order for an oilseed ingredient to be named as an antioxidant for a food product, the ingredient when used at low levels should be able to provide substantial antioxidant protection to the product.
We became interested in antioxidant activity of oilseed products in meat systems in the late 1970's, primarily because of: (a) the research results from Dr. Pratt's laboratory showing that the accumulation of 2-thiobarbituric acid (TBA)-reactive substances in cooked-stored beef slices were substantially reduced when the slices were covered with aqueous extracts (6.7% in soy material) of soybeans and soybean food ingredients rather than plain water (12) and (b) a review by Hayes et al. (13) of antioxidant activity of soybean products. Following the report of Pratt (12) on water-soluble antioxidant activity of soybeans, there was another key study that interested us. Sato et al. (14) added a textured, defatted soy flour product or liquid cyclone-processed, defatted cottonseed flour (i.e., pigment glands-removed, defatted flour from glanded cottonseeds) — at a 4% meat substitution

level — to ground beef loaves. They found that both oilseed ingredients decreased lipid oxidation and warmed-over flavor development in the cooked product during refrigerated storage; however, much greater antioxidant effect was shown by the cottonseed flour.

Use of Oilseed Ingredients as Antioxidants

Model System Studies. Food ingredients prepared from different oilseeds were compared for their antioxidant activity in model systems. Aqueous (15) and methanolic (16) extracts were prepared from defatted flours, protein concentrates and protein isolates from soybean, peanut and glandless cottonseed. Glandless cottonseed is different from the regular glanded cottonseed in that it is essentially devoid of the pigment gland containing gossypol which is toxic to monogastric animals including humans. Glandless cottonseed kernels from the field may contain insignificant amounts (within the safe limit) of total gossypol (17). In our model system studies, linoleate was reacted with metmyoglobin (MetMb), Fe^{2+}-EDTA (in 1:1 molar ratio) or raw beef homogenates, with oxygen uptake measured to determine the rate of linoleate oxidation. The antioxidant activity of the methanolic extract of each oilseed ingredient was also determined by its inhibitory effect on autoxidation of safflower oil (16). With aqueous extracts, there were no overall distinctive differences in antioxidant activity among the three oilseed types (15). With methanolic extracts, however, glandless cottonseed ingredients showed uniquely and consistently higher antioxidant activity than peanut and soy ingredients (16; Table I). Furthermore, the total phenolic content was also markedly higher for the alcoholic extracts of glandless cottonseed ingredients, and antioxidant activities of methanol extracts of defatted flours and protein concentrates from the three oilseed types were significantly correlated with their total phenolic concentrations (16).

More recently, a model meat system consisting of water-extracted beef muscle residue plus MetMb, MetMb-H_2O_2, nonheme iron (Fe^{2+}), or NADPH-dependent enzymic lipid peroxidation system (18,19) was used to obtain additional information on the antioxidative nature of glandless cottonseed ingredients, specifically defatted flour. Defatted glandless cottonseed flour inhibited the oxidation of lipids in the washed beef muscle regardless of the catalytic component or system (20; Table II). However, the magnitude of inhibition by the cottonseed flour varied depending on the thermal status (raw or cooked) of the system as well as catalytic components.

Addition of Oilseed Ingredients to Meat. Frequently, model system results are different, qualitatively and/or quantitatively, from results of actual food products because various constituents in food products can interact

Table I. Effects of methanolic extracts of oilseed
ingredients on linoleate oxidation catalyzed by
Metmyoglobin (MetMb), Fe^{2+}-EDTA and fresh beef
homogenates and on autoxidation of safflower oil

Oilseed protein ingredient[a]	% Inhibition of linoleate oxidation catalyzed by[b]			% Inhibition of autoxidation of safflower oil
	MetMb	Fe2+-EDTA	Beef homog- enates	
Defatted flours				
Cottonseed, glandless	67a	88a	51a	92
Peanut	43b	57b	43b	81
Soybean	35b	61b	46b	87
Concentrates				
Cottonseed, glandless	49a	91a	51a	82
Peanut	37b	47b	28b	74
Soybean	24c	31c	32b	68
Isolates				
Cottonseed, glandless	53a	58a	53a	60
Peanut	35b	32c	51a	25
Soybean	23c	47b	48a	52

SOURCE: Reprinted with permission from ref. 16. Copyright 1981.
[a]Extract equivalent to 0.1% (w/v) oilseed ingredient in the
linoleate oxidation system or 1.2g/10g safflower oil. [b]The
data represent the means of two independent series
(duplicate determinations in each series) of experiments
for each oilseed ingredient and catalytic system. Means in
the same column within an ingredient form (defatted flour,
concentrate or isolate) which are not followed by the same
letter are significantly different ($p<0.05$).

with test variables and thus influence test results.
Therefore, we wanted to find out whether the various
oilseed ingredients that we evaluated in model systems
would perform similarly when used in actual meat products.
The first meat product vehicle employed was cooked ground
beef patties (21). Each oilseed ingredient was used in the
rehydrated state (1:2 dry ingredient-water) to replace 10%
of the meat in raw ground beef patties. In other words,
the experimental patties contained 3.3% dry oilseed
ingredient. The fat content of both experimental (oilseed
ingredient-containing) and control (100% beef) patties was
adjusted to the same percentage (20%). Patties were cooked
to different internal temperatures and stored at 4°C prior
to determination of lipid oxidation by a TBA procedure
which minimizes further lipid oxidation during the assay

Table II. Effect of defatted glandless cottonseed flour on lipid oxidation in water-extracted beef muscle residue catalyzed by MetMb, MetMb-H_2O_2, Fe^{2+} and components of enzymic lipid peroxidation: Raw and cooked systems

Catalytic additive[a]	DGCF addition[b]	TBA no.[c] after storage at 4°C[d]		% Inhibition by DGCF	
		0 day	3 days	0 day	3 days
Raw system					
MetMb	w/o	0.49	0.60		
	w/	0.33	0.39	33	35
MetMb + H_2O_2 (1:0.1)[e]	w/o	1.78	1.16		
	w/	0.38	0.33	79	72
MetMb + H_2O_2 (1:0.25)	w/o	2.63	3.35		
	w/	0.39	0.42	85	88
MetMb + H_2O_2 (1:0.5)	w/o	0.63	2.06		
	w/	0.36	0.41	43	80
MetMb + H_2O_2 (1:1)	w/o	0.87	2.10		
	w/	0.36	0.40	59	81
Fe^{2+}	w/o	1.13	1.23		
	w/	0.61	0.49	46	60
Fe^{2+} + NADPH + ADP	w/o	1.80	2.43		
	w/	1.22	1.37	32	56
Cooked system					
MetMb	w/o	0.89	1.50		
	w/	0.56	0.75	37	50
MetMb + H_2O_2 (1:0.1)[e]	w/o	1.23	1.77		
	w/	0.55	0.70	55	61
MetMb + H_2O_2 (1:0.25)	w/o	1.36	2.26		
	w/	0.62	0.68	54	70
MetMb + H_2O_2 (1:0.5)[b]	w/o	1.13	2.64		
	w/	0.51	0.70	45	74
MetMb + H_2O_2 (1:1)	w/o	2.34	3.97		
	w/	0.57	0.62	76	84
Fe^{2+}	w/o	2.09	3.15		
	w/	0.62	0.76	70	76
Fe^{2+} + NADPH + ADP	w/o	2.19	3.54		
	w/	1.26	1.40	43	61

[a]Levels of catalytic additives: 4 mg MetMb/g; 3 µg Fe^{2+}/g; 0.3 µmol NADPH/g; 0.3 µmol ADP/g. [b]W/o: without DGCF; w/: with DGCF (26.5 mg DGCF/g). [c]mg malonaldehyde/kg sample. [d]TBA data for samples without DGCF were those reported previously (*18*). [e]Molar ratio of MetMb to H_2O_2.

(22). Differences in antioxidative effectiveness among the three oilseed types were dependent on the degree of freshness of the raw meat material. With freshly ground beef, all of the oilseed ingredients tested were effective in preventing lipid oxidation in cooked patties during 5 days of refrigeration, irrespective of the cooking end-point internal temperature. However, when ground beef stored for 3 days at 4°C was used as the raw material, glandless cottonseed ingredients were superior to soy or peanut ingredients in antioxidative effectiveness; those patties containing any type of glandless cottonseed showed no increase in TBA values even after 6 days of refrigeration (Table III). This confirmed what we had found in model systems with methanolic extracts of these oilseed ingredients (16), that is, glandless cottonseed products have the greatest antioxidative effect. Sensory evaluation of the patties prepared from freshly ground beef, which was conducted immediately after cooking (to an internal temperature of 74°C), revealed that patties containing oilseed ingredients were not significantly

Table III. TBA values of all-beef and oilseed ingredient-containing patties prepared from ground beef stored at 4°C for 3 days

Product[a]	Cooked to 70°C			Cooked to 74°C		
	0	3	6	0	3	6
	(Days at 4°C after cooking)					
All-beef	4.1	5.8	7.6	4.2	7.6	10.0
Beef + defat. flour[b]						
Cottonseed, glandless	2.5	2.6	2.5	3.0	2.3	2.4
Peanut	3.5	4.1	4.9	3.2	2.7	4.5
Soy	3.9	4.1	4.6	3.7	4.6	5.6
Soy-textured	3.1	4.0	3.4	2.5	2.6	3.2
Beef + concentrate[b]						
Cottonseed, glandless	2.5	2.4	2.5	2.7	2.5	2.2
Peanut	2.6	2.6	2.5	2.7	2.7	4.4
Soy	3.4	3.6	4.1	2.5	2.1	3.0
Beef + isolate[b]						
Cottonseed, glandless	2.6	2.5	2.6	2.8	2.1	2.4
Peanut	3.2	3.6	5.1	3.3	3.7	4.7
Soy	3.2	3.3	3.6	3.3	4.4	5.0

SOURCE: Adapted from Ziprin et al. (21).
[a]All raw products were formulated to contain 20% fat.
[b]10% as rehydrated product (3.3% dry oilseed ingredient).

different from all-beef patties in all sensory attributes evaluated. It needs to be emphasized that a nonmeat ingredient to be used in a meat product should be compatible sensorially with that meat product and should not adversely affect the eating quality of the meat product at the level used.

Since our study on cooked beef patties confirmed that glandless cottonseed ingredients provided far superior antioxidant protection compared to peanut or soy ingredients, we decided to test defatted glandless cottonseed flour (DGCF; the least expensive glandless cottonseed protein ingredient) — at 1-3% levels based on the final product weight — in restructured beef roasts (*23, 24*). When restructured beef roasts made with added NaCl (0.75%), polyphosphate (0.5%) and water (3%), with or without DGCF, were precooked and refrigerated, both TBA values and sensory warmed-over flavor scores were much lower for samples containing DGCF than those without DGCF. The use of DGCF at 1% and 2% levels had no adverse effect on palatability of the restructured beef product; however, when used at 3%, it adversely affected texture and flavor of the product. A restructured meat product is made from lower-valued muscles, and salt (NaCl), polyphosphate(s) and water are usually added in the restructuring process.

We also tested DGCF in raw ground beef containing different amounts of NaCl, 0-2% (*25, 26*). The NaCl level was included as a variable because NaCl promotes lipid oxidation (*19*) and processed meat products contain varying amounts of NaCl. At each NaCl level, DGCF (2% or 3%) reduced lipid oxidation and discoloration in raw ground beef patties during storage at both 4° and -20°C. The lipid- and color-protecting effects of DGCF were observed in beef patties containing relatively small amounts of fat, 8% (*25*) and 10% (*26*), as well as in beef patties containing a much higher amount of fat, 20% (*26*).

At this point, we conducted a study to determine if the DGCF would have any adverse effect on microbial growth in ground beef stored under retail display conditions (*27*). Ground beef with or without 3% added DGCF was over-wrapped with oxygen-permeable polyvinyl chloride film, as commonly done for retail ground beef, and displayed up to 6 days in a retail case. In contrast to the findings of many studies showing that meat extenders derived from soybeans, usually used in ground meat at 10-30% (rehydrated basis) levels, often increase bacterial counts of raw ground beef (*28*), DGCF (not rehydrated) added to ground beef at 3% levels did not adversely affect the display life in terms of total aerobic plate counts. The effect of DGCF on the composition of microorganisms (especially dominant microorganisms) during the display was similar to the vacuum packaging effect for fresh beef cuts, i.e., *Lactobacillus* spp. constituted a major or dominant part of the microflora of ground beef with added DGCF while *Pseudomonas* spp. or a combination of *Pseudomonas* and heterofermentative *Lactobacillus* spp. dominated the microflora of ground beef with no DGCF.

Use of Oilseed Ingredients in Meat Accompaniments.
The antioxidant property of oilseed ingredients can also be
indirectly utilized. The ingredients can be incorporated
into liquid media (such as gravy and sauce) for precooked
meat products. Many frozen meat entrees (e.g., Salisbury
steaks, roast beef slices with gravy, meat loaf slices with
sauce, etc.) used in many foodservice establishments and
those available at retail outlets have some types of liquid
cover. We used glandless cottonseed, peanut and soy
ingredients (defatted flours, protein concentrates, and
protein isolates), in conjunction with wheat flour, to make
gravies (29). When precooked beef patties (100% beef) were
covered with gravy preparations with or without oilseed
ingredients, to simulate Salisbury steaks, and stored at
4°C, TBA values were much lower for the patties covered
with the gravy preparations containing oilseed ingredients
than those without. No consistent differences among the
oilseed types were observed. In addition, hot water
extracts of the oilseed ingredients were tested as a cover
liquid (6% in dry oilseed ingredient) for refrigerated
roast beef slices (29). TBA values were markedly lower for
the beef slices covered with oilseed ingredient extracts
than for those covered with plain water, regardless of the
oilseed type or ingredient form (Table IV).

Table IV. TBA values of roast beef slices covered with
aqueous extracts of oilseed food ingredients and
stored at 4°C for 3 days

Extract	TBA no.
Defatted flour	
H_2O (control)	4.4
Cottonseed, glandless	1.3
Peanut	1.5
Soybean	1.7
Concentrate	
H_2O (control)	4.1
Cottonseed, glandless	1.0
Peanut	2.9
Soybean	1.5
Isolate	
H_2O (control)	4.6
Cottonseed, glandless	2.0
Peanut	1.8
Soybean	1.9

SOURCE: Reprinted with permission from ref. 29. Copyright 1981.

Use of Oilseed Ingredients in Batter-Breading. More
recently, we became interested in controlling oxidative
flavor deterioration in batter-breaded meat products using
oilseed food ingredients. A variety of precooked/prepared,

batter-breaded meat products are currently available in
retail outlets, and their flavor freshness is not always
predictable. In addition, raw-frozen, batter-breaded
muscle food products are widely used in foodservice. It
was not certain if an antioxidative oilseed ingredient
incorporated into the batter-breading system (rather than
into the meat formulation) would significantly protect the
meat lipids from oxidation. Initially, DGCF and sunflower
seed protein isolate were incorporated into batter used to
coat beef patties(30). Ground beef patties coated with
either batter made of wheat flour alone (control) or batter
made with 1:1 blends of wheat flour and DGCF (or sunflower
seed protein isolate) were fried in commercial cooking oil
(partially hydrogenated soybean oil). TBA values of the
meat part of fried patties stored at 4°C for 5 days were
reduced by more than 50% by the DGCF or sunflower isolate
present in the coating. While presence of DGCF in batter
had no adverse effect on the crust color of fried patties,
sunflower isolate discolored the crust because of its dark
greenish color.   In the ensuing studies, DGCF was
incorporated either into predust plus batter or into
restructured pork muscle nuggets (31, 32). Pork shoulders
were restructured into nuggets (1.25-cm thick) with 0.5%
NaCl, 0.25% polyphosphate and 3% water or with the three
additives plus 1.5% DGCF, and restructured nuggets were
batter-breaded with either commercial coating ingredients
or commercial coating ingredients plus DGCF (40% and 20%
DGCF in predust and dry batter mix, respectively). The
batter-breaded nuggets were pre-fried in vegetable oil
without preservatives and stored at either 4°C (31, 32) or
-20°C (32). TBA values and sensory warmed-over flavor
scores determined on the precooked-refrigerated products
(31, 32) and TBA values on raw-frozen products (32)
indicated that oxidative flavor deterioration in batter-
breaded pork nuggets can be substantially reduced or
minimized by incorporating the oilseed ingredient into
either the coating system or the meat. Data representing
typical results of refrigerated-reheated products are shown
in Table V.

**Heating Oilseed Protein Ingredients with Reducing
Sugars to Enhance Antioxidant Effect**. Many studies
have shown that the nonenzymatic browning (Maillard)
reaction produces antioxidative substances (11). However,
in the great majority of these studies, products resulting
from the reaction of an amino acid or a secondary amine
with a reducing sugar were tested for antioxidant activity.
We were interested in documenting whether or not
nonenzymatic browning between oilseed protein ingredients
and a reducing sugar could enhance the antioxidant value of
oilseed ingredients, and the extent of the enhancement, if
it does. Defatted flours (54-63% protein) and protein
isolates (89-93% protein) prepared from glandless
cottonseed, peanut and soybean were mixed with glucose in a
ratio of 1:1 by weight and heated at 100°C for up to 6
hours. Antioxidant activity of ethanolic extract of the

Table V. TBA values and sensory warmed-over flavor (WOF)
scores for precooked, batter-breaded pork nuggets made
with or without defatted glandless cottonseed
flour (DGCF) in the meat or coating that were
reheated after 0 or 9 days of storage 4°C

Product[e] (Meat/ coating)	TBA no.				WOF score[f]	
	Whole nugget		Meat alone		Whole nugget	
	0	9	0	9	0	9
	Days		Days		Days	
C/C	1.3[a]	2.7[a]	1.1[a]	2.7[a]	5.0[a]	3.9[b]
C/E	1.2[a]	1.5[b]	0.8[b]	1.4[b]	4.7[a]	4.4[a]
E/C	1.0[b]	1.1[c]	0.6[c]	1.2[c]	4.9[a]	4.7[a]
E/E	0.9[b]	1.1[c]	0.5[d]	0.9[d]	4.8[a]	4.5[a]

SOURCE: Adapted from Rhee et al. (31).
[a-d]Means within the same column which are not followed by
the same superscript letter are significantly different
(p<0.05). [e]C=control (without DGCF); E=experimental (with
DGCF). [f]Based on a 5-point scale (1=very pronounced WOF;
5=no WOF).

heated mixtures was determined against autoxidation of
safflower oil.   Antioxidant activity of the mixtures
increased with heating time, as did the extent of
nonenzymatic browning, and the rate of increase in
antioxidant activity was greater for those mixtures (peanut
flour-glucose and peanut isolate-glucose) having lower
original (0-hour) antioxidant activity (33). The percent
antioxidant activity increase after 2-hour heating ranged
from 11% to 71% depending upon the oilseed ingredient used,
whereas that after 6-hour heating ranged from 33% to 257%.
If small amounts of these heated mixtures of oilseed
ingredients plus glucose (or other reducing sugars) are
used in sensorially compatible meat products, they may
possibly decrease warmed-over flavor development more than
do oilseed ingredients alone. This possibility needs to be
tested in actual meat products of different composition.

**Antioxidant   Components   in   Glandless   Cottonseed
Ingredients**.   Phenolic compounds, such as flavonoids,
phenolic acids, tocopherols and phenolic terphenes, seem to
be the major antioxidant compounds in food materials of
plant origin (including most commonly consumed oilseed food
ingredients), with phosphatides, amino acids and peptides
also partially accounting for the antioxidant effect
observed with the materials (13, 34).   Also, possible
antioxidant contribution of phytic acid (35) as well as
that of minute amounts of gossypol (a phenolic compound)
which may be present in glandless cottonseed food

ingredients   can not be totally ruled out.   The superior antioxidant effect of glandless cottonseed ingredients, compared to peanut and soy ingredients, in a meat product (*21*) and model systems (*16*) appears to be due primarily to higher total phenolic concentrations in conjunction with the presence of the flavonoid species of higher antioxidant activity (flavones rather than isoflavones) in cottonseed ingredients (*10*).

Further research.   In addition to oilseed ingredients, there are other natural antioxidants that have been tested, or may have antioxidant potential, in meat products (10). Comparisons of oilseed ingredients with other natural antioxidants for effectiveness in meat products would be useful.   Additionally, the efficacy and performance of antioxidative oilseed ingredients when used in combination with other natural antioxidants need to be investigated.

With the increasing consumers' demand for low-fat meat products due to diet/health concerns, antioxidative oilseed ingredients may be used in meat products for an additional purpose, i.e., to improve the texture of low-fat meat products.   Low-fat meat products generally are less juicy, tender and/or flavorous while more springy or firm.   The texture improving function of certain soy isolate products in lean ground beef patties has been documented (36).   The potential dual functions of some oilseed ingredients — as antioxidants and texture modifiers — need to be investigated, particularly for precooked low-fat meat products targeted at foodservice uses and sales at the retail level.   Research studies along this line are ongoing in our laboratory.

Literature   Cited

1.   Hutchins, B. K.; Liu, T. H. P.; Watts, B. M.   *J. Food Sci.* **1967**, *32*, 214.
2.   Lin, T. S.; Hultin, H. O.   *J. Food Sci.* **1977**, *42*, 136.
3.   Govindarajan, S.; Hultin, H. O.; Kotula, A. W.   *J. Food Sci.* **1977**, *42*, 571.
4.   Koizumi, C.; Nonaka, J.; Brown, W. D.   *J. Food Sci.* **1978**, *38*, 813.
5.   *Warmed-Over Flavor of Meat*; St. Angelo, A. J.; Bailey, M. E., Eds.; Academic Press: Orlando, FL, 1987.
6.   Johnson, P. B.; Civille, G. V.   *J. Sens. Stud.* **1986**, *1*, 99.
7.   Love, *J. Food Technol.* **1988**, *42(6)*, 140.
8.   Tims, M. J.; Watts, B.M.   *Food Technol.* **1958**, *12*, 240.
9.   Pearson, A. M.; Love, J. D.; Shorland, F.B. In Advances in Food Research; Chichester, C. O.; Mrak, E. M.; Stewart, G. F., Eds.; Academic Press: New York, NY, 1977; Vol. 23; pp 1-74.
10.   Rhee, K. S. In *Warmed-Over Flavor of Meat*; St. Angelo, A. J.; Bailey, M. E., Eds.; Academic Press: Orlando, FL, 1987; pp 267-289.

11. Bailey, M. E. *Food Technol.* **1988**, *42(6)*, 123.
12 Pratt, D. E. *J. Food Sci.* **1972**, *37*, 322.
13. Hayes, R. E.; Bookwalter, G. N.; Bagley, E. B. *J. Food Sci.*, **1977**, *42*, 1527.
14. Sato, K.; Hegarty, G. R.; Herring, H. K. *J. Food Sci.* **1973**, *38*, 398.
15. Rhee, K. S.; Ziprin, Y. A.; Rhee, K. C. *J. Food Sci.* **1979**, *44*, 1132.
16. Rhee, K. S.; Ziprin, Y. A.; Rhee, K. C. *J. Food Sci.* **1981**, *46*, 75.
17. Fisher, G. S.; Frank, A. W.; Cherry, J. P. *J. Agric. Food Chem.* **1988**, *36*, 42.
18. Rhee, K. S.; Ziprin, Y. A.; Ordonez, G. *J. Agric. Food Chem.* **1987**, *35*, 1013.
19. Rhee, K. S. *Food Technol.* **1988**, *42(6)*, 127.
20. Rhee, K. S., Texas A&M University, College Station, TX; unpublished data.
21. Ziprin, Y. A.; Rhee, K. S.; Carpenter, Z. L.; Hostetler, R. L.; Terrell, R. N.; Rhee, K. C. *J. Food Sci.*, **1981**, *46*, 58.
22. Rhee, K. S. *J. Food Sci.* **1977**, *43*, 1776.
23. Leu, J.-P. R. *Study of Factors Affecting Oxidative Stability of Precooked Chunked and Formed Beef Roasts*; Ph.D. Dissertation, Texas A&M University, College Station, TX, 1986.
24. Leu, R.; Keeton, J. T.; Rhee, K. S.; Bohac, J. J.; Cross, H. R. **1986**, Presented at the 46th Ann. Meeting of the Inst. of Food Technol., Dallas, TX, June 15-18.
25. Rhee, K. S.; Smith, G. C.; Rhee, K. C. *J. Food Sci.* **1983**, *48*, 351.
26. Rhee, K. S.; Smith, G. C. *J. Food Protect.* **1983**, *46*, 787.
27. Rhee, K. S.; Vanderzant, C.; Keeton, J. T.; Ehlers, J. G.; Leu, R. *J. Food Sci.*, **1985**, *50*, 1388.
28. Draughon, F. A. *Food Technol.* **1980**, *34(10)*, 69.
29. Rhee, K. S.; Ziprin, Y. A. *J. Food Protect.* **1981**, *44*, 254.
30. Rhee, K. S. *J. Food Sci.* **1984**, *49*, 1224.
31. Rhee, K. S.; Leu, R.; Ziprin, Y. A.; Keeton, J. T.; Bohac, J. J.; Cross, H. R. *J. Food Sci.*, **1988**, *53*, 388.
32. Rhee, K. S.; Keeton, J. T.; Ziprin, Y. A.; Leu, R.; Bohac, J. J. *J. Food Sci.* **1988**, *53*, 1047.
33. Rhee, K. S.; Rhee, K. C. *J. Food Protect.* **1982**, *45*, 452.
34. Pratt, D. E. In *Autoxidation in Food and Biological Systems*; Simic, M. G.; Karel, M., Ed.; Plenum Press: New York, NY, 1980; pp 283-293.
35. Graf, E.; Empson, K. L; Eaton, J. W. *J. Biol. Chem.*, **1987**, *262*, 11647.
36. Liu, M. N.; Huffman, D. L.; Egbert, W. R.; McCaskey, T. A.; Liu, C. W. *J. Food Sci.*, **1991**, *56*, 906.

RECEIVED November 7, 1991

CHEMICAL PROPERTIES
OF PHENOLIC COMPOUNDS

Chapter 19

Tannin–Protein Interactions

Ann E. Hagerman

Department of Chemistry, Miami University, Oxford, OH 45056

Over the last 20 years, information about the diverse compounds
classified as "tannins" has rapidly accumulated. Structural studies have
revealed that "tannins" belong to one of three groups of compounds:
proanthocyanidins, gallotannins or ellagitannins. Each of the three
types of tannin interacts strongly with protein, and a variety of
approaches have been used to probe this characteristic reaction. The
interaction is influenced by characteristics of the protein (including size,
amino acid composition, pI, and extent of posttranslational
modification), characteristics of the tannin (size, structure,
heterogeneity of the preparation), and conditions of the reaction (pH,
temperature, solvent composition, time). Under some conditions an
insoluble complex is formed, but soluble complexes may also form.
Both soluble and insoluble complexes are stabilized by reversible,
noncovalent bonds between tannin and protein.

The "tannins" are a diverse group of plant phenolics which are unified by a single
common property: the ability to precipitate protein from aqueous solution.
Tannins are polymeric phenolics. Molecular weights as high as 20,000 have been
reported for condensed tannins (1). The molecular weights of hydrolyzable
tannins range from 500 to 5,000 (2). Lower molecular weight phenolics
including phenolic acids and simple flavonoids may bind protein but cannot
crosslink the complexes as is required for precipitation. Lignin, a highly
methoxylated phenolic polymer, does not precipitate protein and is not a tannin
(3).

Tannins have been known for centuries. The extracts of bark traditionally
used to convert animal skins to leather are rich in tannins (4). Red wine,
beer, and many fruits contain tannins which react with proteins in the mouth
causing the sensation of astringency (5). Many herbal medicines contain

0097–6156/92/0506–0236$06.00/0
© 1992 American Chemical Society

tannin (*6*). More recently, tannins have been under active investigation as anticancer agents (*7*) and because of their role in herbivory (*8*). The nutritional consequences of ingesting tannin have been reviewed (*9*).

Tannin Chemistry

Tannins are divided into two major classes: condensed tannins, which are flavanoid-based polymers; and hydrolyzable tannins, which are polygalloyl esters. Higher plants may produce any combination of condensed and hydrolyzable tannins, although some chemotaxonomic patterns have been noted (*2*). Brown algae contain phloroglucinol-based phlorotannins, which have not been well characterized (*10*). Chemistry and analysis of the tannins have been reviewed recently (*11*) and will only be briefly considered here.

The condensed tannins are a structurally diverse group of flavonoid-based compounds (Figure 1) (*12*). There are several variations in the pattern of hydroxylation on the A and B rings of the flavanoid monomers. The most common condensed tannins, the procyanidins, are hydroxylated on carbons 5 and 7 of the A ring and on carbons 3' and 4' of the B ring. There are three chiral centers on the heterocyclic ring of the flavonoid, and many of the possible stereoisomers of the condensed tannins have been identified. Several interflavanoid bonds have been described, yielding linear or branched structures. Derivitization of the flavanoids is possible, particularly at the nonphenolic hydroxyl on carbon 3.

The condensed tannins do not undergo hydrolysis in acid or base. However, in hot alcohol the flavanoid polymer is oxidatively cleaved to yield colored anthocyanidins; thus the condensed tannins are often called "proanthocyanidins". This reaction in hot alcohol is the basis for the most common of the analytical methods for condensed tannins (*13*). Alternative methods are based on the reaction of the flavanoid with aromatic aldehydes such as vanillin (*14*)(*15*). NMR has been a powerful tool for elucidating the structures of the condensed tannins (*11*).

The hydrolyzable tannins are hydrolyzed in acid, in base or by esterases (tannase) to yield the parent polyol and the phenolic acids. Glucose is the most common alcohol, but hydrolyzable tannins containing diverse alcohols such as hamamelose and quinic acid have been reported (*11*). If gallic acid is the only phenolic acid produced upon hydrolysis, the tannin is a gallotannin (Figure 2). The gallic acid can be esterified directly to the core polyol, or can be esterified to other gallic acid residues to form the so-called depside chains, which can be up to five gallic acid residues long.

Ellagitannins (Figure 2) produce hexahydroxydiphenic acid (HHDP) upon hydrolysis; the free acid spontaneously lactonizes to yield ellagic acid. HHDP is thought to form by oxidative coupling of adjacent galloyl residues on gallotannins. Further oxidation or hydration reactions can subsequently occur to form the complex hydrolyzable tannins such as casuariin and castalagin (*11*).

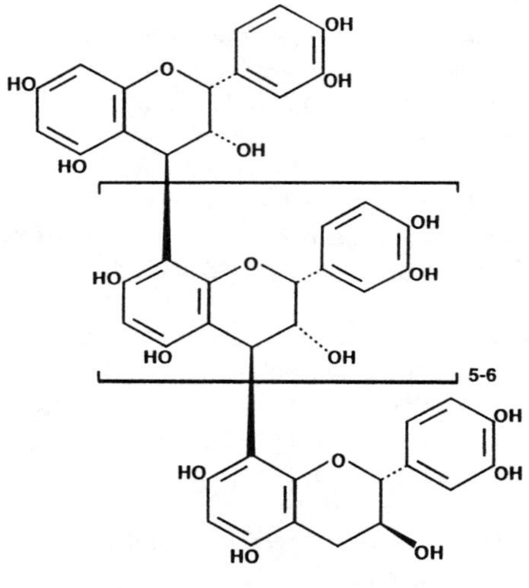

Figure 1. Condensed tannin from <u>Sorghum</u> grain.

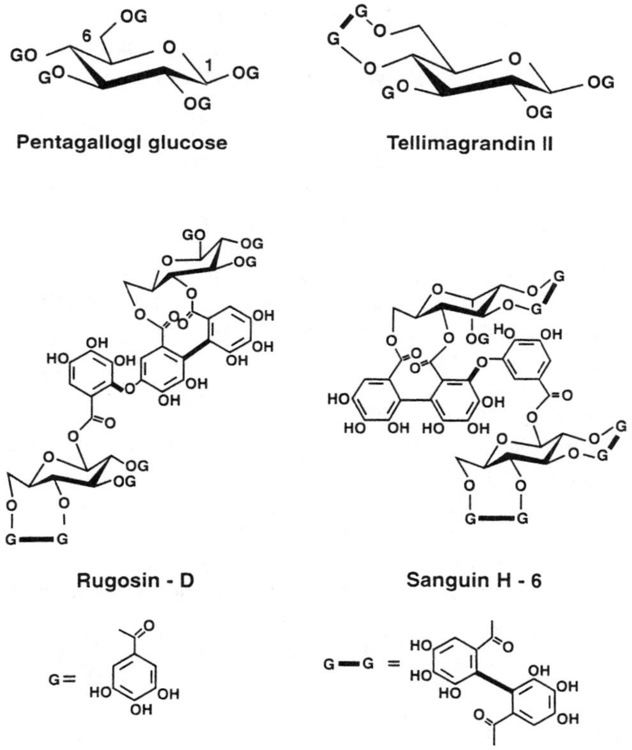

Figure 2. Typical hydrolyzable tannins. A simple gallotannin, pentagalloyl glucose, is shown with several of the more complex ellagitannins.

Gallotannins can be determined by measuring gallic acid after hydrolysis (*16*), and ellagitannins by analyzing ellagic acid after hydrolysis (*17*). Structural determinations are largely based on NMR (*11*).

The commercially available tannins include a condensed tannin, quebracho, which is a polymer of a 5-deoxy flavonoid and is thus a profisetinidin (*12*). Tannic acid is a mixture of esters ranging from tri- to dodecagalloylglucose (*18*). Chestnut tannin is a poorly characterized ellagitannin which is commercially available (*19*). Any tannin standard must be characterized before it is used (*20*), since there are large differences in the purity and compostiion of the various commercial preparations. As much as 50% of crude quebracho tannin may be nontannin phenolics (*21*). The mixture of esters present in tannic acid varies considerably depending on commercial supplier and lot (*22*).

In addition to the specific chemical methods for establishing the structural features of the tannins, numerous methods for determining the ability of tannins to interact with proteins have been described (*20*). In general, the tannin-containing sample is mixed with protein, and the resulting insoluble complex is isolated by centrifugation. If the protein was labeled either with an isotope (*23*) or with a chromophore (*21*)(*24*), precipitated protein can be measured. Tannin in the precipitate or supernatant can be determined spectroscopically (*25*) or chromatographically (*22*). The various types of tannin have quite different abilities to precipitate protein, so it is essential to use an appropriate standard when using a precipitation method to determine tannin levels in plant extracts (*20*).

Precipitation methods are limited to determining insoluble complexes, but under some conditions tannin-protein complexes may be soluble. Several approaches, including electrophoretic analysis of proteins after treatment with tannins (*26*), and competitive binding assays (*27*) have been used to characterize soluble complexes. Equilibrium dialysis has been used to probe tannin-protein interactions (*28*); potential problems result from interaction between the tannin and the dialysis membrane, and from the difficulties of using Scatchard analysis for multivalent, cooperative interactions. Tannin-protein and tannin-caffeine interactions have been examined using microcalorimetry (*29*).

Tannin-protein Interactions

There are four potential types of interaction between phenolics and protein: hydrogen bonding, hydrophobic, ionic, and covalent. Under most conditions, only the first two types of bonding are involved in the formation of tannin-protein complexes.

The phenolic hydroxyl group is an excellent hydrogen bond donor, and forms strong hydrogen bonds with the amide carbonyl of the peptide backbone. Tannins also interact with synthetic polyamides such as nylon (*30*). Hydrogen bonding solvents such as acetone and formamide inhibit tannin-protein interactions. For the series of formamides, N, N-dimethylformamide more effectively inhibits precipitation of bovine serum albumin by tannin than does the weaker hydrogen bond acceptor N-methylformamide, and the unsubstituted formamide is the least effective inhibitor (27). At high pH values, where the phenolic hydroxyl group is ionized, tannins do not interact with protein (25), supporting the contention that the phenolic hydroxyl is a hydrogen bond donor.

Hydrophobic forces are also involved in tannin-protein interactions (31). For example, nonpolar solvents and detergents such as dioxane or sodium lauryl sulfate inhibit the interaction (27). Haslam's group has recently developed a model system for studying tannin-protein interactions in which the alkaloid caffeine is substituted for protein (29). Caffeine and the galloyl groups of gallotannins stack to form parallel layers stabilized by hydrophobic bonds; hydrogen bonds between the caffeine amide and nearby galloyl groups crosslink individual stacks and cause precipitation (29). Based on this work, Haslam has concluded that hydrophobic interactions are more important than hydrogen bonding interactions for stabilizing the tannin-caffeine complex. It is not clear that this model, based on the small planar caffeine molecule, can be directly applied to systems based on large, conformationally complex proteins. The relative importance of hydrogen bonding and hydrophobic forces to the formation of tannin-protein complexes cannot yet be assessed.

It is clear that neither ionic nor covalent bonds are important in stabilizing tannin-protein complexes. Tannins do not interact with proteins at pH values above 10 (25), indicating that the phenolate anion is not involved in complex formation. Under oxidizing conditions low molecular weight phenolics do form a variety of covalent adducts with free amino acids or proteins (32). However, the complexes formed between tannin and protein can be dissociated by mild treatments with detergents (23), caffeine (29), or protein denaturants such as phenol (33). Covalent complexes would be far more difficult to disrupt.

It has been suggested that tannin-protein complexes may be more stable at low temperatures (34). Haslam did find that a variety of phenolics and tannins had higher affinities for caffeine at 45°C than at 60°C (29). He attributed the temperature dependence to the entropy changes associated with exclusion of solvent from the surface during the initial formation of hydrophobic bonds (29). In other studies, it has been reported that the precipitation of bovine serum albumin by tannin was unaffected by temperatures over the range 4°C to 44°C; at higher temperatures protein coagulation interfered with the assay (35).

The complexes that form between tannin and protein may be either soluble or insoluble. Formation of insoluble complexes is favored at pH values near the isoelectric point of the protein, where electrostatic repulsions that prevent aggregation of protein are minimized (25). Formation of insoluble complexes is also favored when tannin is present in excess. When protein is present in excess, the complexes remain soluble because each protein molecule is bound by only a few phenolic ligands (36). Soluble complexes are more difficult to detect than insoluble complexes, and only a few attempts have been made to characterize them (26)(37).

Tannin and protein can be released from complexes by a variety of mild treatments, but in the absence of specific denaturing agents the complexes may be quite stable. Haslam reported that addition of excess protein to a tannin-protein precipitate caused dissolution of the precipitate (28), consistent with the tendency of tannin to form soluble complexes in the presence of excess protein. However, we have noted that when a precipitate containing tannin and labeled protein was suspended in a large excess of unlabeled protein, there was no exchange of unlabeled protein with labeled protein and no decrease in the amount of precipitate present (38). Presumably the multivalent, crosslinked nature of the complex makes it difficult for exchange to occur.

Tannin Structure

The interaction between tannin and protein is influenced by the number of phenolic groups available for interaction and by the arrangement of the groups that are available. The importance of the number of phenolic groups has been most clearly demonstrated by comparisons of simple gallotannins. Pentagalloyl glucose (Figure 2) has a higher affinity for bovine serum albumin than tri- or tetragalloyl glucose (28). Haslam suggested the addition of despside-linked galloyl groups to pentagalloyl glucose would not enhance the interaction of gallotannin with protein (2). That proposal is not supported by experiments that show that the affinity of galloyl polyesters for protein increases with size for components of tannic acid ranging from trigalloyl glucose to at least dodecagalloyl glucose (18)(22).

The importance of the availability of the galloyl groups for interaction with protein is further illustrated by comparison of ellagitannins to gallotannins. Dimerization of the galloyl residues to form HHDP is accompanied by a reduction in affinity for protein apparently because of reduced flexibility of the ellagitannin. Thus the affinity of the ellagitannin Tellimagrandin II (Figure 2) for bovine serum albumin is much lower than the affinity of pentagalloyl glucose, and only slightly higher than the affinity of tetragalloyl glucose (28). Similarly, the affinity of Sanguin H-6 (Figure 2) for bovine serum albumin is far less than the affinity of Rugosin-D (Figure 2) (28). Two of the pairs of galloyl groups found in Rugosin-D are dimerized in Sanguin H-6, making this large ellagitannin have an affinity for bovine serum albumin similar to that of tetragalloyl glucose.

The importance of the arrangement of the phenolic hydroxyls has been demonstrated in studies comparing the simple gallotannin pentagalloyl glucose with the dimeric ellagitannin Rugosin-D. Like pentagalloyl glucose, Rugosin-D has five free galloyl groups, but Rugosin-D has a much higher affinity for bovine serum albumin than pentagalloyl glucose (28). In Rugosin-D, three of the free galloyl groups are esterified to one glucose, and the other two are esterified to the other glucose. Separation of the galloyl groups must improve the "fit" between tannin and protein.

There are only a few studies of the influence of tannin structure on the affinity of condensed tannin for protein (38), in part because obtaining condensed tannins of well defined structure has been difficult. The affinity for protein apparently increases as the degree of polymerization of the tannin increases (39). It has been suggested that the pattern of linkage between flavanoid subunits may have a role in determining affinity for protein (40), but additional studies of this question are justified. The stereochemistry and pattern of hydroxylation on either the A or the B rings of the condensed tannin influence the interaction with protein. The prodelphinidins, with three ortho hydroxy groups on the B ring, have higher affinity than procyanidins, which have only two ortho hydroxy groups on the B ring (*41*). Quebracho tannin, which is a profisetinidin and has hydroxyl groups on carbons 7, 3' and 4', has a lower affinity for protein than Sorghum tannin, which is a procyanidin (5, 7, 3', 4' hydroxylation) (*42*).

There are only a few direct comparisons of the affinities for protein of hydrolyzable tannins and condensed tannins. Similar amounts of purified procyanidin and purified gallotannin are required to precipitate hemoglobin or bovine serum albumin (*39*)(*43*). However, pentagalloyl glucose was six-fold more efficient as a precipitant of beta-glucosidase than was a procyanidin (*44*). Haslam has found that procyanidins have much lower affinity constants for caffeine than do gallotannins, and he suggests similar relative affinities for protein (*29*).

In less direct studies comparing condensed and hydrolyzable tannins, underivatized flavanoids have been compared to flavanoids esterified with gallic acid on carbon 3. Esterification increases the affinity of caffeine for the flavanoid, suggesting that gallic acid binds caffeine more effectively than the flavanoid (*29*). Galloylation at carbon 3 of several condensed tannins increased their activity in cytotoxicity and antiviral assays (*45*), suggesting that the added galloyl group significantly increased the affinity of the tannin for protein. However, further direct comparisons of the affinities of condensed and hydrolyzable tannins for protein would be fruitful.

Protein Structure

Tannins are quite specific protein precipitating agents. The procyanidin from Sorghum grain, a condensed tannin, binds a salivary proline-rich protein selectively in the presence of a 100-fold molar excess of bovine serum albumin (*46*). Similar specificity has been observed for gallotannins (*43*) and for several other condensed tannins (*42*), with the affinities of the tannins for the proteins ranging over four orders of magnitude.

The affinity of tannin for protein is a function of the amino acid composition and the flexibility of the protein. Proteins and polypeptides rich in proline have particularly high affinity for tannin, in part because the bis-alkyl substituted amide nitrogen of the proline-containing peptide bonds enhances hydrogen bonding of the peptide carbonyl (*27*). Polysarcosine (poly N-methylglycine) and polyvinylpyrrolidone, which both have bis-alkyl substituted amide nitrogens, also have high affinity for tannin (*27*).

Protein secondary and tertiary structure also influence the interaction with tannin. Proline obstructs the formation of secondary structures such as alpha-helices. The resulting random coil structures have peptide backbones which are accessible to tannin. Proteins with compact globular structures, including ribonuclease, lysozyme and cytochrome C, have low affinity for tannin because of the inaccessibility of the peptide backbone (27). Polyhydroxyproline and poly(proline-glycine-proline) have low affinity for tannin despite their high proline content. These polymers form interchain hydrogen bonds, making the peptide backbone inaccessible to tannin (27).

Amino acid composition is not the only determinant of affinity of proteins for tannin. For example, in a survey for proline-rich proteins, several samples contained large amounts of amide-bound proline, but there was no evidence for the presence of proteins with high affinity for tannin (47). Other features that may determine the affinity of tannin for protein are the molecular weight of the protein and posttranslational processing. Larger proteins and polymers have higher affinity for tannin than do smaller polymers, but small proline-rich polymers are preferred over large proline-poor polymers (27). Posttranslational modifications such as glycosylation can significantly alter the affinity of a protein for tannin. It has been suggested that in general glycoproteins do not bind tannin as efficiently as unglycosylated proteins (48). However, glycosylation substantially increased the affinity of a salivary proline-rich protein for tannin (49). The glycoprotein-tannin complexes that formed were soluble, so could only be detected with competitive binding assays rather than with simple precipitation assays (20).

Tannin-protein Interactions in Complex Systems

Tannin-protein interactions are of particular interest to food chemists and nutritionists interested in maintaining desirable flavor and quality of tannin-containing foods. Only preliminary work has been done with the complex systems of interest to food scientists. Several types of studies have shown that the behavior of heterogeneous mixtures of tannins and proteins requires further examination.

It has frequently been assumed that targets for dietary tannin include digestive enzymes. Direct in vivo tests of that hypothesis have yielded contradictory results. In some animals, dietary tannin does inhibit enzymes including proteases, lipases and glycosidases (50). In other organisms, tannin increases enzyme activity or induces enzyme synthesis (51).

Although in simple in vitro systems it seems clear that tannins inhibit enzymes (52)(53), as more complex systems are studied the results become less straghtforward. Tannins sometimes increase the activity of proteases by interacting with and denaturing the substrate protein (54). Butler has shown that the solubilized forms of some membrane-bound proteins interact strongly with tannin, but the native proteins do not interact with tannin (55). Apparently the membrane prevents the tannin from binding to the protein. Tannic acid inhibits alpha amylase activity unless the tannin is preincubated with a bean lectin (56). Apparently a lectin-tannin complex formed, preventing interaction of the tannin with the enzyme. The lectin is also inactivated by the reaction with tannin.

The mixtures of tannins found in plants may behave quite differently from purified samples of tannin. For example, the precipitation of purified proteins by purified condensed tannin or purified gallotannin is always complete within 15 minutes of mixing the reactants (*25*). However, when crude plant extracts are used, additional precipitate may accumulate for up to 14 hours (*36*). The kinetics of the reaction must be far more complex for mixtures of tannins than for homogeneous preparations. All of these studies demonstrate that the behavior of tannins in complex systems cannot be accurately predicted based on our current understanding of tannin-protein interactions.

Conclusions

During the last twenty years, substantial progress has been made in establishing the nature of tannin-protein interactions. As the structures of tannins have been elucidated, it has become possible to undertake detailed studies of how tannin and protein structure influence the interaction. Recent studies have revealed that it may be difficult to predict the behavior of tannins in multicomponent mixtures like those found in vivo; further studies of tannin-protein interactions in these complex mixtures are clearly needed.

Literature Cited

1.   Williams, V. M.; Porter, L. J.; Hemingway, R. W. *Phytochemistry* **1983**, 22, 569-572.
2.   Haslam, E. *Plant polyphenols. Vegetable Tannins Revisited*; Cambridge University Press: Cambridge, U.K., **1989**.
3.   Hagerman, A. E.; Butler, L. G. In *Herbivores: Their Interactions with Secondary Plant Metabolites*; Rosenthal, G. A.; Berenbaum, M. R. Eds.; Academic Press: NY, New York, **1992**.
4.   Bliss, E. D. In *Chemistry and Significance of Condensed Tannins*; Hemingway, R. W.; Karchesy, J. J. Eds.; Plenum Press: New York, NY, **1989**.
5.   Pierpoint, W. S. In *Flavonoids in biology and medicine III. Current issues in flavonoids research*; Das, N. P. Ed.; National University of Singapore: Singapore, **1990**.
6.   Haslam, E.; Lilley, T. H.; Cai, Y.; Martin, R.; Magnolato, D. *Planta Medica* **1989**, 55, 1-8.
7.   Athar, M.; Khan, W. A.; Mukhtar, H. *Cancer Research* **1989**, 49, 5784-5788.
8.   McArthur, C.; Hagerman, A. E.; Robbins, C. T. In *Plant Defenses Against Mammalian Herbivory*; Palo, R. T.; Robbins, C. T. Eds.; CRC Press: Boca Raton, FL, **1991**.
9.   Butler, L. G. and Rogler, J. C. This volume.
10.   Steinberg, P. D. In *Ecological Roles for Marine Secondary Metabolites*; Paul, V. J. Ed.; Comstock Purblishing, Ithaca, NY, **1991**.
11.   Porter, L. J. In *Methods in Plant Biochemistry. Plant Phenolics*; Harborne, J. B. Ed.; Academic Press, New York, NY, **1989**.

12.   Hemingway, R. W. In *Chemistry and Significance of Condensed Tannins*;
       Hemingway, R. W.; Karchesy, J. J. Eds.; Plenum Press, New York, NY,
       1989.
13.   Porter, L. J.; Hrstich, L. N.; Chan, B. C. *Phytochemistry* **1986**, 25,
       223-230.
14.   Price, M. L.; Van Scoyoc, S.; Butler, L. G. *J. Agric. Food Chem.* **1978**,
       26, 1214-1218.
15.   Putnam, L.; Butler, L. G. *J. Chromatog.* **1985**, 318, 85-93.
16.   Inoue, K. H.; Hagerman, A. E. *Anal. Biochem.* **1988**, 169, 363-369.
17.   Wilson, T. C.; Hagerman, A. E. *J. Agric. Food. Chem.* **1990**, 38,
       1678-1683.
18.   Verzele, M.; Delahaye, P.; Damme, F. V. *J. Chromatog.* **1986**, 362,
       363-374.
19.   Hagerman, A. E. and Robbins, C. T. unpublished observations **1991**.
20.   Hagerman, A. E.; Butler, L. G. *J. Chem. Ecol.* **1989**, 15, 1795-1810.
21.   Asquith, T. N.; Butler, L. G. *J. Chem. Ecol.* **1985**, 11, 1535-1544.
22.   Hagerman, A. E.; Robbins, C. T.; Weerasuriya, Y.; Wilson, T. C.;
       McArthur, C. *J. Range Manag.* **1992** in press.
23.   Hagerman, A. E.; Butler, L. G. *J. Agric. Food Chem.* **1980**, 28,
       944-947
24.   Bate-Smith, E. C. *Phytochemistry* **1973**, 12, 970-912.
25.   Hagerman, A. E.; Butler, L. G. *J. Agric. Food Chem.* **1978**, 26,
       809-812.
26.   Austin, P. J.; Suchar, L. A.; Robbins, C. T.; Hagerman, A. E. *J.·
       Chem. Ecol.* **1989**, 15, 1335-1347.
27.   Hagerman, A. E.; Butler, L. G. *J. Biol. Chem.* **1981**, 256, 4494-4497.
28.   McManus, J. P.; Davis, K. G.; Beart, J. E.; Gaffney, S. H.; Lilley, T.
       H.; Haslam, E. *J. Chem. Soc. Perk. Trans. II* **1985**, 1429-1438.
29.   Cai, Y.; Gaffney, S. H.; Lilley, T. H.; Magnolato, D.; Martin, R.;
       Spencer, C. M.; Haslam, E. *J. Chem. Soc. Perk. Trans. II* **1990**,
       2197-2209.
30.   Loomis, W. D.; Battaile, J. *Phytochemistry* **1966**, 5, 423-438.
31.   Oh, H. I.; Hoff, J. E.; Armstrong, G. S.; Haff, L. A. *J. Agric. Food
       Chem.* **1980**, 28, 394-398.
32.   Pierpoint, W. S. *Biochem. J.* **1969**, 112, 619-629.
33.   Hagerman, A. E.; Butler, L. G. *J. Agric. Food Chem.* **1980**, 28,
       947-952.
34.   Jones, W. W.; Mangan, J. L. *J. Sci. Food Agric.* **1977**, 28, 126-136.
35.   Hagerman, A. E. *Condensed Tannin of Sorghum Grain: Purification and
       Interactions with Proteins*. Ph. D. Dissertation. Purdue University: W.
       Lafayette, IN, **1980**.
36.   Hagerman, A. E.; Robbins, C. T. *J. Chem. Ecol.* **1987**, 13, 1243-1259.
37.   Calderon, P.; Van Buren, J.; Robinson, W. B. *J. Agric. Food Chem.*
       1968, 16, 479-482.

38. Hagerman, A. E. In *Chemistry and Significance of Condensed Tannins*; Hemingway, R. W.; Karchesy, J. J., Eds.; Plenum Publishing: New York, NY **1989**.
39. Porter, L. J.; Woodruffe, J. *Phytochemistry* **1984**, 23, 1255-1256.
40. Ezaki-Furuichi, E.; Nonaka, G.; Nishioka, I.; Hayashi, K. *Agric. Biol. Chem.* **1987**, 51, 115-120.
41. Asano, K.; Ohtsu, K.; Shinagawa, K.; Hashimoto, N. *Agric. Biol. Chem.* **1984**, 48, 1139-1146.
42. Asquith, T. N.; Butler, L. G. *Phytochemistry* **1986**, 25, 1591-1593.
43. Hagerman, A. E.; Klucher, K. M. In *Plant Flavonoids in Biology and Medicine: Biochemical, Pharmacological and Structure Activity Relationships*; Cody, V.; Middleton, E.; Harborne, J. Eds.; New York: Alan R. Liss, Inc.: New York, NY, **1986**.
44. Haslam, E. *Biochem. J.* **1974**, 139, 285-288.
45. Takechi, M.; Tanaka, Y.; Takehara, M.; Nonaka, G. I.; Nishioka, I. *Phytochemistry* **1985**, 24, 2245-2250.
46. Mehansho, H.; Hagerman, A.; Clements, S.; Butler, L.; Rogler, J.; Carlson, D. M. *Proc. Nat. Acad. Sci. U.S.A.* **1983**, 80, 3948-3952.
47. Mole, S.; Butler, L. G.; Iason, G. *Biochem. Syst. Ecol.* **1990**, 18, 287-293.
48. Strumeyer, D. H.; Malin, M. J. *Biochem. J.* **1970**, 118, 899-900.
49. Asquith, T. N.; Uhlig, J.; Mehansho, H.; Putnam, L.; Carlson, D. M.; Butler, L. *J. Agric. Food Chem.* **1987**, 35, 331-334.
50. Longstaff, M.; McNab, J. M. *Brit. J. Nut.* **1991**, 65, 199-216.
51. Ahmed, A. E.; Smithard, R.; Ellis, M. *Brit. J. Nut.* **1991**, 65, 189-197.
52. Tamir, M.; Alumot, E. *J. Sci. Food Agric.* **1969**, 20, 199-202.
53. Griffiths, D. W. *J. Sci. Food Agric.* **1979**, 30, 458-462.
54. Mole, S.; Waterman, P. G. *Phytochemistry* **1987**, 26, 99-102.
55. Blytt, H. J.; Guscar, T. K.; Butler, L. G. *J. Chem. Ecol.* **1988**, 14, 1455-1465.
56. Fish, B. C.; Thompson, L. U. *J. Agric. Food Chem.* **1991**, 39, 727-731.

RECEIVED December 17, 1991

Chapter 20

Implication of Phenolic Acids as Texturizing Agents During Extrusion of Cereals

Suzanne M. Gibson and George Strauss

Department of Chemistry, Rutgers, The State University of New Jersey, New Brunswick, NJ 08904

Ferulic and other phenolic acids in corn meal and other cereals were recognized as the components that, by their immobilization during extrusion cooking, caused an increase in fluorescence anisotropy. Fractionation of uncooked corn meal and extrudates by successive extractions showed that extrusion-cooking caused phenolic acids to accumulate in one of the fractions. Such extrusion resulted in the formation of high-molecular weight water-insoluble complex consisting of carbohydrate, protein, and phenolic acids, in amounts that progressively increased with harsher extrusion conditions. The texture and cohesiveness of extrudates is considered to arise from a network of this complex with starch that is formed during the extrusion process.

The formation of textured products from cereal flours and meals by extrusion cooking is a highly complex process. In a typical screw extruder the feed material is mixed with water, then pushed forward, kneaded, and sheared under high pressure at gradually increasing temperatures that may reach 200°C. Finally the cooked dough is extruded through a narrow die that imparts further shear. The discrete cell components such as starch granules, protein bodies, and cell-wall fragments are homogenized, physically disrupted, and caused to react chemically. The resulting product is cohesive, and has a texture that can vary from brittle to rubbery, depending on type of feed material and processing conditions.

The purpose of the present investigation, made with corn meal, was to identify the specific chemical changes responsible for the development of cohesiveness and texture during extrusion. In earlier work (1) we found that a fluoresent component of corn meal undergoes an increase in its fluorescent anisotropy as a result of extrusion. This indicated that a fluorophore had become less mobile. The anisotropy increases resulting from extrusion correlated well with the degree of harshness of extrusion conditions (temperature, shear rate, etc.) (2), and also correlated with product properties such as compressive strength, density, expansion ratio, and taste and aroma ratings. These close correlations suggested a direct involvement of the fluorescent component in the reaction that creates texture.

0097–6156/92/0506–0248$06.00/0
© 1992 American Chemical Society

As described here, this fluorescent component of corn meal has now been identified as ferulic acid, a member of the class of phenolic acids. Corn meal is known to contain both soluble and insoluble phenolics, with ferulic acid comprising 85 % of the total phenolics (3-5). It is primarily bound by its carboxyl group to an insoluble grain component by an ester linkage. Phenolic acids with their carboxyl and hydroxyl groups are capable of combining with proteins and polysaccharides by hydrogen bonds, by chelation, or by covalent bonds, forming bridges or crosslinks (6).

Phenolic acids occur mainly in the cell walls which are rich in several types of pentosans. One group, the water-soluble pentosans, are non-starch polysaccharides of xylose or galactose, with randomly attached side groups of single or double arabinose units that in turn are esterified with phenolic acids. Another type, the water-soluble arabinogalactans, are branched and contain no phenolic acids, but are covalently bound to a peptide, probably via its hydroxyproline. A further type, the water-insoluble pentosans, lack the arabinose moiety (7).

Pentosan-phenolic acid esters when extracted from cereals form gels in the presence of oxidizing agents, such as H_2O_2/peroxidase (8). The gels contain ca. 25% protein and 75% arabinoxylan, with trace amounts of ferulic and diferulic acid (9).

Several mechanisms have been proposed for the gelling reaction:
(1) Oxidative phenolic coupling of ferulic acid residues on adjacent pentosan chains forming crosslinks;
(2) oxidative phenolic coupling of tyrosine residues on a protein (presumably from the arabinogalactans present in the mixture of water-soluble pentosans) with ferulic acid residues, also forming crosslinks;
(3) oxidative coupling of cysteine sulfhydryl groups with the double bond (rather than with the phenolic ring) of ferulic acid (10).

Such crosslinking reactions could account for the observed texturization of extruded cereal meals, provided that the crosslinking process can be shown also to involve the bulk starch and protein constituents. If the cell-wall material (the main source of pentosan-phenolis acid esters), amounting to only 2-3% of the cereal, were to react only with itself then only minor overall physical changes could be expected.

In view of the above considerations, we investigated changes in the binding of phenolic acids to the starch and protein components as a result of extrusion. Several fractions were isolated by successive extractions, and phenolics were liberated from them by alkaline or enzymatic hydrolyses. These data, obtained for uncooked corn meals and for extrudates produced from them, could then be compared and correlated the corresponding fluorescence anisotropies.

Materials and Methods

Materials. Two types of corn meal were used. CC400 corn meal from the Lauhoff Grain Co. contained 7% protein. Its starch was 70% amylopectin and 30% amylose. Pure n Thick corn meal from the National Starch Co. contained 11% protein. Its starch was 100% amylopectin. Extrudates were produced on a ZSK-20 Werner & Pfleiderer twin-screw extruder. Extrudates labeled #8 and #3, from CC400 corn meal, had 25% feed moisture and were extruded at a screw speed of 300 rpm at 100°C and 200°C, respectively. Extrudate #32, from Pure n Thick corn meal, had 20% feed moisture and was extruded at 500 rpm and 140°C. The α-amylase, from aspergillus oryzae, was obtained from Sigma Chemical Co.

Fluorescence Anisotropy. This parameter is a measure of the retention of polarization of the emitted fluorescence when a sample is irradiated with plane polarized light. It was measured on solid samples by a front-face technique developed in this laboratory, as described eleswhere (1). Briefly, a cuvette containing the powdered sample was positioned off-center in the sample compartment of a conventional fluorescence spectrometer so that the cuvette face was irradiated obliquely. The instrument was fitted with polarizers in the excitation and emission light paths. Fluorescence intensities were measured with the planes of polarization in the two paths parallel, and again when they were normal to each other. The anisotropy was proportional to the difference of these two measurements.

Fluorescence Excitation Spectra. Using the same front-face technique as above, fluorescence spectra of solid samples were obtained by scanning through excitation wavelengths while keeping the emission wavelength constant. In this way, spectra equivalent to absorption spectra - unobtainable directly on solids - were recorded.

Fractionation by Successive Extractions. The overall extraction scheme used and the labeling of fractions are given in Figure 1.

STEP 1: Uncooked corn meal and extrudates were extracted by warm 70% aqueous ethanol and centrifuged three times after which no further material could be removed. The pooled supernatants are labeled **1** in the extraction scheme. Their absorption at 320 nm, with the use of a standard solution as reference, gave the total extractable phenolics.

STEP 2: The solid residue was extracted with water and centrifuged three times at 1000 rpm. This gave a solid residue collecting at the bottom and a turbid colloidal supernatant, **2** in the scheme, that did not clear during centrifugation. A known aliquot of the supernatant was evaporated to dryness in vacuo and weighed, giving its weight fraction. This material was treated with 1 N KOH overnight to hydrolyze esterified phenolics. UV absorption then gave the total phenolics content of this fraction. Another aliquot of the supernatant was treated with a-amylase for 24 hours. This resulted in a clear supernatant, **2A**, and a quick-settling white precipitate, **2B** .

STEP 3: The solid residue from the centrifugation after water extraction was suspended in dimethyl sulfoxide (DMSO), sonicated briefly, stirred for 16 hours, and then centrifuged. This was repeated three times. The opalescent supernatant, **3**, was mixed with ethanol. The resulting precipitate was washed with ethanol, dried and weighed. Treatment with KOH and measurement of the UV absorption, as above, gave the phenolics content of this fraction.

STEP 4: The residue, **4**, from the DMSO extraction was washed with water and with ethanol (three times each), dried and weighed. Again, KOH treatment and UV absorption gave the phenolics content.

Protein Determination. Protein contents of the several fractions were measured using the Bradford coomassie blue assay. Ethanolic fractions to be analyzed were first vacuum dried, then redissolved in DMSO and mixed with an equal volume of water; aqueous fractions were mixed with an equal volume of DMSO; those fractions originally in DMSO were mixed with an equal volume of water. 0.2 ml of the 50% DMSO soultions thus produced in each case were then mixed with 1 ml DMSO and 3 ml of the coomassie blue/phosphoric acid reagent and allowed to stand for 10 minutes before measuring their absorbance at 595 nm against a BSA standard (11).

Identification of Phenolic Acids. Ethanol extracts of corn meal were evaporated to dryness, then taken up in 1 N KOH (room temperature, 24 hours) to liberate any phenolics present as esters. Following acidification to pH 3 with HCl, the samples were extracted with ethyl acetate and spotted on TLC plates (silica gel G soft-layer plates containing a fluorescent indicator, from Fisher Scientific Co.). Residues from the ethanol extractions were also treated with KOH, acidified, extracted, and spotted as above. The solvent was benzene-methanol-acetic acid (20:4:1). All components were visualized on the plates as dark spots under short-wave (254 nm) UV light. The phenolic acids appeared as blue fluorescent spots under long-wave (360 nm) light.

Identification of Carbohydrates. The sample was hydrolyzed with concentrated H_2SO_4, then neutralized with saturated $Ba(OH)_2$, concentrated, and spotted on TLC plates. The solvent used was ethyl acetate-methanol-acetic acid-water (60:15:15:10). Following the procedure of Dubois et al. (12), the saccharides were visualized as brown spots after spraying the plates with a reagent consisting of 3 g phenol and 5 ml H_2SO_4 in 95 ml ethanol, then heating them for 15 minutes at 110°C.

Molecular Weight Determination. A sample of the supernatant 2 was mixed with DMSO to a final DMSO concentration of 85%. The resulting solution was centrifuged at 12,000 G for 15 minutes in an Eppendorf microcentrifuge to remove dust and any traces of undissolved material, then analyzed in a Nicomp Model 200 dynamic light scattering photometer-autocorrelator. The resulting particle size distribution data were converted to molecular weights, using polystyrene latex suspensions as standards.

Results

Identification of Fluorescent Components. Front-face fluorescence excitation spectra of solid corn meal and extrudate samples revealed excitation and emission maxima at 360 and 420 nm, respectively. Addition of a 0.1 M solution of sodium carbonate caused these peaks to shift to longer wavelengths (384/480 nm), indicating that the fluorescence was due to the presence of phenolic compounds (13).

Certain extrudates, extruded under extremely harsh conditions (low moisture, very high temperature) exhibited browning. Such samples gave excitation spectra with longer wavelengths (380/460 nm) than extrudates formed under milder conditions. Addition of base caused a shift to even longer wavelengths (400/480 nm), showing that the 380/460 nm peaks still were due to phenolic compounds. The red shift (relative to milder extrusion), may be due to a longer conjugation chain in the molecule resulting from a crosslinking reaction, as suggested by Amado and Neukom (7).

TLC was performed on the ethanol extract 1 of corn meal (see extraction scheme, Figure 1), and on the residue from this extraction, as described under Methods. The resulting chromatograms are shown in Figure 2, together with those of ferulic and coumaric acid. The ethanol extract contained ferulic acid as the principal solute, with lesser amounts of coumaric acid and minor other compounds not identified. The residue from the ethanol extraction showed only ferulic acid.

Quantitation of Free and Complexed Phenolic Acid and Protein Constituents. Changes in the association of phenolics and proteins with starch as a result of extrusion were investigated by fractionating corn meals and extrudates by successive extractions as diagrammed in Figure 1 and further described under Methods. Fractions were analyzed for weight, phenolic acid content, and protein content.

Weight distributions of the fractions 1, 2, 3, and 4 for each sample are shown as a histogram in Figure 3. The colloidal aqueous extract (fraction 2) increased dramatically as a result of extrusion, going from 6% to 67% for the CC400 corn meal, and from 1% to 88% for Pure n Thick corn meal. These increases occurred mostly at the expense of the DMSO-soluble fraction 3. This latter fraction appears to be mostly unreacted starch and protein.

As mentioned above, the fluorescence anisotropy of corn meal invariably increases upon extrusion, with the increase depending on the harshness of extrusion conditions. In Figure 7A the amount of carbohydrate in fraction 2 is plotted against fluorescence

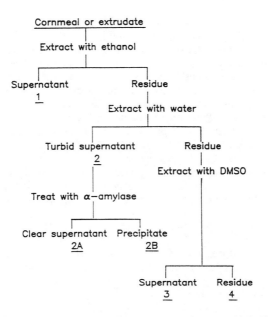

Figure 1. Fractionation scheme by successive extractions, and labeling of fractions.

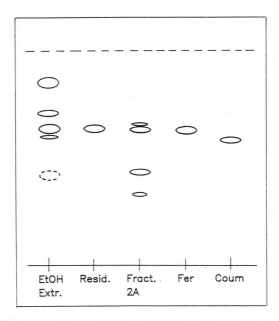

Figure 2. Identification of phenolic acids of corn meal in fraction **1**, the residue, and fraction **2A** by thin-layer chromatography.

anisotropy for CC400 corn meal and two extrudates made from it under mild and harsh conditions, respectively. The data show excellent correlation.

The distribution of phenolic acids among the four fractions is shown in Figure 4. Extrusion produced large increases in the percentage of the total phenolic acids found in the colloidal fraction **2**, paralleling the weight increase of that fraction. The additional phenolics going into fraction **2** are seen to arise from the ethanol-soluble fraction **1** and from the completely insoluble fraction **4**. The amount of phenolics in fraction **2** vs. fluorescence anisotropy, plotted in Figure 7C, shows very large increases as a result of extrusion that, like the carbohydrate content, also correlate with the anisotropy.

The distribution of protein is shown in Figure 5. Fraction **4** contained a negligible percentage of the total protein present and so is not included. For uncooked corn meal CC400, fraction **2** was devoid of protein. On extrusion, up to 25% of the total protein was found in this fraction. These changes, as a function of fluorescence anisotropy, are plotted in Figure 7B.

The ethanolic supernatants **1**, in addition to their phenolic acid content determined quantitatively, also had UV absorption spectra with peaks at 280 nm, characteristic of tyrosine and thus showing protein, and also a triplet of peaks at 420-480 nm, typical of carotenoids.

The turbid supernatant **2**, evidently a colloidal suspension, remained unchanged, with no material settling out in 1-2 days. Visually, much more suspended material was observed with extrudates than with uncooked corn meal. The Pure n Thick corn meal and the extrudate made from it, where the starch is 100% amylopectin, gave turbid supernatants when extracted with cold water. The CC400 corn meal and its extrudates, however, containing 30% amylose, produced turbid supernatants only when extracted with hot water at 90°C. All supernatants **2** gave an intense blue color with iodine, showing a high concentration of starch.

The fluorescence anisotropy of the colloidal supernatant was 0.204; this compares with an anisotropies of 0.134 for this particular extrudate (measured before fractionation), and 0.05-0.06 for uncooked corn meals.

Molecular weight determinations by dynamic light scattering were made on aliquots of supernatant **2** dissolved in 85% DMSO. The average molecular weights were approximately 1.6×10^6 before extrusion, and up to 11×10^6 after extrusion. Figure 7D shows that these molecular weight increases paralleled those of the anisotropy .

Following hydrolysis by α-amylase at pH 6.9 for 24 hours, a heavy precipitate **2B** settled out from the starch hydrolysate, whereas nothing settled out from the intact colloidal starch suspension. The precipitate tested negative with iodine, showing it to be a non-starch saccharide. Its contribution to the total solutes in fraction **2** increased from about 1% to 9% as a result of extrusion. This precipitate had a very high molecular weight, estimated at $< 20 \times 10^6$. The clear fraction **2A** was precipitated by addition of ethanol, thus separating it from the large amount of glucose generated by the a-amylase digestion. It was then acid-hydrolyzed, and subjected to TLC analysis for the presence of saccharides as described above. Figure 6 shows the chromatograms for extrudates #8 and #3, together with those of several saccharides as reference. The main components in this fraction were glucose, xylose, arabinose, and galactose. The amounts of these saccharides were distinctly higher for extrudate #3 which had been extracted at a higher temperature than #8. A similar treatment and TLC analysis of fraction **2B** also showed the presence of pentoses.

The residue from the water extraction was in turn extracted with DMSO as described under Methods. This extract (fraction **3**) apparently was unreacted material since it

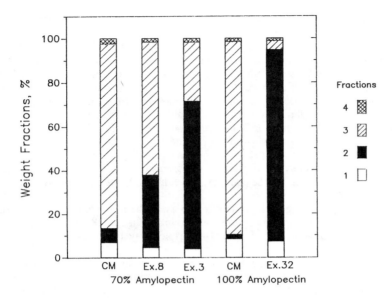

Figure 3. Weight distribution among the fractions isolated by successive extractions from corn meals and extrudates.

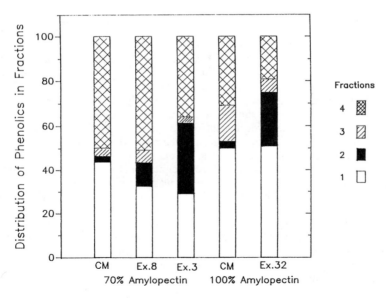

Figure 4. Distribution of phenolic acids among the fractions from corn meals and extrudates.

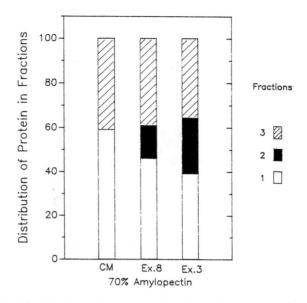

Figure 5. Distribution of protein among the fractions from corn meal and extrudates.

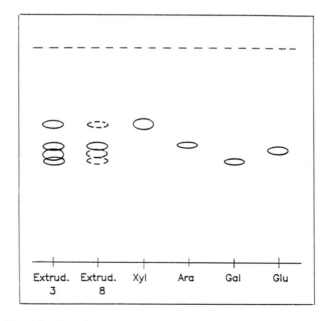

Figure 6. Identification of saccharides in the fractions **2A** isolated from extrudates by thin-layer chromatography.

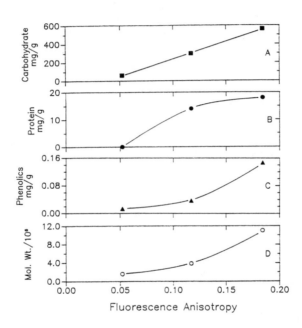

Figure 7. Characteristics of the colloidal fraction **2** (see Figure 1) isolated from corn meal and two extrudates, as a function of fluorescence anisotropy. A, B, C: Carbohydrate, protein, and phenolics contents; D: Molecular weights.

represented most of the mass for uncooked corn meal, but much smaller fractions for extrudates.

The residue from the DMSO extraction (fraction **4**), was very high in phenolic acids. Being insoluble also in ethanol and water, it appears to be unreacted cell wall material known to be present in corn meal (14).

Discussion

The phenolic acids identified in corn meal by chromatographic and spectroscopic techniques had the same absorption and emission peaks as those observed in fluorescence anisotropy measurements. This shows that the anisotropy increases observed as a result of extrusion are due to the immobilization of phenolic acids. Such immobilization, by itself, does not indicate whether phenolic acids take part in the chemical change leading to texturization since they may simply become physically entrapped in a rigid matrix.

Clarification of this point has been provided by the fractionation of corn meals and extrudates. The distributions among four fractions for total mass (Figure 3), phenolic acids (Figure 4), and protein (Figure 5) showed systematic increases in each of these measurements for the colloidal supernatant (fraction **2**) when progressing from uncooked corn meal to extrudates formed under conditions of increasing harshness. The redistribution of fractional mass very clearly shows that unprocessed starch and protein (fraction **3**) is converted into a hydrocolloid material (fraction **2**). The transfer of phenolic acids from fractions **1** and **4** into fraction **2**, and the transfer of protein from fraction **1** into **2** demonstrates that these components, too, are being incorporated into the colloidal fraction. All these transfers occurred to different extents, depending on processing conditions. There was also an almost 10-fold higher mean molecular weight for solutes in **2** for an extrudate, compared to uncooked corn meal (Figure 7D).

The presence of phenolic acids and pentose saccharides in this material makes it likely that it consists of pentosans such as FAXX (ferulate ester of an arabinoxylan) (15) that have been identified in cereal cell walls where they provide stiffness by crosslink reactions.

Taken together, these results show conclusively that extrusion results in the formation of a high-molecular weight water-insoluble complex consisting of carbohydrate, protein, and phenolic acids, in amounts that increase with increasing harshness of extrusion. This complex appears to entrap unreacted starch but without making it inaccessible to a-amylase. The starch may, therefore, be entrapped non-covalently by a wide-meshed network. The formation of gels formed from flour extracts has been ascribed to such a network (7), in which the insoluble fraction, i.e. the polymer, amounted to only 0.2% based on the flour. This is analogous to a swollen polymer network forming a gel by entrapping many times its own mass of another substance (16). The complex in fraction **2** can be regarded as the crucial reaction product that, despite its small contribution to the total mass, provides texture and cohesiveness for the whole extrudate structure. A further indication for the existence of a polymeric network in extrudates is their variation in elastic properties that range from rubbery to brittle. As shown by Flory (16), these properties depend on the density of crosslinks, i.e. the mesh size. The observed range of extrudate properties may thus be the result of different extents of the crosslinking reaction.

The progressive increase in fluorescence anisotropy with progressively harsher extrusion conditions can now be accounted for in the light of the observed redistribution of phenolic acids as a result of extrusion. They are seen to move from fractions where they are not complexed and thus mobile, with a low anisotropy, into fraction **2** where they are immobilized, possibly due to crosslinking, and hence have a high anisotropy. The measured anisotropy in a sample then is an average value that is a measure of the relative amounts of immobilized and mobile phenolic acids present.

The results reported here are consistent with the following model of the changes occurring during extrusion of corn meal:

1. The initial well-known steps are the hydration, swelling, and gelatinization of starch, accompanied by loss of crystallinity of starch granules and expulsion of amylose (if present) from them.

2. High-molecular weight polysaccharides are degraded into shorter fragments, exposing new binding sites for subsequent reactions.

3. Proteins are denatured, also providing new binding sites.

4. Starch and protein chains are intermixed with cell wall fragments containing phenolic acids esterified with pentosans.

5. Crosslinking reactions occur between polymeric pentosans and/or between pentosans and proteins or starch, resulting in a high-molecular weight network of carbohydrate and polymer chains. Such a network constitutes less than 10% of the total mass but can entrap a much larger mass of starch. The crosslinked network is considered the essential structure that provides texture and cohesiveness.

Acknowledgments

This is Publication No. 10544-17-90 of the New Jersey Agricultural Experiment Station supported by State Funds and the Center for Advanced Food Technology. The Center for Advanced Food Technology is a New Jersey Commission on Science and Technology Center.

Literature Cited

1. Gibson, S.M.; Strauss, G. Cereal Chem. **1989**, 310.
2. Strauss, G.; Gibson, S.M.; Adachi, J.D. In Applied Food Extrusion Studies, Kokini, J.L.; Ho, C.-T.; Karwe, M.V., Eds.; Marcel Dekker: New York, NY, 1991.
3. Harris, P.J.; Hartley, R.D. Nature 1976, 259, 508.
4. Sosulski, F.; Krygier, K.; Hogge, L.J. Agric. Food Chem. **1982**, 30, 337.
5. Delcour, J.A.; Vinkx, C.J.A.; Vanhamel, S. J. Chromatogr. **1989**, 467, 149.
6. Fry, S.C. Biochem. J. **1982**, 203, 493.
7. Amado, R.; Neukom, H. In New Approaches to Research on Cereal Carbohydrates; Hill, R.D.; Munck, L., Eds.; Progress in Biotechnology; Elsevier: Amsterdam, The Netherlands, 1985; Vol. 1, 241-251.
8. Ciacco, C.F.; D'Appolonia, B.L. Cereal Chem. **1982**, 59, 96.
9. Neukom, H.; Markwalder, H.U. Cereal Foods World **1978**, 23, 374.
10. Hoseney, R.C.; Faubion, J.M. Cereal Chem. **1981**, 58, 421.
11. Bradford, M.M. Anal. Biochem. **1976**, 72, 143.
12. Dubois, M.; Gilles, K.A.; Hamilton, J.K.; Rebers, P.A.; Smith, F. Anal. Chem. **1956**, 28, 350.
13. Schnabl, H.; Weissenbock, G.; Sachs, G.; Scharf, H. J. Plant Physiol. **1989**, 135, 249.
14. Selvendran, R.R.; DuPont, M.S. Cereal Chem. **1980**, 57, 278.
15. Mueller-Harvey, I.; Hartley, R.D. Carbohydr. Res. **1986**, 148, 71.
16. Flory, P.J. Principles of Polymer Chemistry; Cornell Univ. Press: Ithaca, NY, 1953.

RECEIVED November 7, 1991

Chapter 21

Red Raspberry Phenolic

Influences of Processing, Variety, and Environmental Factors

A. Rommel, R. E. Wrolstad, and R. W. Durst

Department of Food Science and Technology, Oregon State University, Corvallis, OR 97331–6602

Flavonols, ellagic acid and other phenolics were characterized and quantified in pilot-plant processed red raspberry juices (n=46) and commercial juice concentrates (n=9) by HPLC/diode array spectral techniques. Samples were prepared using mini-columns packed with Polyamide-6; a fraction eluted with methanol contained ≤ 8 quercetin-glycosides, quercetin and kaempferol. A second fraction eluted subsequently with 0.5% ammonia in methanol contained 3 flavonol-glucuronides, 2 other flavonol-forms, 2 aglycons, ellagic acid, and $\leq$ 16 forms of ellagic acid; 36 additional flavonol-forms were measured in trace amounts in the two fractions. The mean total concentrations of flavonols and ellagic acid forms in pilot-plant juices were 122 ppm and 28 ppm, respectively. After acid hydrolysis, quercetin, kaempferol, ellagic acid (3 forms), 2 catechins, 3 hydroxycinnamic and 3 hydroxybenzoic acids were characterized. The varieties Willamette and Heritage contained the most flavonols, and Willamette and Meeker the most ellagic acid and derivatives. Juice produced by centrifugation had the highest flavonol contents, while diffusion extraction and standard processed juices contained the most ellagic acid. Mold decreased the concentrations of flavonol-glycosides greatly, but had little effect on ellagic acid and its derivatives.

The compositional data base for red raspberry juice needs to be expanded to help regulatory agencies and the food industry to monitor quality and detect adulteration. Raspberry juices and concentrates are targets for adulteration as red raspberries are very expensive. Of raspberry constituents, secondary plant metabolites (e.g. flavonoids and other phenolics) are particularly suitable as authenticity indicators as they are present within a finite range (1). Anthocyanin pigment analyses have been useful for determining the authenticity of red raspberry juice concentrates (2), however, they do not always provide sufficient evidence for adulteration. Additional phenolic authenticity indicators such as non-anthocyanin flavonoids (e.g. flavonols, catechins), benzoic and cinnamic acids would complement anthocyanins very well. Non-anthocyanin flavonoids would be very suitable for detecting adulteration by cheaper fruit juices, not red in color (e.g. apple, pear), which are potential

0097–6156/92/0506–0259$08.00/0
© 1992 American Chemical Society

adulterants. Most non-anthocyanin flavonoids are not commercially available and so cannot be added to hide adulteration.

Raspberry juice composition has also become of increasing interest because of the potential health effects of its phenolic constituents, in particular, quercetin and ellagic acid. Of the flavonols, mainly quercetin but also kaempferol and the quercetin-glycoside rutin have been shown to have anticarcinogenic effects in mammals (e.g. 3-7). Quercetin is formed in the human mouth and gastrointestinal tract via bacterial hydrolysis of quercetin-glycosides and glucuronides (3-5, 8, 9). Although quercetin has been demonstrated to be mutagenic in bacterial assay systems, it has been found not to be carcinogenic or teratogenic *in vitro* and *in vivo* by many research groups (3). Rather, quercetin has been found to be a potent anticarcinogen against skin, colon and mammary cancers in rodents (e.g. 3-7). Quercetin may also inhibit the induction and progression of human cancers (4, 5, 10). Epidemiological evidence supports the theory that flavonols and other phenolics have anticarcinogenic effects in humans, as there is an inverse correlation between individuals who consume a diet rich in fruits and vegetables and their risk of developing cancer (6, 11-13).

Ellagic acid has anticarcinogenic effects against a wide range of carcinogens in several tissues. Significant inhibition of colon, esophageal, liver, lung, tongue, and skin cancers has been shown in rats and mice by *in vitro* and *in vivo* mutagenicity and carcinogenicty investigations; both topical application and feeding were used in *in vivo* studies with mice and rats (e.g. 14-23). Ellagic acid is a possible chemopreventative agent also in human carcinogenisis (23-25). Nonetheless, there is still some conflicting evidence as to the anticarcinogenic effectiveness of ellagic acid (19). Furthermore, it has not been resolved how much ellagic acid and quercetin are absorbed into the body from dietary sources (3, 5, 19, 26).

In previous studies of raspberry juice anthocyanins, sugars and acids, a small number of samples were analyzed, and the influences of processing techniques and environmental factors were not investigated (2). Only fresh raspberries and a few varieties have been studied for their contents of non-anthocyanin flavonoids (flavonols: 1, 27-29; catechins: 30), ellagic acid (15, 31) and benzoic and cinnamic acids (32, 33). There is little information about the influence of juice concentrate processing on the phenolic composition of red raspberry juices.

Raspberries contain glycosides of the flavonols quercetin and kaempferol (Figure 1). Ellagic acid is released from cell-walls through hydrolysis of ellagitannins to hexahydroxydiphenic acid which forms an inner dilactone spontaneously, called ellagic acid (31, 34, 35; Figure 2). Numerous derivatives of ellagic acid exist, formed through methylation, glycosylation and methoxylation (19).

It was the objective of our study to create an expanded compositional data-base for red raspberry juice consisting of flavonoids (anthocyanins, flavonols, catechins) and other phenolics, sugars, and organic acids. This report is restricted to ellagic acid, non-anthocyanin flavonoids, and benzoic and cinnamic acids in 55 red raspberry juices. It was the objective of this part of our study to identify and quantify such phenolics so that they can be used a) as supplementary authenticity indicators together with anthocyanins, sugars and organic acids and b) as a data-base for evaluating the effects of raspberry juice on health. The anthocyanin composition of these same raspberry juices is reported by Boyles (36).

Samples

Fifty-five different juices were investigated, 45 were produced in the pilot plant of the Department of Food Science and Technology at Oregon State University and at the Agriculture Canada Research Station, in Summerland, British Columbia, from berries grown from 1988-1990 inclusive in Canada, Poland and the United States

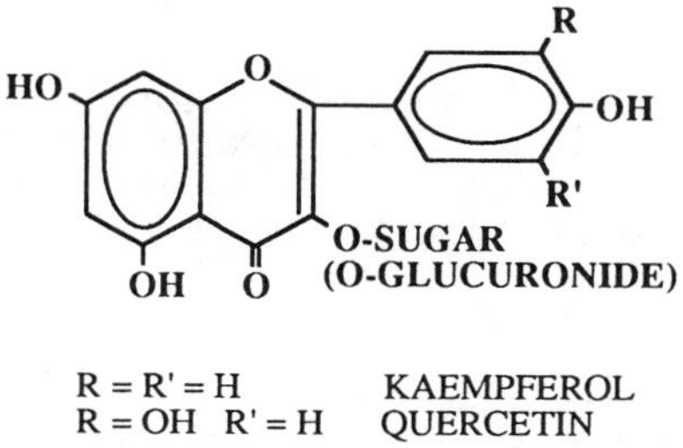
R
HO
O
OH
R'
O-SUGAR
(O-GLUCURONIDE)
OH
O
R = R' = H KAEMPFEROL
R = OH R' = H QUERCETIN

Figure 1. Structures of quercetin and kaempferol-glycosides and glucuronides.

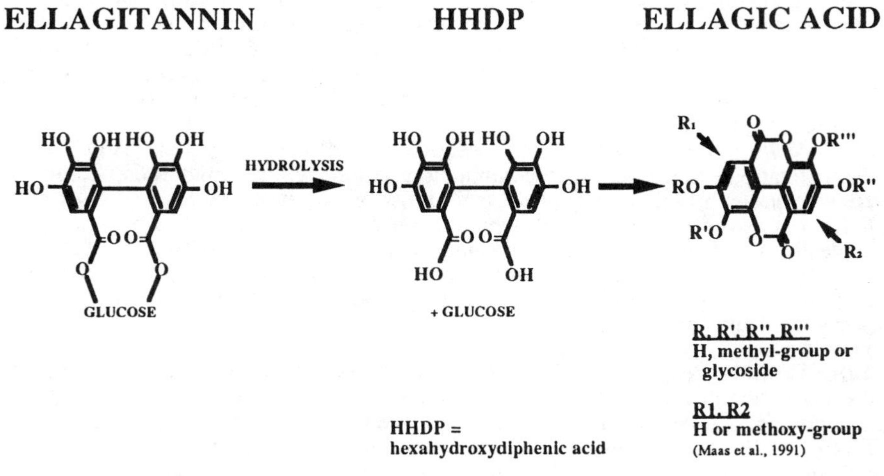
ELLAGITANNIN
HHDP
ELLAGIC ACID
HYDROLYSIS
GLUCOSE
+ GLUCOSE
R, R', R'', R'''
H, methyl-group or glycoside
R1, R2
H or methoxy-group
(Maas et al., 1991)
HHDP =
hexahydroxydiphenic acid

Figure 2. Structures of ellagic acid and derivatives and their precursors in plant cell walls.

(experimental samples; Table I). Nine commercial raspberry juice concentrates were provided by seven companies.

Experimental juices were made from 10 different varieties representing the principal varieties grown commercially in the United States, Canada and Eastern Europe. In the United States most raspberries are grown in Oregon and Washington. Many of the raspberries imported to the United States are of Eastern European origin (e.g. Poland). Two lesser known varieties were included in this study for comparison to the common ones; they were Heritage, a variety that ripens in September, and Golden, a yellow-colored raspberry.

The following environmental factors were evaluated: geographic origin (United States, Canada, Poland), maturity (underripe, ripe, overripe), harvesting-technique (machine vs. hand-picking), and mold-contamination. One batch of very moldy ripe berries, Meeker variety, was provided which was partially fermented as evident by a low content of soluble solids of the berries.

Experimental Procedures

Juice Processing Methods. A 'standard' process, which is typical of industrial processes, was used to produce most juices in our pilot plant (Figure 3a). Where not enough berries could be provided for juice production in the pilot plant (in the cases of the Polish, Heritage and Golden varieties), a standard juice making process was simulated in the laboratory. Single-strength (i.e. unconcentrated) juices produced by the standard process were stored frozen.

The effect of concentration was investigated for some samples using a centrifugal film evaporator (Centri-Therm, model CT-1B; vacuum, -0.85 kg/cm^3; temperature, 60°C). In addition, concentration via a direct osmotic process was also evaluated for its effects on phenolic concentrations (*37, 38, 39*). In this process, juice separated by a membrane from high fructose corn syrup was concentrated by direct osmosis to a desired concentration (up to 45-50 °Brix), at room (26°C) or chilled (8°C) temperatures.

Two alternative processing methods were also investigated. High-speed centrifugation (*40*), alone or in combination with different commercial pectolytic enzyme preparations, was applied to aliquots of one batch of fruit (Figure 3b). Diffusion extraction (Figure 3c), which has been used successfully for producing apple and pear juices (*41, 42*), was conducted on another aliquot of the same batch of fruit in one experimental trial. In this process, water at 63°C was used counter-currently to extract soluble components; the extract was subsequently depectinized, pasteurized and concentrated.

Juice Sample Preparation for HPLC Analysis. Preliminary fractionation of the phenolics in red raspberry juice was required to achieve satisfactory HPLC separations. Conventional isolation techniques (C$_{18}$-cartridges, Sephadex LH-20, PVPP, cation and anion-exchangers) were unsuccessful. This difficulty was attributed to the complexity and number of flavonol and ellagic-acid compounds present in red raspberries. We obtained good preparations, however, by modifying the procedure of Wald and Galensa (*43*) and Henning and Herrmann (*44*), using TLC-grade Polyamide-6 in minicolumns (Figure 4). We recovered a methanol fraction containing flavonol-glycosides and aglycons and an ammonia/methanol fraction containing flavonol-glucuronides, acylated flavonol-glycosides, aglycons, and ellagic acid and derivatives. Sample-preparation was replicated for each of the 55 juices analyzed.

Materials Used. TLC-grade Polyamide-6, particle size < 100 µm (J.C.Baker Inc., Phillipsburg, NJ 08865); Bio-Rad minicolumns, 10 cm length (Bio-Rad Laboratories, Richmond, CA); glass beads, 212-300 µm size (Sigma

Table I. Experimental Raspberry Juice Samples

Variety	Origin	UR	R	OR	Mo	Std.	Std.& Conc.	Centr.	Enz.& Centr.	Diff. Extr.	Std.& Osm.
Meeker	USA (OR)	X				X					
Meeker	USA (OR)	X				X					
Meeker	USA (OR)		X			X					
Meeker	USA (OR)			X		X					
Meeker	USA (OR)			X		X					
Meeker	USA (OR)			X		X					
Meeker	USA (OR)			X		X					
Meeker	USA (OR)		X		X	X					
Willamette	USA (OR)	X				X					
Willamette	USA (OR)	X				X					
Willamette	USA (OR)			X		X					
Willamette	USA (OR)			X		X					
Willamette	USA (OR)			X		X					
Willamette	USA (OR)			X		X					
Willamette	USA (OR)			X		X					
Willamette	USA (OR)			X		X					
Willamette	USA (OR)		X			X					
Willamette	USA (OR)		X			X					
Willamette	USA (OR)		X				(X)*				
Willamette	USA (OR)		X				(X)				
Willamette	USA (OR)		X				X				
Willamette	USA (OR)		X					X			
Willamette	USA (OR)		X								XmA 8
Willamette	USA (OR)		X								XmA26
Willamette	USA (OR)		X								XmB26
Willamette	Canada (BC)		X					X			
Willamette	Canada (BC)		X						X SP		
Willamette	Canada (BC)		X						X Be		
Willamette	Canada (BC)		X							X	
Chilcotin	Canada (BC)		X				X				
Skeena	Canada (BC)		X				X				
Mall. Promise	Poland		X				(X)				
Mall. Promise	Poland		X				(X)				
Mall. Promise	Poland		X				(X)				
Mall. Seedling	Poland		X				(X)				
Mall. Seedling	Poland		X				(X)				
Mall. Seedling	Poland		X				(X)				
Noma	Poland		X				(X)				
Noma	Poland		X				(X)				
Noma	Poland		X				(X)				
Veten	Poland		X				(X)				
Veten	Poland		X				(X)				
Veten	Poland		X				(X)				
Heritage	USA (OR)		X				(X)				
Heritage	USA (OR)			X			(X)				
Golden	USA (OR)	X					(X)				

UR underripe, R ripe, OR overripe, Mo moldy, Std. standard, Conc. concentration, Centr. centrifugation, Enz. enzyme, Diff. Extr. diffusion extraction, Osm. osmosis, BC British Columbia, OR Oregon, *simulation of a standard juice making process in the laboratory, Mall. Malling, SP Pectinex Ultra SP-enzyme, BE Pectinex BE-enzyme, mA8 membrane A at 8°C, mB26 membrane B at 26°C.

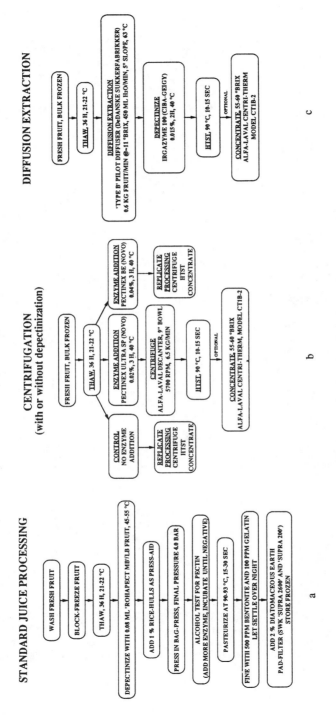

Figure 3. Flow diagrams for juices processing by a) a standard procedure, b) centrifugation (with and without depectinization) and c) diffusion extraction.

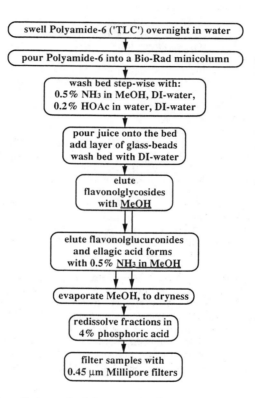

Figure 4. Flow diagram for juice sample preparation for HPLC analysis.

Chemical Co., St. Louis, MO 63178);   filters, 0.45 μm pore size, type HA
(Millipore Corp., Bedford, MA 01730).

Acid and Base Hydrolysis. Juice samples were hydrolyzed a) in 2N HCl for 30
min in boiling water as described by Hong and Wrolstad (45) and b) in 2N NaOH
for 2 hours at room temperature in the dark in nitrogen atmosphere, followed by 30
min hydrolysis in 2N HCl in boiling water, as described by Markham (46).

HPLC Analysis.   We tested many columns and solvent-systems before
establishing a procedure to separate all the components in the two juice fractions.
Columns commonly used for separating flavonoids, i.e. C_{18}-columns with high
carbon-load and end-capped with methanol, or polymer columns, were not
successful. A similar C_8-column, recommended for separating flavonol-glycosides
(47-49), did not work for our raspberry juices. A C_{18}-column, not end-capped with
low carbon-load, however, gave a good separation: 'Spherisorb ODS-1' 5μ, 250
mm length, 4.6 mm ID (Alltech Associates, Inc., Deerfield, Il).

   Solvents: A--100% acetonitrile, B--1% acetic acid in deionized water; flow
rate: 0.6 ml/min. Phenolics were detected at the following wavelengths (nm):
ellagic and benzoic acids (260), gallic acid and catechins (280), cinnamic acids (320),
flavonols (360), anthocyanins (520).

   HPLC programs: 1) Methanol and ammonia eluates--5 min at 16% A; to
19% A in 30 min; 5 min at 19% A; to 30% A in 7 min; to 50% A in 10 min; to
100% A in 5 min; 5 min at 100% A; return to initial conditions in 5 min (total run
time: 77 min). 2) Acid and base-hydrolyzed samples--from 5 to 15% A in 35 min;
to 25% A in 20 min; to 55% A in 18 min; to 100% A in 5 min; 5 min at 100% A;
return to initial conditions in 7 min (total run time: 90 min).

   Instrumentation: A Perkin-Elmer liquid chromatograph (series 400, The
Perkin-Elmer Corp., Norwalk, CT) was used, equipped with a Hewlett-Packard
diode-array detector (model 104A) and data station (series 9000, Hewlett-Packard
Co., Palo Alto, CA), and a Beckman autosampler (model 501, Beckman
Instruments, Inc., San Ramon, CA).

Peak Characterization. 1) by UV-spectra, which are very characteristic of
different classes of compounds and in some instances, compounds within classes;
e.g. quercetin and kaempferol have slightly different absorption maxima; 2) by
comparison to standards: standards were separated by HPLC either by themselves or
mixed with juices; 3) by comparison of raspberry flavonol-chromatograms to those
of other authentic fruits of known composition (e.g. blackberry, cherry, currants;
50); 4) through relative retention-times of known and unknown peaks; 5) through
the elution-order of glycosides of the same aglycon in cases where standards were
not available.

Standards Used for Peak Characterization. Flavonols: quercetin-3-
glucoside, quercetin-3-galactoside, quercetin-3-xylosylglucuronide, kaempferol-3-
glucoside, kaempferol-3-xylosylglucoside, and kaempferol-3-glucuronide were
provided by Prof. Dr. Herrmann (University of Hannover, Germany); quercetin-3-
glucoside and quercetin-3-arabinoside (Carl Roth GmbH & Co., 7500-Karlsruhe 21,
Germany); quercetin-3-rutinoside (rutin), kaempferol, quercetin and quercetin-3-L-
rhamnoside (quercitrin) (Sigma Chemical Co., St. Louis, MO 63178). Ellagic acid
(Sigma Chemical Co.). Hydroxybenzoic acids: gallic, protocatechuic, and p-
hydroxybenzoic acids (Sigma Chemical Co.). Hydroxycinnamic acids: caffeic, p-
coumaric, and ferulic acids (Sigma Chemical Co.)

Quantification method. Flavonolglycosides, aglycons, and ellagic acid and
derivatives were quantified via internal and external standards. **Internal**

Standards. Naringin (for the methanol fraction) and 4-methylumbelliferyl-ß-D-glucuronide (for the ammonia/methanol fraction) (both obtained from Sigma Chemical Co.). Half a ml of a naringin stock solution (500 ppm naringin in DI-water) and a 4-methylumbelliferyl-ß-D-glucuronide stock-solution (250 ppm in DI-water), respectively, were added to 11 ml of single-strength red raspberry juice. Because juices were concentrated during sample preparation, less internal standard was added to the juice samples than to external standards. Juices spiked with internal standards were separated into fractions as described in Figure 6. For quantification, peak areas were normalized to the appropriate internal standards.

      External standards. Rutin (for the methanol fraction) and ellagic acid (for the ammonia/ methanol fraction) (both obtained from Sigma Chemical Co.). A set of external standards for each fraction, consisting of the external standard at four different concentrations, was run alternately with the juice samples throughout analysis. Separate sets of standards had to be used as purified ellagic acid (Sigma Chemical Co.) is only soluble in > 80% ethanol in water. Set 1: rutin at 1.5, 25, 50, 150 ppm (in DI-water); 100 ppm naringin and 50 ppm 4-methylumbelliferyl-ß-D-glucuronide were added to each aliquot. Set 2: ellagic acid at 0.6, 10, 20, 60 ppm (in 80% ethanol); 100 ppm rutin (as internal standard for ellagic acid) was added to each aliquot. For each set of standards a standard curve was fitted by linear regression (peak area vs. concentration in ppm). The concentration (c) of each individual flavonol or ellagic acid form was calculated from measured peak area (a) using the equation

$$c = I + S\,a$$

where I and S were the intercept and slope of the fitted line for the corresponding external standard. Adjustments were made for differences in the concentrations of the internal standards in the juice samples compared to the sets of standards. The concentrations of flavonols and ellagic acid and its forms were normalized to standard single-strength juice °Brix.

Limitations of Quantification. 1) The internal standard naringin degraded in the stock-solution stored in the freezer over time. A cubic function was determined to estimate a corrected naringin peak area for the time at which each juice sample was analyzed. 2) The ellagic acid used as an external standard is only soluble in > 80% ethanol, resulting in decreased peak sharpness. 3) The influence of Polyamide-6 on reproducibility of sample preparation is presently unknown. 4) Benzoic and cinnamic acids could not be quantified as no reproducible procedure for sample preparation could be developed. The variable influences of these factors account for a significant portion of the variance observed between replicate analyses.

Statistical analysis. To evaluate the reproducibility of the data, we determined a) the standard errors of means of phenolics and b) the percentages of variances due to differences between samples and replicate sample preparations with analysis of variance (ANOVA). Standard errors of means are listed rather than standard deviations. There was considerable variation between replications. While this limited the conclusions that can be drawn from treatment effects, the quantitative estimates are still useful for evaluating trends and providing an overall perspective.

Results & Discussion

Examples of HPLC chromatograms of methanol fractions of juices made from the varieties Willamette and Norna are shown in Figures 5a and b. Both juices were made from ripe berries by the standard process. The HPLC profiles of the two varieties were qualitatively quite similar, however, quantitatively very different in the

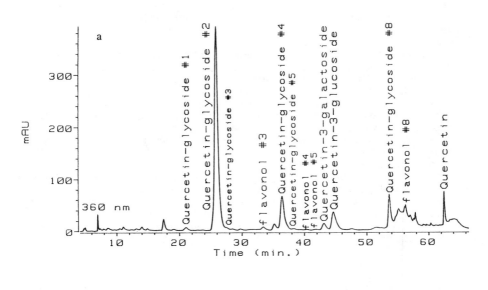

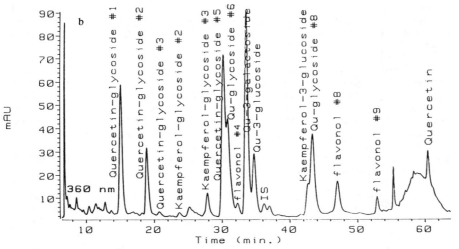

Figure 5. HPLC chromatograms of methanol fractions (separated by Polyamide-6) of red raspberry juices made from a) ripe Willamette variety and b) ripe Norna variety.

glycosides present (e.g. quercetin-glycosides #1 and #2, quercetin-3-galactoside and quercetin-3-glucoside). Some trace peaks (e.g. 3 kaempferol-glycosides) were detected in Norna variety but not in Willamette. In the case of the ammonia fractions of the same Willamette and Norna juices (Figures 6a and b) the differences in compositional profiles were mainly quantitative, e.g. the ratio of ellagic acid to other ellagic acid forms was much greater in Willamette variety than Norna variety.

Ellagic Acid and Ellagic Acid Forms. Concentrations of ellagic acid and its derivatives for all red raspberry juices samples (excluding the extremely moldy sample) are summarized in Table II. The percentage of total variance due to differences between replicate sample preparations was high for some compounds. Ellagic acid was detected in all juice samples with a mean concentration of 10.1 ppm for 45 experimental samples and up to 52.1 ppm in the 9 commercial samples. As many as 16 additional ellagic acid forms were detected in the juices investigated, with individual concentrations ranging from traces to 3 ppm (means) in experimental juices and between traces and 23.5 ppm in commercial samples. The total concentration of ellagic acid and derivatives in experimental juices was 28.2 ppm (means) and in commercial juices ranged from 22.4 to 80.4 ppm.

The concentrations of summed ellagic acid forms in juices made from 10 different varieties are shown in Figure 7a. Figure 7b compares the concentrations of ellagic acid in the same juices. The varietal distribution of ellagic acid largely reflected that of summed ellagic acid forms. Varieties differed greatly in their contents of ellagic acid and summed forms. Willamette and Meeker varieties contained most summed ellagic acid forms (between ca. 7 and 36 ppm) and ellagic acid (up to ca. 20 ppm); Veten and Norna contained medium amounts of ellagic acid and summed forms. The concentrations of ellagic acid and summed forms decreased with increasing ripeness in Meeker variety, while Willamette juice made from ripe fruit had the highest contents and juice made from underripe fruit had the lowest. For other varieties, there was no consistent relationship between ellagic acid concentrations and ripeness.

Flavonols. Concentrations of quercetin and kaempferol-glycosides, glucuronides and other forms detected in both fractions are summarized in Table III. Quercetin was present almost entirely as glycosides and glucuronides in all raspberry juices. The total mean concentration of flavonols in 45 experimental samples was 122 ppm, that of quercetin-forms was 118 ppm, and that of kaempferol-forms was 3.6 ppm. Eight quercetin-glycosides, 3 flavonol-glucuronides and 2 other flavonol-forms were detected and measured in most samples. Quercetin-3-glucuronide was the major flavonol with a mean concentration of 55 ppm for the experimental samples; it ranged from 9.7 to 89 ppm in the commercial samples.

Second in quantity was quercetin-glycoside #2, which was present in experimental samples with a mean of 29.2 ppm and a range from 1.4 to 67.9 ppm in commercial samples. Presence of such an early eluting flavonol-glycoside has not been reported previously in raspberries. We speculate that this glycoside may be quercetin-3-sophoroside as: diglycosides have earlier retention times than monoglycosides; cyanidin-3-sophoroside is the major raspberry anthocyanin (all flavonols are glycosylated by the same enzymes; *51*); quercetin and kaempferol are the only flavonol aglycons present in raspberries (because the hydroxylation patterns of anthocyanins and flavonols are the same; *52*); glycosides of cyanidin and pelargonidin are present in these juices in about the same ratio (*36*) as those of quercetin and kaempferol. We confirmed the presence of 6 of the flavonol-glycosides and glucuronides reported previously for raspberries (*27, 29*).

Table IV summarizes 36 additional quercetin and kaempferol-forms which were present in trace amounts (i.e. < 1 ppm) to varying degrees in the juices

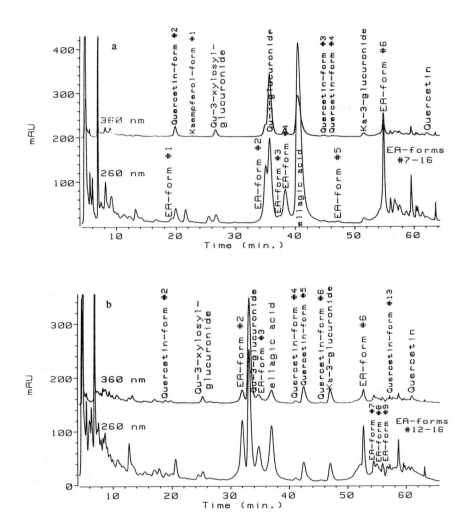

Figure 6. HPLC chromatograms of ammonia/methanol fractions (separated by Polyamide-6) of red raspberry juices made from a) ripe Willamette variety and b) ripe Norna variety.

Table II. Ellagic Acid and Ellagic Acid Forms in Experimental and Commercial Raspberry Juices

Ellagic Acid Compound	n	Concentration in Experimental Juices (ppm)			Concentration-Range in Commercial Juices (ppm; n=9)
		Mean ± Std. Error	% Variance due to Difference		
			Among Juices	Between Repl.	
Ellagic Acid (EA)*	45	10.12 ± 1.53	71.7	28.3	5.48 - 52.14
EA-form #2	44	2.95 ± 0.27	70.6	29.4	0.87 - 7.19
EA-form #6	44	2.13 ± 0.20	65.9	34.1	0.78 - 3.78
EA-form #3	44	2.12 ± 0.20	25.5	74.5	trace - 1.88
EA-form #4	13	1.65 ± 0.25	22.0	78.0	n.d. - 2.40
EA-form #1	20	1.52 ± 0.44	13.1	86.9	2.06 - 23.51
EA-form #12	44	1.12 ± 0.10	83.3	16.7	trace - 1.48
EA-form #8	32	0.77 ± 0.08	38.4	61.6	trace - 2.44
EA-form #9	32	0.74 ± 0.05	1.6	98.4	n.d. - 2.46
EA-form #7	37	0.71 ± 0.04	23.5	76.5	n.d. - 1.49
EA-form #10	26	0.69 ± 0.04	12.7	87.3	n.d. - 0.85
EA-form #11	31	0.62 ± 0.01	32.9	67.1	trace
EA-form #15	32	0.62 ± 0.01	30.4	69.6	n.d. - 0.99
EA-form #13	38	0.61 ± 0.01	31.8	68.2	trace
EA-form #14	35	0.61 ± 0.00	32.6	64.4	n.d. - 0.71
EA-form #16	44	0.61 ± 0.01	24.3	75.7	trace - 0.77
EA-form #5	34	trace• ± 0.00	33.3	66.7	n.d. or trace
EA + 16 EA-forms	45	28.17 ± 2.34	69.9	30.0	22.36 - 80.36

*reported previously for fresh raspberries by Bate-Smith (1959) and Daniel et al. (1989); •$\leq$ 0.6 ppm.

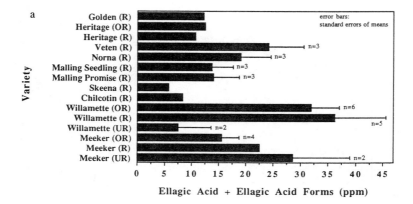

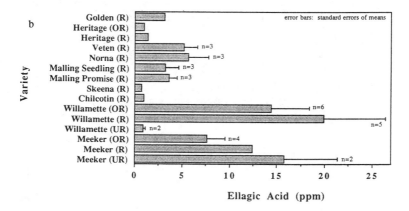

Figure 7. a) Total concentrations of ellagic acid forms and b) concentrations of ellagic acid, in juices made from different raspberry varieties (UR-underripe, R-ripe, OR-overripe).

Table III. Quercetin and Kaempferol-Glycosides, Glucuronides, and Other Forms in Experimental and Commercial Raspberry Juices

Flavonol Compound	n	Concentration in Experimental Juices (ppm)			Concentration-Range in Commercial Juices (ppm; n=9)
		Mean ± Std. Error	% Variance due to Difference Among Juices	Between Repl.	
Quercetin-3-glucuronide*	45	54.37 ± 4.48	64.2	35.8	9.72 - 88.51
Quercetin-glycoside #2 (3-sophoroside?)	44	29.21 ± 4.88	88.1	11.9	1.35 - 67.88
Quercetin-glycoside #4 (MeOH-fraction)	30	7.78 ± 1.32	75.7	24.3	n.d. - 14.76
Quercetin-form #2 (NH3-fraction)	43	3.86 ± 0.54	57.9	42.1	trace• - 8.96
Quercetin-glycoside #6 (MeOH-fraction)	8	3.41 ± 1.23	13.2	86.8	n.d. or trace
Quercetin-glycoside #5 (MeOH-fraction)	10	3.33 ± 0.97	64.0	36.0	n.d. or trace
Quercetin-3-glucoside*	45	2.99 ± 0.55	86.1	13.9	trace - 8.94
Quercetin-glycoside #8 (MeOH-fraction)	42	2.55 ± 0.33	58.3	41.7	trace - 6.46
Quercetin-glycoside #1 (MeOH-fraction)	26	2.48 ± 0.49	55.6	44.4	nd - 6.18
Quercetin-3-xylosylglucuronide*	39	2.43 ± 0.25	56.8	43.2	trace - 5.72
Kaempferol-3-glucuronide*	45	2.31 ± 0.48	96.4	3.6	trace - 4.50
Quercetin-3-galactoside*	43	2.03 ± 0.38	77.0	23.0	trace - 6.18
Quercetin-form #5 (NH3-fraction)	36	1.90 ± 0.46	93.9	6.1	n.d. - 2.37
Quercetin* (sum, both fractions)	45	1.75 ± 0.23	73.8	26.2	1.89 - 12.59
Kaempferol* (sum, both fractions)	39	1.25 ± 0.20	3.0	97.0	trace - 2.00
Sum, all Above Quercetin-forms	45	118.09 ± 8.47	77.1	22.9	31.15 - 211.31
Sum, all Above Kaempferol-forms	45	3.55 ± 0.64	79.2	20.8	2.00 - 6.00
Sum, all Above Flavonols	45	121.64			

*reported previously for fresh berries by Ryan and Coffin (1971) and/or Henning (1981); •≤1ppm.

Table IV. Quercetin and Kaempferol-forms Present in Trace Amounts in Raspberry Juices in Both Fractions•

Quercetin-glycoside #3 (M)	5 experimental samples, 1 commercial sample
Quercetin-glycoside #7 (M)	2 commercial samples
Quercetin-form #1 (A)	1 Willamette (overripe) sample
Quercetin-form #3 (A)	4 Willamette, 2 (of 3) Malling Seedling samples
Quercetin-form #6 (A)	Golden (4.4 ppm); centri. (1.7 ppm); 24 exp., 6 com. spls.
Quercetin-form #7 (A)	all Heritage, Golden, 3 Willamette, 1 commercial sample
Quercetin-form #8 (A)	Golden, 2 commercial
Quercetin-form #9 (A)	16 experimental, 3 commercial samples
Quercetin-form #10 (A)	Golden
Quercetin-form #11 (A)	Golden
Quercetin-form #12 (A)	Golden
Quercetin-form #13 (A)	32 experimental, 4 commercial samples
Quercetin-form #14 (A)	Golden, 1 Willamette, 1 Heritage (overripe) sample
Kaempferol-3-glucoside* (M)	4 Meeker, all Noma, all Veten, Skeena, 1 osmosis sample
Kaempferol-glycoside #1 (M)	1 commercial sample
Kaempferol-glycoside #2 (M)	22 experimental, 2 commercial samples
Kaempferol-glycoside #3 (M)	6 commercial samples
Kaempferol-glycoside #4 (M)	6 (of 7) Meeker, 1 commercial sample
Kaempferol-glycoside #5 (M)	1 Noma sample
Kaempferol-glycoside #6 (M)	Golden, 3 Malling, 1 OR-Heritage, 1 Will., 1 com. sample
Kaempferol-form #1 (A)	all centri., 1 Meeker (OR), 2 Will. samples, 1 com. sample
Kaempferol-form #2 (A)	all centrifugation, all Heritage samples
Kaempferol-form #3 (A)	centrifugation + SP-enzyme (1.4 ppm); centr.+ BE-enzyme
Kaempferol-form #4 (A)	1 Veten sample (1.5 ppm), Chilcotin
Kaempferol-form #5 (A)	1 Meeker, 1 Willamette, 1 Malling Seedling sample
Kaempferol-form #6 (A)	Golden, 1 Noma, 1 Veten sample
Flavonolglycoside #1 (M)	2 commercial samples
Flavonolglycoside #2 (M)	1 commercial sample
Flavonolglycoside #3 (M)	29 experimental, 6 commercial samples
Flavonolglycoside #4 (M)	29 experimental, all commercial samples
Flavonolglycoside #5 (M)	24 experimental, 6 commercial samples
Flavonolglycoside #6 (M)	1 osmosis sample; diffusion extraction; 1 com. sample
Flavonolglycoside #7 (M)	all Heritage samples
Flavonolglycoside #8 (M)	26 experimental, 8 commercial samples
Flavonolglycoside #9 (M)	22 experimental, 4 commercial samples
Flavonolglycoside #10 (M)	all Heritage, 2 Malling samples, 1 commercial sample

*reported previously for fresh raspberries by Henning (1981), •≤ 1 ppm, M methanol fraction, A ammonia/methanol fraction.

investigated. Only certain varieties seemed to contain some of these forms and certain processes appeared to enhance their concentrations.

The sum of all quercetin-forms (Figure 8a) ranged from a few to ca. 160 ppm among experimental varieties, with Heritage, Willamette and Norna containing the highest concentrations. For quercetin-3-glucuronide the concentration pattern was quite similar (Figure 8b), Heritage, Golden, Malling Promise, and Norna varieties having the greatest quantities (between ca. 75 to 125 ppm). Quercetin-glycoside #2 (Figure 8c) was present in much greater amounts in Willamette than other varieties and was not detected in Golden. Quercetin-3-glucoside (Figure 8d) was present in all varieties at 5 ppm or less, Heritage, Willamette and Malling Promise having the highest concentrations.

There were great differences in the concentrations of total kaempferol-forms among varieties (Figure 9), Heritage and Norna containing between ca. 11 and 20 ppm, while all others contained only a few ppm or traces. The ratio of quercetin-3-glucuronide to ellagic acid also differed greatly among varieties (Figure 10).

Influences of Environmental Factors. There was no apparent correlation between ripeness and the concentrations of total quercetin forms (Figure 8a), quercetin-3-glucuronide (Figure 8b) and quercetin-3-glucoside (Figure 8d) in juices made from the varieties Meeker and Willamette. Quercetin-glycoside #2 (Figure 8c) increased in Willamette juice with increasing ripeness. Mold decreased the contents of quercetin-glycosides and glucuronides considerably in the variety Meeker as shown by the total concentration of quercetin-forms (Figure 11a). Mold did not, however, have an effect on total ellagic acid forms, as investigated in the same variety (Figure 11b). With the juice samples available for this study we could not determine whether geographic origin of the berries or harveting method influenced the concentrations of flavonols and ellagic acid.

Influences of Processing. Diffusion extraction and centrifugation processed juices were made from aliquots of the same batch of ripe Willamette raspberries, grown in British Columbia (Canada). Standard processed juices (control), with or without concentration (by vacuum or osmosis), were made from equal aliquots of another batch of ripe Willamette berries, grown in Oregon. Juices made by the standard process from small batches of underripe and overripe Willamette berries, grown in Oregon, were produced for comparison. The influences of these juice processing techniques on concentrations of total ellagic acid and quercetin-forms are shown in Figure 12. Centrifugation processed juice contained much more total quercetin-forms (ca. 280 ppm) than all other juices. The influence of pectinases considerably decreased the concentrations of total quercetin forms (to ca. 190 ppm; Figure 12b) and total ellagic acid forms (from ca. 37 ppm to 17 ppm; Figure 12a) compared to centrifugation alone. Diffusion extraction juice contained even less total quercetin forms (ca. 130 ppm), however, it produced a juice with far greater total contents of ellagic acid forms than other juices (ca. 70 ppm; Figure 12a).

The concentration of total ellagic acid forms in standard processed juice (single-strength, large batch) was as high as that in juice produced by diffusion extraction; however, these juices were made from different batches of berries. All concentration techniques decreased total ellagic forms considerably compared to the control; vacuum evaporation and osmotic concentration at 26°C (membrane A) had the greatest decreasing effects. There were fewer differences in the concentrations of summed quercetin forms among juices produced by the standard process, with or without concentration. Vacuum concentration decreased total quercetin-forms compared to unconcentrated juice, however, standard single-strength juice made from a smaller batch of berries contained less total quercetin and ellagic acid forms than vacuum concentrated juice.

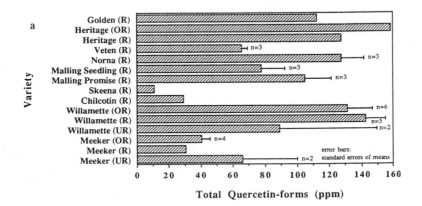

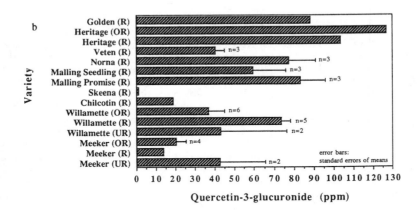

Figure 8. a) Total concentrations of quercetin-forms and concentration of b) quercetin-3-glucuronide in juices made from different raspberry varieties (UR-underripe, R-ripe, OR-overripe).

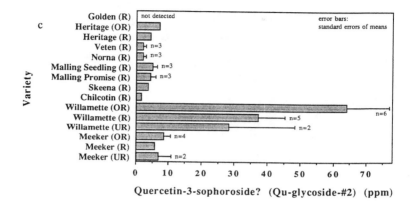

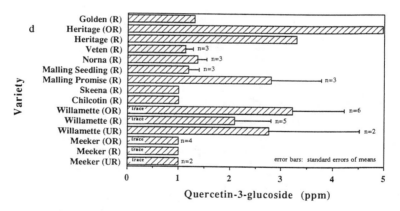

Figure 8. Concentrations of c) quercetin-glycoside #2, and d) quercetin-3-glucoside in juices made from different raspberry varieties (UR-underripe, R-ripe, OR-overripe).

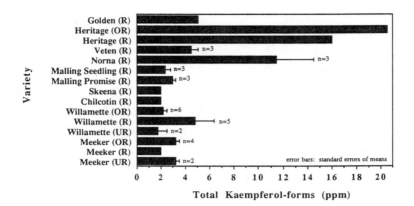

Figure 9. Total concentrations of kaempferol-forms in juices made from different raspberry varieties (UR-underripe, R-ripe, OR-overripe).

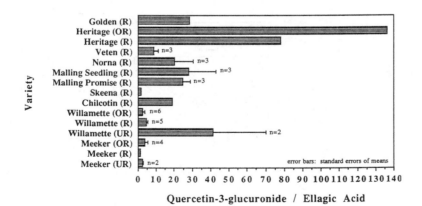

Figure 10. Ratios of quercetin-3-glucuronide to ellagic acid for juices made from different raspberry varieties (UR-underripe, R-ripe, OR-overripe).

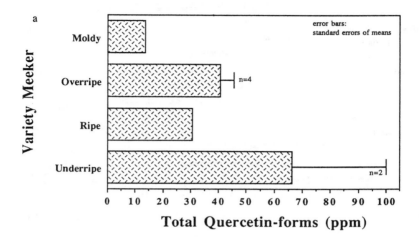

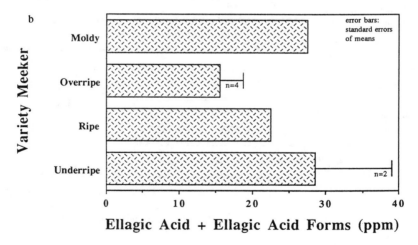

Figure 11. Effect of mold-contamination on total concentration of a) quercetin-forms and b) ellagic acid forms in raspberry juice made from the variety Meeker.

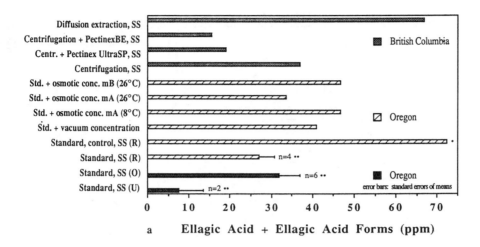

a   **Ellagic Acid + Ellagic Acid Forms (ppm)**

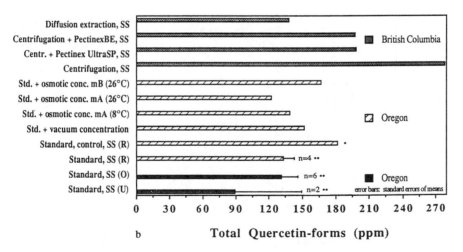

b   **Total Quercetin-forms (ppm)**

Figure 12. Effects of different processing techniques on total concentration of a) ellagic acid forms and b) quercetin-forms in raspberry juices made from the variety Meeker (m-membrane, SS-single-strength, U -underripe, R-ripe, O -overripe, •large batch, ••small batch).

Comparison of Experimental and Commercial Samples. In general, commercial samples contained more ellagic acid and ellagic acid derivatives than experimental samples (mean of 52.6 ppm for commercial samples compared to a mean of 17.9 ppm for experimental samples; Figure 13a). There was, however, considerable variation in ellagic acid and quercetin contents (Figure 13b) among the commercial samples, which could have resulted from varietal and/or processing differences. Three commercial samples had particularly high contents of summed quercetin-forms (ca. 150-210 ppm).

Hydrolyzed Juice Samples. An example of a HPLC chromatogram of ripe Willamette juice, produced by the standard process and acid hydrolyzed in 2N HCl for 30 min, is shown in Figure 14a. Ellagic acid, 2 ellagic acid forms, quercetin, kaempferol, 2 flavan-3-ols ((+)-catechin and (-)-epicatchin), 3 hydroxycinnamic acids (caffeic, p-coumaric and ferulic), as well as 3 hydroxybenzoic acids (gallic, protocatechuic, and p-hydroxybenzoic) were characterized by comparison with standards. The same Willamette juice, hydrolyzed with 2N NaOH for 2 hours followed by 2N HCl for 30 min (Figure 14b) had a very similar HPLC pattern of compounds, except that the ratio of ellagic acid to the first ellagic acid form was greater and that the second ellagic acid form was absent.

Conclusions

The chromatographic profiles of flavonols and ellagic acid were qualitatively very similar for all varieties studied, however, there were great quantitative differences due to variety, processing technique and environmental factors. Ellagic acid was present in all samples studied; 16 additional ellagic acid derivatives were detected. Commercial samples contained more ellagic acid than experimental juices. Quercetin was the primary flavonol aglycon in all raspberry juices, present almost entirely in glycosylated form. Three flavonol-glucuronides, 8 quercetin-glycosides, 2 other flavonol-forms, quercetin and kaempferol were characterized and quantified in most samples; 36 additional flavonol-forms were detected in trace amounts. Quercetin-3-glucuronide was the major flavonol-glycoside with an unidentified quercetin-glycoside (#2, quercetin-3-sophoroside?) being second in concentration. While the quantitative data for the flavonols and ellagic acid derivatives have limitations because of the variation between sample replications, the estimated concentrations are useful for giving a general picture of the amounts of these compounds in red raspberry products. Red raspberry flavonol and ellagic acid profiles should be a useful auxiliary technique for detecting adulteration, e.g. in cases where qualitative differences are found.

The varieties Heritage, Willamette and Norna had the highest concentrations of total quercetin and kaempferol forms; Heritage, Golden, Malling Promise and Norna contained the most quercetin-3-glucuronide. Quercetin-glycoside-#2 was present in Willamette variety in much higher concentrations than in other varieties. The pattern of concentration of ellagic acid for different varieties was similar to that of summed ellagic acid forms, Meeker and Willamette containing the most. Diffusion extraction and the standard process produced juices with by far the greatest concentrations of ellagic acid and its forms, while centrifugation processed juice contained much more summed quercetin-forms than juices made by other techniques. Depectinization and concentration (by vacuum or osmosis) decreased total ellagic acid and quercetin forms considerably. Mold-contamination decreased the contents of quercetin-glycosides and glucuronides greatly, while it had little effect on ellagic acid and its derivatives. Varieties grown in the same region showed much variation in both ellagic acid and flavonol contents. There was no trend for ellagic acid and flavonol concentrations with increasing ripeness, nor could the influence of harvesting method be determined with the samples available.

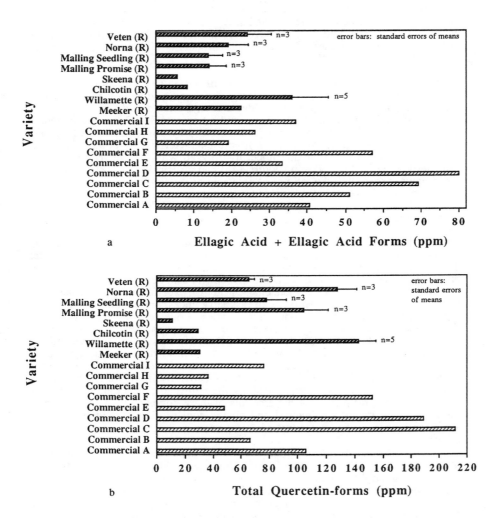

Figure 13. Total concentrations of a) ellagic acid forms and b) quercetin-forms in experimental (single-strength) and commercial (diluted from concentrates) raspberry juices.

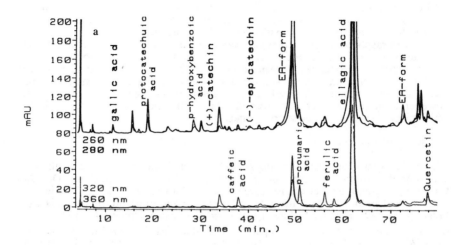

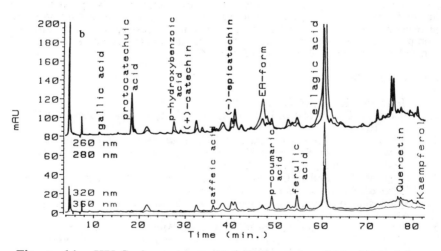

Figure 14. HPLC chromatograms of Willamette red raspberry juice hydrolyzed in a) 2N hydrochloric acid for 30 min and b) 2N sodium hydroxide for 2 hours and 2N hydrochloric acid for 30 min.

In acid hydrolyzed juices, quercetin, kaempferol, ellagic acid, two ellagic acid forms, (+)-catechin, (-)-epicatchin, and caffeic, p-coumaric, ferulic, gallic, p-hydroxybenzoic, and protocatechuic acids were identified. Base hydrolyzed samples contained the same phenolic compounds except for one of the ellagic acid forms. Hydrolysis simplifies the HPLC chromatographic profile of the phenolics in raspberry juice dramatically and would therefore be a useful tool for detecting adulteration, particularly if these phenolics could be quantified reliably.

Red raspberry juices contain considerable amounts of flavonols and ellagic acid derivatives. As quercetin and ellagic acid have been shown to have anticarcinogenic properties, the possible beneficial effects of consuming red raspberry products, particularly in combination with other phenolics-containing fruits and vegetables, cannot be ignored. Before dietary recommendations can be made, however, more research needs to be conducted on absorption and the metabolic and anticarcinogenic effects of these compounds.

Acknowledgments

We would like to thank the following persons and institutions for their contributions and support, which made this project possible: Tom Beveridge and Dr. Dan Cummins, Agriculture Canada Research Station, Summerland, B.C., Canada; Dr. Witold Plocharski, Research Institute of Pomology and Floriculture, Skierniewice, Poland; Prof. Dr. Herrmann, Institute of Food Chemistry, University of Hannover, Germany; Dr. David Thomas, Department of Statistics, Oregon State University; Lucy Wisniewski, Salem, Oregon; Brian Yorgey, Research Assistant, Department of Food Science and Technology, Oregon State University; North Willamette Research and Extension Center, Aurora, Oregon; Oregon Caneberry Commission, Salem, Oregon. We would also like to thank the following companies for providing juice concentrates: Clermont Inc.; Endurance Fruit Processing, Inc.; J.M. Smucker Company; Kerr Concentrates Inc.; Milne Fruit Products; Osmotek Inc.; Sanofi Bio-Industries; Rudolf Wild GmbH & Co. KG.

Literature Cited

1.   Herrmann, K. *J. Fd. Technol.* **1976**, *11*, 433-448.
2.   Spanos, A.S.; Wrolstad, R.E. *J. Assoc. Off. Anal. Chem.* **1987**, *70*(6), 1036-1046.
3.   Deschner, E.E. Presented at the *202nd National Meeting of the American Chemical Society*, New York, NY, Symposium on Phenolic Compounds in Foods and Health, 1991; AGFD Abstract No. 114.
4.   Leighton, T.; Ginther, C.; Nortario, V.; Harter, W. Presented at the *202nd National Meeting of the American Chemical Society*, New York, NY, Symposium on Phenolic Compounds in Foods and Health, 1991; AGFD Abstract No. 116.
5.   Verma, A.K. Presented at the *202nd National Meeting of the American Chemical Society*, New York, NY, Symposium on Phenolic Compounds in Foods and Health, 1991; AGFD Abstract No. 113.
6.   Weisburger, J.H. Presented at the *202nd National Meeting of the American \Chemical Society*, New York, NY, Symposium on Phenolic Compounds in Foods and Health, 1991; AGFD Abstract No. 86.
7.   Yasukawa, K.; Takido, M.; Takeuchi, M.; Nitta, K. In *Plant Flavonoids in Biology and Medicine II. Biochemical, Cellular, and Medical Properties*, Cody, V.; Middleton, Jr.E.; Harborne, J.B.; Beretz, A. Eds.; Progress in Clinical and Biological Research; Alan R. Liss, Inc.: New York, NY, 1988, Vol. 280; pp 247-250.

8. Bokkenheuser, V.D.; Winter, J. In *Plant Flavonoids in Biology and Medicine II. Biochemical, Cellular, and Medical Properties*; Cody, V.; Middleton, Jr.E.; Harborne, J.B.; Beretz, A. Eds.; Progress in Clinical and Biological Research; Alan R. Liss, Inc.: New York, NY, 1988, Vol. 280; pp 143-146.

9. Shillitoe, E.J.; Hoover, C.I.; Fisher, S.J.; Abdel-Salam, M.; Greenspan, J.S. *J. Natl. Cancer Inst.* **1984**, *73*(3), 673-678.

10. Yoshida, M.; Sakai, T.; Hosokawa, N.; Marui, N.; Matsumoto, K.; Fujioka, A.; Nishino, H.; Aoike, A. *FEBS Lett.* **1990**, *260*(1), 10-13.

11. Committee on Diet, Nutrition, and Cancer. *Diet, Nutrition, and Cancer*, National Academy Press: Washington, D.C., 1982.

12. Stich, H.F.; Rosin, M.P. In *Nutritional and Toxicological Aspects of Food Safety*; Friedman, W. Ed.; Advances in Experimental Medicine and Biology; Plenum Press: New York, NY, 1984, Vol. 177; pp 1-29.

13. Stoner, G.D. *Proceedings of the Annual Meeting of the North American Strawberry Growers Association*; Grand Rapids, MI, 1989.

14. Chang, R.L.; Huang, M.T.; Wood, A.W.; Wong, C.Q.; Newmark, H.L.; Yagi, H.; Sayer, J.M.; Conney, A. *Carcinogenesis* **1985**, *6*, 1127-1133.

15. Daniel, E.M.; Krupnick, A.S.; Heur, Y.-H.; Blinzler, J.A.; Nims, R.W.; Stoner, G.D. *J. Food Comp. Anal.* **1989**, *2*, 338-349.

16. Das, M.; Bickers, D.R.; Mukhtar, H. *Carcinogenesis* **1985**, *6*(10), 1409-1413.

17. Del Tito, B.J. Jr.; Mukhtar, H.; Bickers, D.R. *Biochem. Biophys. Res. Commun.* **1983**, *114*(1), 388-394.

18. Lesca, P. *Carcinogenesis* **1983**, *4*, 1651-1653.

19. Maas, J.L.; Galletta, G.J.; Stoner, G.D. *HortScience* **1991a**, *26*(1), 10-14.

20. Mandal, S.; Stoner, G.D. *Carcinogenisis* **1990**, *11*(1), 55-61.

21. Mukhtar, H.; Das, M.; Del Tito, B.J.; Bickers, D.R. *Biochem. Biophys. Res. Commun.* **1984**, *119*, 751-757.

22. Tanaka, T.; Iwata, H.; Niwa, K, Mori, Y.; Mori, H. *Japan J. Cancer Res.* **1988**, *79*, 1297-1303.

23. Tanaka, T.; Yoshimi, N.; Sugie, S.; Mori, H. Presented at the *202nd National Meeting of the American Chemical Society*, New York, NY, Symposium on Phenolic Compounds in Foods and Health, 1991; AGFD Abstract No. 90.

24. Lee, K.-H. Presented at the *202nd National Meeting of the American Chemical Society*, New York, NY, Symposium on Phenolic Compounds in Foods and Health, 1991; AGFD Abstract No. 92.

25. Teel, R.W.; Babcock, M.S.; Dixit, R.; Stoner, G.D. *Cell Biol. Toxicol.* **1986**, *2*(1), 53-62.

26. Stavric, B; Matula, T.I.; Klassen, R.; Downie, R.H. Presented at the *202nd National Meeting of the American Chemical Society*, New York, NY, Symposium on Phenolic Compounds in Foods and Health, 1991; AGFD Abstract No. 116.

27. Henning, W. *Z. Lebensm. Unters. Forsch.* **1981**, *173*, 180-187.

28. Herrmann, K. *Ernaehrungs-Umschau* **1974**, *21*(6), 177-181.

29. Ryan, J.J.; Coffin, D.E. *Phytochem.* **1971**, *10*, 1675-1677.

30. Mosel, H.D.; Herrmann, K. *Z. Lebensm. Unters. Forsch.* **1974**, *154*, 324-327.

31. Bate-Smith, E.C. In *The Pharmacology of Plant Phenolics*; Fairbairn, J.W., Ed.; Academic Press: New York, NY, 1959; pp 133-147.

32. Herrmann, K. *Crit. Rev. Food Sci. Nutr.* **1989**, *28*(4), 315-347.

33. Schuster, B.; Herrmann, K. *Phytochem.* **1985**, *24*(11), 2761-2764.

34. Bate-Smith, E.C. *Phytochem.* **1972**, *11*, 1153-1156.

35. Wilson, T.C.; Hagerman, A.E. *J. Agric. Food Chem.* **1990**, *38,* 1678-1683.
36. Boyles M.J. M.S. Thesis, Oregon State University, 1991.
37. Beaudry, E.G.; Lampi, K.A. *Food Technol.* **1990a,** *44,* 121.
38. Beaudry, E.G.; Lampi, K.A. *Flussiges Obst.* **1990b,** *57,* 663-664.
39. Wrolstad, R.E.; McDaniel, M.R.; Durst, R.W.; Micheals, N.J.; Lampi, K.A.; Beaudry, E.G. Submitted to *J. Food Sci.*
40. Beveridge, T.; Harrison, J.E.; McKenzie, D.-L. *Can. Inst. Food Sci. Technol. J.* **1988,** *21*(1), 43-49.
41. Schobinger, U.; Dousse, R.; Duerr, P.; Tanner, H. *Fluessiges Obst* (Sonderdruck) **1978,** *6,* 1-5.
42. Luethi, H.R.; Glunk, U. *Fluessiges Obst* **1974,** *41,* 498-505.
43. Wald, B.; Galensa, R. *Z. Lebensm. Unters. Forsch.* **1989,** *188,* 107-114.
44. Henning, W.; Herrmann, K. *Z. Lebensm. Unters. Forsch.* **1980,** *170,* 433-444.
45. Hong, V.; Wrolstad, R.E. *J. Assoc. Off. Anal. Chem.* **1986,** *69*(2), 199-207.
46. Markham, K.R. In *Techniques of Flavonoid Identification;* Academic Press: New York, New York, 1982; pp 59-60.
47. Harborne, J.B.; Boardley, M. *J. Chromatogr.* **1984,** *299,* 377-385.
48. Harborne, J.B.; Boardley, M.; Linder, H.P. *Phytochem.* **1985,** *24*(2), 273-278.
49. Hostettmann, K.; Domon, B.; Schaufelberger, D.; Hostettmann, M. *J. Chromatogr.* **1984,** *283,* 137-147.
50. Macheix, J.-J.; Fleuriet, A.; Billot, J. In *Fruit Phenolics;* CRC Press, Inc.: Boca Raton, Florida, 1990; pp 105-126.
51. Bilyk, A.; Sapers, G. *J. Agric. Food Chem.* **1986,** *34*(4), 585-588.
52. Wildanger, W.; Herrmann, K. *Z. Lebensm. Unters. Forsch.* **1973,** *151,* 103-108.

RECEIVED August 10, 1992

Chapter 22

Changes in Phenolic Compounds During Plum Processing

Hui-Yin Fu[1], Tzou-Chi Huang[2], and Chi-Tang Ho[1]

[1]Department of Food Science, Cook College, Rutgers, The State University
of New Jersey, New Brunswick, NJ 08903
[2]Department of Food Science and Technology, National Pingtung
Polytechnic Institute, Pingtung, Taiwan

Semi-dried pickled plums are the most important processed fruit product in Taiwan. The formation of the red-
brown color on the skin of the product can only be
achieved through a full sun-drying process. This variation in pigment is the most noticeable change that
occurs during the processing. The total quantity of
polyphenols decreased significantly in the sun-dried
sample. A mixture of leucoanthocyanin and (+)-cate-
chin purified from plum flesh by preparative TLC was
used as a model system to study the mechanism for the
formation of the pigment. A red-brown pigment similar to that in the sun-dried plums was isolated and
purified after exposure to tungsten light. Radiation
in the visible portion of sunlight appeared to catalyze the photopolymerization of the red-brown pigment.

Plums (Plunus mume Sieb. et Zucc.) have been cultivated in China for
at least 3000 years. They were introduced to Taiwan and are now
common in the mountainous areas.
    Plum fruits are harvested in the late spring, and are processed
into different commercial products. Among them, semi-dried plums
are the most popular. The annual production is estimated at
5,000,000 tons and is exported exclusively to Japan.
    High salt (ca. 25%) pretreatment for 3 to 6 months was per-
formed traditionally to obtain the characteristic aroma. The pink
color on the skin of the product developed as a result of exposure
to sunshine for a minimal period of two days (7 hours exposure per
day). The water content was reduced from 95% in plum flesh to
about 60% in the final product.
    This paper reports the possible mechanism for the formation of
the desirable red-brown colors in sun-dried pickled plums.

0097–6156/92/0506–0287$06.00/0
© 1992 American Chemical Society

Phenolic Compounds in Plums

An HPLC equipped with a Photodiode Array UV-VIS detector was used to characterize phenolic compounds in plums. Identification was accomplished by comparing their retention times and UV absorption spectra with those of authentic standards.

Phenolic compounds extracted from plums by ethyl acetate were separated into acidic and neutral fractions using a Sep-Pak C18 cartridge according to the method of Lee and Jaworski (1). The acidic fraction contained mainly three isomers of caffeoylquinic acids (structures shown in Figure 1): 3-caffeoylquinic acid (neo-chlorogenic acid) (2.7 mg/g dry matter), 4-caffeoylquinic acid (cryptochlorogenic acid) (1.9 mg/g dry matter) and 5-caffeoylquinic acid (chlorogenic acid) (1.9 mg/g dry matter) as shown in Table I. The high concentration of caffeoylquinic acids in plum flesh is in accord with the results reported for other fruits of the same genus, prunus (2-3).

The neutral fraction contained mainly (-)-epicatechin (0.53 mg/g dry matter) and (+)-catechin (0.4 mg/g dry matter), at about the same level as other fruits of the same genus.

Table I.   Phenolic Compounds in Plum Flesh

Compound	Concentration (mg/g dry matter)
Acidic Fraction	
3-caffeoylquinic acid	2.7
4-caffeoylquinic acid	1.9
5-caffeoylquinic acid	1.9
Neutral Fraction	
(+)-catechin	0.45
(-)-epicatechin	0.53
procyanidins	
B-1	0.06
B-2	0.09
B-3	0.08
B-4	0.05

Changes of Phenolic Compounds during Sun-Drying Processing

During the processing of semi-dried pickled plums, significant color changes were observed. The original green color of plum flesh changed to a yellowish color after curing, to a light reddish color during the early stages of the sun-drying, and to a red-brown color after sun-drying. Interestingly, the red-brown color on the skin of the product could only be formed through the full sun-drying process. For oven-dried products, a black-brown instead of a red-brown color developed on the skin.

According to data previously reported by us (4), it was found that the content of both total chlorophyll and carotenoids decreased gradually through the whole process. Total chlorophyll was reduced from 14.0 mg/g dry matter in the flesh to 1.04 mg/g dry matter after being kept in brine for 4 weeks, and dropped to a negligible amount at the end of the two-day sun-drying period. A similar trend was observed for the total carotenoids. They decreased from 2.64 mg/g dry matter in the flesh plums to 1.05 mg/g dry matter after being kept in brine, and to 0.34 mg/dry matter after being sun-dried.

Quantitative changes in the major phenolic compounds at different stages of the processing of semi-dried pickled plum are shown in Figures 2 and 3. HPLC was used to monitor the degradation of phenolic compounds. Among the neutral phenolics in plums (-)-epicatechin degraded most drastically, followed by (+)-catechin. It was estimated that 47% of (-)-epicatechin and 78% of (+)-catechin remained in the cured sample, and only 1.2% and 15.6%, respectively, were left in the sun-dried sample. It was noticeable that (-)-epicatechin degraded almost completely during the curing and sun-drying processes. Decrease of (-)-epicatechin in curing may be due to the high specific activity of polyphenol oxidase for this phenolic as compared with that of (+)-catechin (4). For acidic phenolics, there were 44% of 3-caffeoylquinic acid, 37% of 4-caffeoylquinic acid and 35% of 5-caffeoylquinic acid remaining in cured plums, and only 15%, 16% and 10%, respectively, left in the sun-dried plums.

All three isomers of caffeoylquinic acid degraded at similar rates. This probably indicated that the oxidation of o-dihydroxy groups of the caffeoylquinic acids was the main mechanism for their degradation.

Properties of Red-Brown Pigment

The formation of a red-brown pigment was the most noticeable change occurring during the processing of sun-dried pickled plums. The red-brown pigments were first extracted with cold acetone from the fully sun-dried plums. Addition of one and a half volumes of chloroform to the aqueous filtrate resulted in a biphasic separation. The aqueous phase was concentrated and then subjected to gel-filtration on a Sephadex G-25 column according to the method of Somers (5). Two main fractions were obtained which, from their absorption at 420 nm, indicated separation between polymeric proanthocyanin (Fraction I; elution volume 10-40 mL) and oligomeric flavan (Fraction II; elution volume 40-100 mL) as shown in Figure 4.

Fraction I was a polymeric polyphenol with a minimum molecular weight of 2,000 estimated for complete exclusion from this particular gel. This fraction was immobilized on a cellulose TLC plate and developed in a butanol-acetic acid-water (BAW; 4:1:1) mobil phase. A tan-colored band at the origin was extracted with hot water, and UV absorption maximum at 280 nm was read after suitable dilution in the same solvent (Figure 5). In the treatment with butanol-hydrochloric acid (100:5) at 100°C, Fraction I gave mainly cyanindin. Cyanidin was liberated and characterized by HPLC.

The color of Fraction II was yellow-brown. HPLC analysis showed that the second fraction was a mixture of oxidation products derived from plum polyphenols. At least 5 peaks were observed in the HPLC profile. The UV absorption spectrum of one of the major peaks was very similar to that of the caffeoylquinic acid oxidation product (6). Both compounds showed a maximum absorbance near 320 nm.

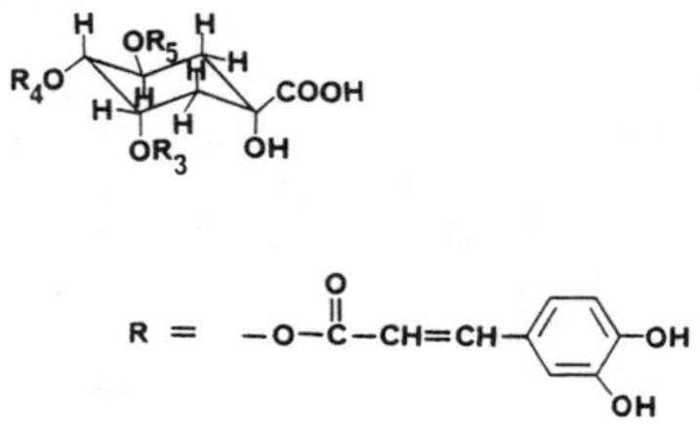

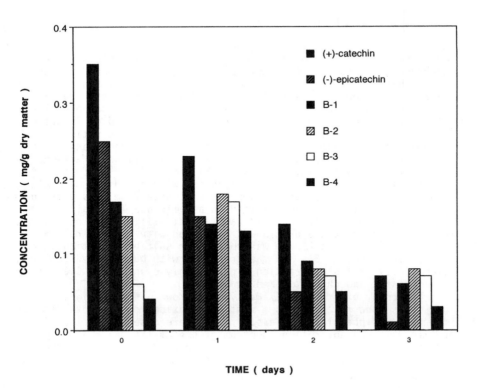

3-Caffeoylquinic acid (Neochlorogenic acid):
R₃ = R;   R₄ = R₅ = H

4-Caffeoylquinic acid (Cryptochlorogenic acid):
R₄ = R;   R₃ = R₅ = H

5-Caffeoylquinic acid (Chlorogenic acid):
R₅ = R;   R₃ = R₄ = H

Figure 1.   Structures of caffeoylquinic acids.

Figure 2.   Evolution of neutral phenolics in plum during sun-drying.

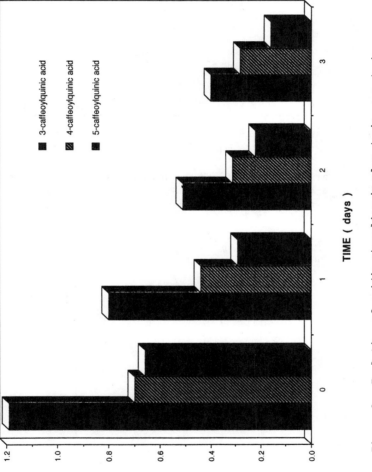

Figure 3. Evolution of acidic phenolics in plum during sun-drying.

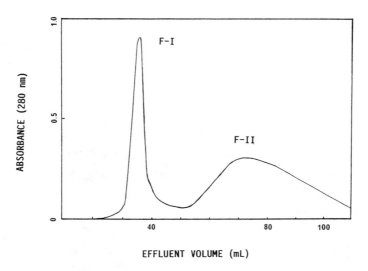

Figure 4. Gel filtration of red-brown pigments from sun-dried plum on Sephadex G-25.

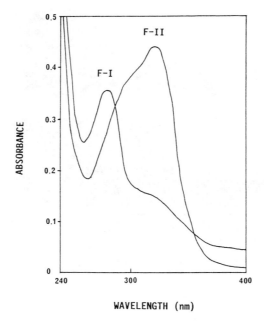

Figure 5. UV absorption spectra of red-brown pigments extracted from sun-dried plum.

Formation of Red-Brown Pigment

A mixture of (+)-catechin and procyanidins (20:1, by weight) purified from plum flesh by preparative cellulose TLC was set up as a liquid model system to study the formation mechanism of the pigment (7). A red-brown pigment was isolated and purified after exposure to radiation from a tungsten lamp at 40°C. The sample exposed to sunlight produced a similar pigment. Both absorption spectrum of the main fraction, and the effluent volume of the red-brown pigment agreed well with that found in the sun-dried plums as shown in Figure 5.

A model system composed of (-)-epicatechin was set up to confirm the occurrence of photopolymerization during the sun-drying process. The rate of red-brown pigment formation was found to be directly related to the reaction temperature. Absorbance at 420 nm increased as the heating temperature increased. The browning reaction of (-)-epicatechin was also found to be pH-dependent. A minimum pH value of 5 at 45°C is required for the red-brown pigment formation as shown in Figure 6. No significant browning reaction was observed below pH 4 after heating for 6 hours.

Addition of procyanodin to the (-)-epicatechin solution increased the rate of the browning (Figure 7). A photo-sensitive constituent, riboflavin, doubled the rate of the browning reaction, and pheophytin further enhanced the reaction. The red-brown pigment formation was monitored by Sephadex G-25 column chromatography. The elution patterns of (-)-epicatechin solution samples heated for 2.4 and 6 hours, respectively, are shown in Figure 8. A molecular weight of about 2000, and a UV absorption maximum at 280 nm were characterized for the first fraction.

Cellulose powder coated with (+)-catechin and leucoanthocyanin was used as the solid model system. A red-brown pigment was produced in the samples exposed to radiation from a tungsten lamp as well as the sample exposed to sunlight.

It was concluded that (+)-catechin, (-)-epicatechin and procyanidins are the major constituents that participate in the formation of the red-brown pigment. Radiation in the visible portion of the spectrum appears to catalyze the photopolymerization of the red-brown pigment. The red-brown pigments, the flavalans, may be formed by linkages between C-4 of one (+)-catechin or (-)-epicatechin, and C-8 of another unit through the oxidative or photosensitive polymerization reaction postulated by Creasy and Swain (8). Leucocyanidin may also condense with (+)-catechin or (-)-epicatechin to generate procyanidins, B1-B4 and, consequently, to the polymeric red-brown pigment following the mechanism analogous to that proposed by Delcour et al. (9). Cell destruction may take place during the sun-drying process. Overnight tempering treatment (plums in crates covered with plastic film) enhances the diffusion of plasmic polyphenolic compounds. Accumulation of these phenolic acids on the epidermis cell leads to the pronounced oxidation reaction of o-diphenols.

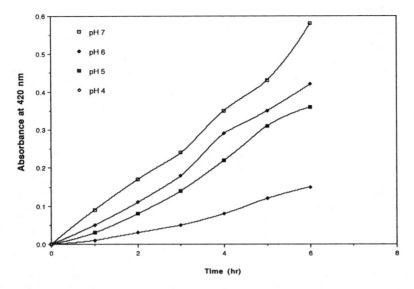

Figure 6.   Effects of pH values on the browning reaction in (-)-epicatechin model system.

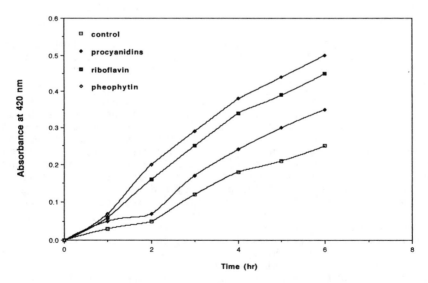

Figure 7.   Effects of photosensitizers on the browning reaction in (-)-epicatechin model system.

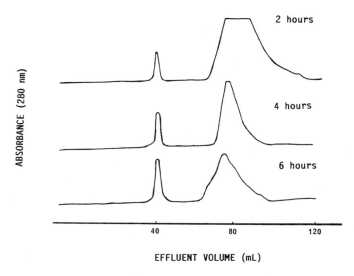

Figure 8. Sephadex G-25 gel-filtration curve of red-brown pigments from heated (+)-catechin model system.

Acknowledgments

This publication, New Jersey Agricultural Experiment Station No. D-10205-2-92, has been supported by State Funds and Hatch Regional Project NE-116. We thank Mrs. Joan B. Shumsky for her secretarial aid.

Literature Cited

1. Lee, C. Y.; Jaworski, A. W. *Amer. J. Enol. Vitic.* **1987**, *38*, 277-281.
2. Raynal, J.; Moutounet, M.; Souguet, J.-M. *J. Agric. Food Chem.* **1989**, *37*, 1046-1050.
3. Herrmann, K. *CRC Crit. Rev. Food Sci. Nutr.* **1989**, *28*, 315-347.
4. Huang, T.-C.; Li, C.-F.; Hwang, L.-S.; Chang, W.-H.; Lee, T-C. Bull. *Nat'l. Pingtung Inst. Agric.* **1982**, *23*, 63-66.
5. Somers, T. C. *Nature,* **1966**, *209*, 368-370.
6. Oszmianski, J.; Lee, C. Y. *J. Agric. Food Chem.* **1990**, *38*, 1202-1204.
7. Huang, T.-C. In *The Role of Chemistry in the Quality of Processed Foods,* Fennema, O.; Chang, W.-H.; Lii, C., Eds.; AVI Publishing Co.: Westport, CT, **1986**, pp. 108-117.
8. Creasy, L. L.; Swain, T. *Nature* **1965**, *208*, 151-153.
9. Delcour, J. A.; Ferreira, D.; Roux,D. G. *J. Chem. Soc. Perkin Trans. I.* **1983**, 1711-1717.

RECEIVED January 13, 1992

BIOCHEMICAL PROPERTIES OF PHENOLIC COMPOUNDS

Chapter 23

Biochemical Mechanisms of the Antinutritional Effects of Tannins

Larry G. Butler[1] and John C. Rogler[2]

Departments of [1]Biochemistry and [2]Animal Science, Purdue University, West Lafayette, IN 47907-1153

In livestock diets, tannins may diminish weight gains, apparent digestibility and feed utilization efficiency. These antinutritional effects have generally been attributed to inhibition by tannins of digestion of dietary proteins. Other effects associated with dietary tannin are systemic, requiring absorption of inhibitory material from the digestive tract into the body. We have been investigating this problem using ^{15}N-labelled protein, ^{14}C-labelled tannins, and drugs such as propranolol which inhibit the secretion of specialized proline-rich tannin-binding salivary proteins that otherwise diminish the antinutritional effects. Dietary tannins do somewhat inhibit the digestion of protein, but the evidence suggests that the protein affected is largely endogenous, not dietary, and that this inhibition accounts for only a small part of the antinutritional effects. Inhibition of digestion seems to be less significant than inhibition of post-digestive metabolism, a systemic effect. Low MW polyphenol components associated with tannins are more readily absorbed from chick diets than are tannins, and may account for the major antinutritional effects.

Tannins are phenol-rich oligomers of flavonoids (condensed tannins) or glucose esters of gallic or ellagic acids (hydrolyzable tannins) which strongly bind proteins. Tannins occur widely throughout the plant kingdom. Their distribution in plant tissues and their numerous functions have been summarized by Chalker-Scott and Krahmer (1). Not unexpectedly for compounds which serve as deterrents to herbivory, consumption of tannins negatively impacts herbivore nutrition.

Structural distinctions between and within condensed and hydrolyzable tannins are being recognized as responsible for somewhat different nutritional effects in the diets of herbivores (2, Mole, S.; Rogler, J. C.; Butler, L. G., in review). The wide range of antinutritional effects reported for condensed

0097–6156/92/0506–0298$06.00/0

© 1992 American Chemical Society

tannins, as well as metabolic defenses of herbivores against these effects, were reviewed by Butler (3). A more recent review addresses nutritional consequences of both condensed and hydrolyzable tannins (4). We summarize here recent advances in our understanding of the mechanism of antinutritional effects of condensed and hydrolyzable tannins.

Similarities and differences between the effects of hydrolyzable and condensed tannins. Relatively few nutritional studies have been done with purified tannins added to tannin-free basal diets. Purification of tannins in sufficient quantities for definitive feeding trials is a formidable undertaking. Most nutritional studies have been done with crude extracts such as quebracho or with intact plant material containing tannin *in situ*. It is important to keep in mind that the observed antinutritional effects observed in these studies are not necessarily due to tannin, but may be due to non-tannin materials which occur exclusively with tannin (5).

Hydrolyzable tannins such as tannic acid have been more readily available in purified forms suitable for feeding trials than have condensed tannins. Unlike condensed tannins, hydrolyzable tannins are subject to enzymatic hydrolysis in the digestive tract. The resulting products include gallic acid, which is readily absorbed and excreted in the urine (6). Jung and Fahey (7) reported that dietary gallate depresses palatability and growth rate in chicks. These effects are presumably due to absorbed gallate and therefore may be considered "systemic" effects (see below).

Protein digestibility by deer and sheep was reduced by feeding plants containing either condensed or hydrolyzable tannins or by supplementation of the diet with condensed (quebracho) tannin (Hagerman, et al, *J. Range Management*, in press). Supplementation at similar levels with a hydrolyzable gallotannin (commercial tannic acid) had no effect on protein digestibility by these ruminants. No tannin (as gallate) was found in feces after feeding tannic acid, suggesting that this hydrolyzable tannin was degraded in the gut and the resulting gallic acid was taken up and excreted in the urine. When fireweed containing gallotannin was fed, 27% of the ingested gallic acid was found in the feces, indicating that the gallotannin was somehow protected against hydrolysis in the gut, presumably by complexation with protein. On the basis of comparison of the molecular weight distributions of the commercial and fireweed gallotannins, Hagerman et al suggested that differences in molecular weight distribution and consequent affinity for proteins were responsible for the differences in the nutritional effects of the two gallotannins.

Biochemical basis for antinutritional effects: Despite their large differences in structure, condensed and hydrolyzable tannins often produce rather similar antinutritional effects. The major effects are diminished weight gains and lower efficiency of utilization of nutrients, particulary protein. There is usually an increase in the level of fecal nitrogen. Food consumption is sometimes, but not always, diminished by dietary tannin, with hydrolyzable tannins usually more effective than condensed tannins. The effects of tannins on food consumption are even less significant when consumption is calculated on a per-weight basis rather than a per-animal basis (8). Tannin-consuming animals are usually smaller than their counterparts, and thus consume less food.

It is often assumed that the similar effects of condensed and hydrolyzable tannins on herbivore nutrition are due to the well-recognized shared capacity of these phenol-rich but otherwise dissimilar materials to bind and coagulate proteins (9,10). This characteristic astringency and associated enzyme inhibition has been proposed to account for various *in vivo* biological activities, including antinutritional properties, of both condensed and hydrolyzable tannins (11).

Tannins differ greatly in their affinity for proteins, and proteins likewise differ greatly in their affinity for various tannins (10,12). But both condensed and hydrolyzable tannins inhibit most *in vitro* enzyme assays, probably because in many of these assays the enzyme is the only material present which is capable of binding the tannin. Because the digestive enzymes are the first, and perhaps only, enzymes to be exposed to dietary tannins when food is consumed, the site of the antinutritional effects of tannin is often assumed to be the digestive tract (13).

The antinutritional effects of tannins have almost always been interpreted in terms of inhibition by tannin of the digestion of dietary protein (13-15). The diminished weight gains of rats and chicks can usually be overcome by an increase in the level of protein in the diet (16). Supplementation with the equivalent amount of free amino acids does not overcome the diminished weight gains due to tannin (16), so inhibition of digestion cannot be the major problem. Inhibition of enzymes *in vitro* is an inadequate model of the *in vivo* effects of tannin, because of the presence in the digestive tract of many other proteins and tannin-binding agents (17). There is a growing body of evidence, some of which is presented below, that this perception of tannins primarily as inhibitors of digestion is overly simplistic and inadequate.

Digestion of dietary vs endogenous (defensive) protein. There was early evidence, partially based on rats fed ^{14}C-casein, that the increased fecal protein induced by dietary tannin is not of dietary origin, but instead is endogenous protein from the lining and secretions of the digestive tract (18,19). The endogenous origin of the tannin-induced increase in fecal protein has now been confirmed by feeding ^{15}N-labelled proteins or [14]C- or ^{15}N-amino acids (to label endogenous proteins) along with tannins from tea (20), beans (21,22) or quebracho and tannic acid (23).

The probable origin of the increased fecal protein induced by dietary tannin in many but not all (24,25) tannin-consuming mammals is proline-rich salivary proteins which have an unusually high affinity for tannin (26). These specialized tannin-binding proteins are virtually absent from the saliva of rats and mice until induced by dietary tannin (27), and both condensed or hydrolyzable tannin are effective inducers (Mehansho, et al., *J. Agric. Food Chem.*, in press). In other animals these salivary proteins appear to be constitutive. In pigs, for example, proline-rich proteins are a major component of the ileal digesta of animals fed diets free of both tannin and protein (28). Proline-rich proteins comprise about 70% of the total proteins in human saliva (26).

The effectiveness of proline-rich salivary proteins as a defense against dietary tannin was tested by feeding tannin-containing diets to rats, with and without 0.05% propranolol, a beta adrenergic antagonist which blocks formation of rat salivary proline-rich proteins (29). Diets containing 2% or more tannin produced lower weight gains and feed utilization efficiencies than corresponding low tannin diets (29). Consumption rates were lower for animals eating the high tannin diets, but these animals were smaller; on an equal rat weight basis the rats consumed more high tannin diet than low tannin diet. Propranolol in tannin-free diets did not have a significant effect on these parameters after a brief (3-5 day) adaptation period. In contrast, rats on high tannin diets with propranolol generally lost weight instead of the modest weight gains made by rats on the corresponding diet with the same amount of tannin but no propranolol. This exacerbation of the effects of dietary tannin by propranolol is presumably due to the rats' loss of their capacity to protect themselves by making proline-rich salivary proteins (29). The results suggest that salivary proline-rich proteins constitute an important defensive system against dietary tannins. Tannin-containing diets fed to hamsters, which are unable to make salivary proline-rich proteins in response to dietary tannin, produced a high mortality rate in a previous study (30). Without the defensive system of proline-rich tannin-binding salivary proteins, the antinutritional effects of dietary tannin would be much more severe.

The strong complex formed between these endogenous proline-rich proteins and dietary tannins is not dissociated by altering the pH (E. Haslam, personal communication), and seems to pass through the digestive tract relatively intact as judged by the high proline content of the fecal protein (13,29,31). Salivary proline-rich proteins typically contain relatively low amounts of essential amino acids (27) so their loss in the feces as a complex with tannin is a useful trade-off if it spares dietary proteins richer in essential amino acids (24).

Digestion vs postdigestive metabolism (systemic effects). Direct absorption of hydrogen-bonding, nondialyzable, protein-precipitating tannin macromolecules seems quite unlikely in the normally functioning animal (11). Tannin macromolecules may occur together with lower MW precursors and structurally related materials more likely to be absorbed from the digestive tract. Purification of tannins requires removal of considerable amounts of non-tannin phenolic materials (32,33).

Several of the effects usually associated with dietary tannin occur in bodily tissue (systemic effects), not just in the intestinal tract. These effects include development of chick leg abnormalities (34), induction of liver enzymes and increased output of urinary glucuronides (35), hypertrophy of the parotid glands (27), diminished urine volume (29), and mortality too rapid to be the result of impaired digestion (30). These effects were observed on feeding "high tannin" sorghums, which contain condensed tannins and associated polyphenolic materials. There were no discernable lesions which could account for apparent uptake of tannins from the digestive tract (36).

Systemic effects, rather than inhibition of digestion, are also apparently mainly responsibile for the diminished growth rate associated with dietary tannin (19). We have examined this problem utilizing a method which allows us to distinguish between direct effects of tannin on digestion and/or absorption, and effects on post-absorptive metabolism (systemic effects) (29). The growth analysis protocol, which was developed for use with insects by Waldbauer (37), expresses the efficiency with which experimental animals (rats in our case) convert ingested food into body substance (ECI) as the product of the approximate digestibility of the diet (AD) and the efficiency with which digested and absorbed food is converted to new body substance (ECD). Rigorous evaluation of AD and ECD in a feeding trial requires that all the component measurements are expressed in the same units; our results are expressed both on a dry weight and a nitrogen basis so that we can compare the effects of tannin on both dry matter and protein digestion and metabolism.

Comparison of the effects of dietary tannin on AD and ECD gave a somewhat unexpected result (29). On a dry weight basis, the approximate digestibility (AD) of low tannin diets was about 88%, and this was only diminished to about 78% by the presence of tannin in the diet. The major effect of tannin was on ECD, the efficiency of conversion to new biomass of food which had already been digested and absorbed. For low tannin diets ECD was about 7%, and this was diminished to 0.7% by the presence of tannin in the diet.

On a nitrogen basis the results were similar but somewhat exaggerated (29). AD(N) was about 75% in the absence of tannin, and was independent of propranolol. Tannin diminished AD(N) to about 45% in the absence of propranolol, and to about 25% in the presence of propranolol. As expected, tannin inhibited digestion of protein to a greater degree than it inhibited the digestion of dry matter. But again, the major effect of tannin was inhibition of ECD(N). In the absence of tannin and propranolol, ECD(N) was about 30%, and this was diminished to virtually zero in the presence of tannin. Propranolol in the absence of tannin caused an increase of ECD(N) to about 48%, and in the presence of tannin and propranolol ECD(N) was strongly negative due to losses of body protein on this diet.

These results show that in the monogastric system studied, the major effect of dietary condensed tannins is not to inhibit food consumption or digestion. Instead, the major effect is on the efficiency with which digested and absorbed nutrients are converted to new body substance. This could be due to a direct inhibitory effect on a key metabolic pathway or an indirect effect due to diversion of metabolism into detoxification of polyphenols and/or their degradation products. In either case, the site affected is not the intestine but is within the body, which requires absorption of the components responsible for these (systemic) effects. Alternatively, the lowered consumption rate and protein digestibility could result in nitrogen starvation, inefficient utilization of non-nitrogenous nutrients, and low ECD values. This mechanism would not require uptake of tannins or associated materials from the intestine. Measurement of tannin and phenol recoveries in the feces do not resolve this problem; recoveries are low but are similarly low when the diet is merely moistened but not fed. Moist conditions favor irreversible complexation of tannin with proteins and possibly other materials so that the tannin can no longer be extracted and assayed.

It is clear that in the case of "high tannin" sorghum, and perhaps for other sources of dietary tannin, toxic materials do get absorbed from the digestive tract. We recently found, using ^{14}C-labeled tannin and lower MW polyphenols purified separately from sorghum grain, that the polymeric condensed tannins were not absorbed from the digestive tract of chicks, but were completely recovered in the feces and intestinal contents. On the other hand, lower MW polyphenols associated with tannin in the sorghum seed were taken up rather efficiently and recovered in serum as well as other tissues (Jimenez-Ramsey, L. M., et al., *Proc. 2nd North American Tannin Conf.*, in press). Like the tannin polymers, these associated and apparently toxic materials are not present in tannin-free sorghums often used as controls in nutritional studies of sorghum tannins.

It seems likely, at least for high tannin sorghum, that the antinutritional effects may not be due to the tannins, but to lower MW, more readily absorbed polyphenols, possibly biosynthetic precursors of tannin.   The less highly polymerized sorghum tannins were reported to more effectively inhibit N utilization than the highly polymerized tannins from this grain (19).   The tentative designation of major antinutritional effects to absorbable tannin-associated materials rather than to the tannins themselves may explain not only the systemic effects mentioned above, but also other observations not necessarily nutritional in nature.   Low MW polyphenols (but not monomeric units such as catechin) from high tannin sorghums are more effective than purified sorghum tannin polymers at inducing production of salivary proline-rich proteins in rats (Mehansho, et al., *J. Agric. Food Chem.*, in press) and at repelling birds (38).

Identification and characterization of the absorbable toxins associated with condensed tannins is underway in this laboratory.

Acknowledgements

Tannin research in this laboratory is partially supported by USAID Grant No. DAN 1254-G-00-0021 through INTSORMIL, the International Sorghum and Millet CRSP and by a Title XII Program Support Grant from USAID.

Literature Cited

1.   Chalker-Scott, L.; and Krahmer, R. L. In *Chemistry and Significance of Condensed Tannins*; Hemingway, R. W.; Karchesy, J. J.; Branham, S. J., Eds.; Plenum Press, New York, 1989; pp 345-368.
2.   Clausen, T. P.; Provenza, F. D.; Burritt, E. A.; Reichardt, P. B.; Bryant, J. P. *J. Chem. Ecology* 1990, *16*, pp 2381-2392.
3.   Butler, L. G. In *Chemistry and Significance of Condensed Tannins*; Hemingway, R. W.; Karchesy, J. J.; Branham, S. J., Eds.; Plenum Press, New York, 1989; pp 391-402.
4.   Salunkhe, D. K.; Chavan, J. K.; Kadam, S. S. *Dietary Tannins: Consequences and Remedies*, CRC Press, Inc., Boca Raton, FL, 1990.
5.   Butler, L. G. In *Toxicants of Plant Origin*; Cheeke, P. R., Ed.; Phenolics; CRC Press, Boca Raton, FL 1989, Vol. IV, pp 95-121.
6.   Booth, A. N.; Masri, M. S.; Robbins, D. J.; Emerson, O. H.; Jones, F. T.; De Eds, F. *J. Biol. Chem.* 1959, *234*, pp 3014-3016.
7.   Jung, H.G.; and Fahey, G.C. *J. Animal Sci.* 1983, *57*, pp 206-219.
8.   Mole, S.; Waterman, P. G. In *Allelochemicals in Agriculture, Forestry, and Ecology*, Waller, G. R., Ed.; Am. Chem. Soc. Symp. Series, Washington, D.C. 1987,; pp 582-587.
9.   Haslam, E.; and Lilley, T. H. *CRC Critical Reviews in Food Science and Nutrition* 1988, *27*, pp 1-40.
10.   Hagerman, A. E. In *Chemistry and Significance of Condensed Tannins*; Hemingway, R. W.; Karchesy, J. J.; Branham, S. J., Eds.; Plenum Press, New York, 1989; pp 323-333.
11.   Singleton, V. L.; Kratzer, F. H. *J. Agric. Food Chem.* 1969, *17*, pp 497-512.
12.   Asquith, T. N.; Butler, L. G. *Phytochem.* 1986, *25*, 1591-1593.
13.   Mitaru, B. N.; Reichert, R. D.; Blair, R. *J. Nutrition* 1984, *114*, pp 1787-1796.

14. Singleton, V. L.; Kratzer, F. H. In *Toxicants Occurring Naturally in Foods*, 2nd ed., National Acad. Sci., New York, 1973, pp 327-345.

15. Deshpande, S.S.; Sathe, S.K.; and Salunkhe, D.K. In *Nutritional and Toxicological Aspects of Food Safety*; Freidman, M., Ed.; Plenum Press, New York, 1984; pp 457-495.

16. Rogler, J. C.; Ganduglia, H. R. R.; Elkin, R. G. *Nutrition Research* **1985**, 5, pp 1143-1151.

17. Blytt, H. J.; Guscar, T. K.; Butler, L. G. *J. Chem. Ecol.* **1988**, *14*, 1455-1465.

18. Glick, Z.; Joslyn, M. A. *J. Nutrition* **1970**, *100*, pp 516-520.

19. Martin-Tanguy, J.; Vermorel, M.; Lenoble, M.; Martin, C. *Ann. Biol. Anim. Bioch. Biophys.* **1976**, *16*, pp 879-890.

20. Shahkhalili, Y., Finot, P. A., Hurrell, R. and Fern, E. *J. Nutrition* **1990**, 120, pp 346-352.

21. Costa de Oliviera, A.; and Sgarbieri, V. C. *J. Nutrition* **1986**, *116*, pp 2387-2392.

22. Marques, U. M. L.; and Lajolo, F. M. *J. Agric. Food Chem.* **1991**, *39*, pp 1211-1215.

23. Mole, S.; Rogler, J. C.; Butler, L. G. *Use of ^{15}N-Labeled Protein to Determine the Effect of Dietary Tannin on the Relative Abundance of Endogenous and Dietary Protein in Feces*, Proc. Groupe Polyphenols XVth International Conference, Strasbourg, July, 1990, pp 121-123.

24. Mole, S.; Butler, L. G.; and Iason, G. *Biochem. System. Ecology* **1990**, 18, pp 287-293.

25. Austin, P. J.; Suchar, L. A.; Robbins, C. T.; Hagerman, A. E. *J. Chem. Ecology* **1989**, *15*; pp 1335-1347.

26. Mehansho, H.; Butler, L. G.; Carlson, D. M. *Annual Review of Nutrition* **1987**, 7, pp 423-440.

27. Mehansho, H.; Hagerman, A. E.; Clements, S.; Butler, L. G.; Rogler, J. C.; Carlson, D. M. *Proc. Natl. Acad. Sci. USA* **1983**, *80*, pp 3948-3952.

28. Zebrowska, T. Les Colloques de l'INRA, #12, pp 225, 1982, presented in Friedman, M., Absorption and Utilization of Amino Acids, Vol III, CRC Press, Boca Raton, FL, 1989, pp 209.

29. Mole, S.; Rogler, J. C.; Morell, C.; Butler, L. G. *Biochem. System. & Ecology* **1990**, 18, pp 183-197.

30. Mehansho, H.; Ann, D. K.; Butler, L. G.; Rogler, J. C.; Carlson, D. M. *J. Biol. Chem.* **1987**, *262*, pp 12344-12350.

31. Eggum, Bj. O.; Christensen, K. D. In *Breeding for Seed Protein Improvement Using Nuclear Techniques*, Intern. Atomic Energy Agency, Vienna, 1975, pp 135-143.

32. Hagerman, A. E.; Butler, L. G. *J. Agric. Food Chem.* **1980**, *28*, pp 947-952.

33. Asquith, T. N.; Butler, L. G. *J. Chem. Ecol.* **1985**, *11*, pp 1535-1544.

34. Elkin, R. G.; Featherton, W. R.; Rogler, J. C. *Poultry Sci.* **1978**, *57*, pp 757-762.

35. Sell, D. R.; Rogler, J. C. *Proc. Soc. Exper. Biol. Med.* **1983** *174*, pp 93-101.

36. Sell, D. R.; Reed, W. M.; Chrisman, C. L.; Rogler, J. C. *Nutr. Reports Intl.* **1985**, *31,* pp 1369-1374.

37. Waldbauer, G. P. *Rec. Adv. Insect Physiol.*. **1964**, *5,* pp 229-241.

38. Bullard, R. W.; Garrison, M. V.; Kilburn, S. R.; York, J.O. *J. Agric. Food Chem.* **1980**, *28,* pp 1006-1011.

RECEIVED August 13, 1992

Chapter 24

Enzymatic Oxidation of Phenolic Compounds in Fruits

Chang Y. Lee

Department of Food Science and Technology, Cornell University,
Geneva, NY 14456

Phenolic compounds are closely associated with the sensory and nutritional quality of fresh and processed plant foods. The enzymatic browning reaction of phenolic compounds, catalyzed by polyphenoloxidase, is of vital importance to fruit processing due to the formation of undesirable color and flavor and the loss of nutrients. This reaction in fruits has often been considered to be a linear function of the phenolic content and polyphenoloxidase activity. However, it was shown that individual phenolic compounds have a wide range of browning with the monomeric catechins and the dimeric procyanidins browning more intensely than other phenolics such as the acidic compounds. In model systems with polyphenol oxidase, phloretin glycosides produced the dark-brown colored reaction products, while quercetin glycosides did not produce brown- colored reaction products. Chlorogenic acid, a common phenolic constituent in most fruits, yielded a color intensity one-third that of the procyanidins. However, mixtures of chlorogenic acid and other phenolic compounds, such as catechin and phloridzin, showed some synergistic effects in that the reaction was accelerated and a dark brown color was produced. It appears that the rate of browning in fruit products is not a linear function of the total phenolic content and that browning of fruits depends on concentration and nature of polyphenol compounds that are co-present.

Polyphenol compounds present in fruits are directly related to sensory qualities such as color, astringency, bitterness, and aroma; indirectly, they participate in the oxidation reactions that occur during postharvest handling and processing of fruits. To food scientists, the major importance of polyphenol compounds is their effects on the

0097–6156/92/0506–0305$06.00/0
© 1992 American Chemical Society

sensory quality of fresh and processed products, due to enzymatic oxidations. The change in color following mechanical or physiological injury of postharvest fruits is due mainly to oxidation of polyphenol compounds by polyphenol oxidase. In fruit processing, such color change is accentuated during preparation for storage, dehydration, freezing, and canning. Formation of a dark brown color causes the product to become unattractive and is accompanied by undesirable changes in flavor, as well as a reduction in the nutritive value of the products. Ultimately, there is a detrimental economical impact as a consequence of this enzymatic browning.

The occurrence of enzyme-catalyzed oxidative browning has long been recognized ever since Lindet (1) first reported in 1895 that the changes in color of fresh apple cider were due to oxidation of tannin by the oxidase present in apple tissue. Since then, numerous papers that relate to oxidizing enzymes and phenolic compounds involved in browning have been published.

In order to evaluate the enzymatic browning of fruits, most research work has focused on the qualitative and/or quantitative relationship between oxidative enzyme (mainly o-diphenyl oxidase, o-DPO) and its substrates, total phenol. Due to the chemical complexity of polyphenol compounds, isolation and identification of individual polyphenol compounds was once difficult. Therefore, the total polyphenol content was used conveniently in the enzymatic browning reactions. However, this traditional method of using total phenols in enzymatic browning has evolved during recent years into the individual polyphenol compound approach, since today, we are able to separate individual polyphenol compounds in fruits more easily than before; furthermore, a wide range of variations in color development among different individual polyphenol compounds is currently known.

The objectives of this study were to:

1. identify the major polyphenol compounds in apples, grapes, and peaches using HPLC,

2. isolate and obtain individual polyphenol compounds in appreciable quantity by using preparative HPLC,

3. study the browning characteristics of individual polyphenol compounds and in combination in model systems, and

4. make correlations between the polyphenols and the browning intensity.

Enzymatic browning in relation to polyphenol oxidase and its substrates.

In the commercial practice of fruit processing, selection of cultivars that display minimum browning has been an important task. Therefore, correlations are frequently sought between browning intensity and the main parameters (polyphenol oxidase and polyphenols) that are responsible. Enzyme activity is known to be an important factor in browning of grape juice and wine, but no direct correlation has been established (2 - 5). Occasionally, we find in the literature that total polyphenol compounds are more closely correlated with browning intensity than is the enzyme (6 - 12). However, the results are often contradictory and there is no simple explanation to the problem. Thus, in order to understand the nature of the problem, we made several attempts using many different cultivars of apples, grapes, and peaches.

We analyzed 15 peach cultivars for polyphenoloxidase (PPO) activity and polyphenol compounds according to the methods published previously(*13,14*). We identified 5 major compounds in peaches (*15*): chlorogenic acid, neochlorogenic acid, procyanidin B3, catechin and caffeic acid (Table I). A large difference in polyphenol content among different cultivars was observed. Chlorogenic acids were the major phenolic compounds in most of the peach cultivars studied, followed by procyanidin B3. PPO activity also varied greatly among different cultivars.

Table I. PPO Activity, Phenolics and the Degree of Browning in Various Peach Cultivars

Cultivar	PPO act., unit/g	Deg of browning (δL)	Phenolics (µg/g)				
			Chlrg	Neochlr	ProcB3	Catech	Caff
1. Deep Early Hale	2635	18.4	7	9	27	15	6
2. Eden	4505	36.3	68	50	41	36	t
3. Elberta	2798	31.8	59	59	85	35	12
4. Harmony	1793	10.2	14	15	30	25	10
5. Hayhaven	2146	27.0	30	18	41	29	10
6. Harken	2105	12.5	28	13	27	17	5
7. La Red	2132	9.1	NA	NA	NA	NA	NA
8. Madison	1923	21.2	38	25	9	6	12
9. Newhaven	1722	19.6	14	17	4	7	6
10. Pekin	2390	24.4	17	16	24	13	t
11. Redhaven	2299	15.9	41	33	28	20	t
12. Reliance	2361	24.8	74	54	18	11	t
13. Triogem	3340	23.1	42	24	40	16	10
14. Veecling	1449	22.8	23	20	11	7	11
15. Velvet	3196	26.3	25	21	27	12	6

NA: Not analyzed. t: Trace amount.   (Some data are from ref.15.)

When we compared the degree of actual browning of individual peach cultivars with its PPO activity, we found that peach cultivars having higher PPO activity had a higher rate of browning (e.g. Eden). Conversely, peaches low in PPO activity, such as Harmony, showed a lower rate of browning. Peach cultivars (e.g. Elberta, Eden) high in total polyphenol compounds showed a high degree of browning and those (e.g. Harken) low in total polyphenol compounds were low in the browning intensity. When we plotted browning against PPO and total polyphenol compounds, there appeared to be some close correlationship between PPO activity and browning (figure not shown), and between the degree of browning and total phenolics (Figure 1).

Enzymatic browning in relation to individual polyphenol compounds.

We carried out a study on grapes in the same manner as we did for peaches, but

we could not obtain any close correlation between browning and polyphenol compounds. We isolated and identified the major phenolic compounds in 15 cultivars of
white grapes *(16)*. We identified 12 phenolic compounds, of which *trans*-caffeoyl tartaric
acid ester was the major component of acidic polyphenols, while catechin, epicatechin,
catechin-gallate, catechin-catechin-gallate and procyanidin B3 were the major constituents of the neutral polyphenol compounds. When we plotted the degree of
browning with either enzyme activity or total polyphenol content, we found a very poor
correlation (r=0.3).

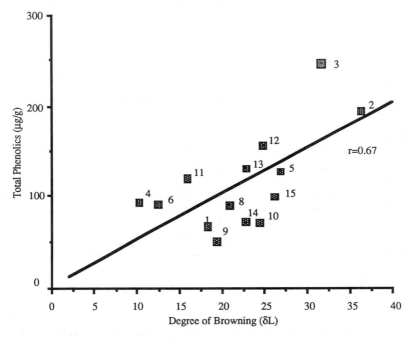

Figure 1. Relationship between actual browning and total polyphenols
among different peach cultivars (numbers refer to cultivars in Table I)
(Some data are from ref.15.)

Numerous attempts have been made by many scientists to correlate the
browning reactions of grape juices in relation to polyphenoloxidase *(4,11,14,)* and
polyphenol compounds *(2,11,17,18,19,20)*. However, there is no single report that
has demonstrated that any one of either enzyme or polyphenol compounds is directly
correlated to the oxidative enzymatic browning of grapes.

Therefore, an attempt was made to relate the relative browning rate of grapes in terms of individual polyphenol compounds using model systems, and to correlate it to the actual browning of grape juice. The major polyphenol compounds isolated from grapes were exposed to polyphenoloxidase individually in model systems and observed for their rate of browning (*21*). Relative browning measurements of individual polyphenol compounds revealed a wide range of browning rates (Figure 2). Catechin and epicatechin had the fastest initial rates of browning, reaching a maximum within 6 hrs of the reaction, while those of procyanidin B2 and B3 were very slow at the beginning but continously increased, reaching a maximum at 48 hrs. This result indicated that the measurement of browning should be made at least after 24 hrs reaction to derive an accurate result. Procyanidin B3 showed the highest absorbance among the phenolic compounds studied. Indeed, Cheynier et al. (*22,23*) also found that monomeric flavan-3-ols and dimeric proanthocyanidins are potent compounds in browning. Catechin-gallate and catechin-catechin-gallate showed a relatively slow rate of browning, with maximum absorbance at 24 hrs being about 25% of that exhibited by epicatechin. The browning of all acidic polyphenol compounds, including the major component of grapes, *t*-caffeoyl tartrate, was the lowest among phenolics, whose contribution to the browning reactions appeared to be a minor one.

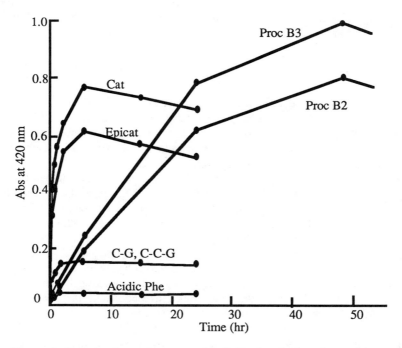

Figure 2. Relative browning rate of individual polyphenol compounds isolated from grapes. (Reproduced with permission from ref. 21. Copyright 1988.)

In order to establish the relationship between polyphenol compounds and the degree of browning, we identified a maximum browning value (the highest absorbance) for each polyphenol from Figure 2 as the browning index (BI). They were procyanidin B3, 0.96; procyanidin B2, 0.82; catechin, 0.72; epicatechin, 0.63; catechin-gallate (and catechin-catechin-gallate), 0.16; and acidic phenolics, 0.05. Based on these BI values and the content of each polyphenol compound in white grapes, the browning potential (BP) of a given grape cultivar was calculated by the following method:

$$BP = (BI_{p1} \times C_{p1}) + (BI_{p2} \times C_{p2}) + \cdots\cdots\cdots + (BI_{pn} \times C_{pn})$$
$$\text{where} \quad p_1, p_2, \cdots\cdots p_n : \text{individual polyphenol compound}$$
$$C: \text{concentration of phenolics in mg/100g.}$$

Accordingly the BP values of 15 selected white grape cultivars were calculated and compared with the observed degree of browning at 24 hrs. When the calculated BP values versus observed actual browning values were plotted, a relatively high correlation (r=0.85) was noted for 12 of 15 cultivars studied. Therefore, for those cultivars that fall into this high correlation, it may be possible to predict the browning potential in grape juice or wine by analyzing for individual polyphenol compounds in fresh grape. However, the other three cultivars, Aurore, Pinot Blanc, and Ravat 34 that did not fall into the majority group appeared to have other factors that influenced the browning reactions greatly.

Enzymatic browning in relation to polyphenol compounds interactions.

In apples, several major polyphenol compounds from the flesh and skin were identified. They were chlorogenic acid, epicatechin, procyanidin B3 and C1, and several quercetin and phloretin glycosides (Table II). The relative amount of polyphenol compounds in the skins was much higher than that in the flesh, and quercetin glycosides were found only in the skins.

When we plotted the degree of browning against polyphenol oxidase or total polyphenol compounds, we found a poor correlation. Therefore, the degree of browning for each polyphenol compound isolated from apples was studied individually in model systems. Catechins, procyanidins and phloretin glycosides resulted in a wide range of color development when individually subjected to polyphenol oxidase. Catechin and epicatechin turned brown rapidly, whereas procyanidins reacted slowly and needed at least 48 hrs to develop the same color intensity as achieved by catechins after 3 hrs. The browning rates of flavan polymers, B2 and C1 were much slower than monomers at the outset. However, during the prolonged reaction time, these polymers developed a full color that was equal to that of monomers in a short time. The browning rate of chlorogenic acid initially was very close to that of catechin and epicatechin, but reached a plateau very quickly within one hour; the color intensity was about one-third that of procyanidins. Chlorogenic acid alone seems to be a less significant factor in color development than flavans. The phloretin glycoside solutions turned brown

Table II. Concentration of Polyphenol Compounds in Flesh and Skins of Apples

Phenolic	R.I. Green.		G.Delicious		Cortland		Monroe		Empire	
	(mg/100g)									
	Skin	Flesh	S	F	S	F	S	F	S	F
Chlorogenic acid	9	15	12	8	8	5	18	9	8	12
Epicatechin	79	11	30	4	40	4	38	7	19	-
Procyanidin B2	100	11	23	4	43	7	18	7	14	2
Quercetin arabinoside	24	-	18	-	20	-	37	-	23	-
Quercetin xyloside	16	-	14	-	9	-	19	-	18	-
Quercetin glucoside	10	-	9	-	13	-	-	-	20	-
Quercetin galactoside	36	-	50	-	28	-	60	-	35	-
Quercetin rhamnoside	34	-	39	-	19	-	26	-	32	-
Phloretin glucoside	10	1	15	1	12	-	11	1	16	1
Phloretin xyloglucoside	27	3	16	2	12	2	19	3	7	1

slowly, but after 3 hrs their absorbances were intermediate between those of catechins and procyanidins, with phloretin xyloglucoside being more readily oxidized than phloretin glucoside. After 24 hrs of reaction, however, the absorbance values of the phloretin glycosides reached a value of 1.2, the highest of all compounds tested. This finding does not support the previous reports suggesting that phloretin glucoside was a poor *o*-DPO substrate (*24*). Phloretin xyloglucoside, whose contribution to browning was not previously reported, was more readily oxidized than phloretin glucoside. However, quercetin glycosides, which were the major polyphenol compounds in apple skins did not react with polyphenol oxidase and produced no brown color. In general, flavonols and their derivatives are abundant in fruits, but they are reported to be very poor substrates for *o*-DPO (*25, 26*).

Since quercetin glycosides did not brown, and also some flavonoids are reported to display an antioxidant activity (*27, 28*), it was our interest to study the role of apple quercetin and phloretin gycosides in the enzymatic browning using a model system. When five quercetin glycosides were each mixed with epicatechin and were exposed to polyphenoloxidase, the mixtures showed a minor change in color compared to those with epicatechin as the sole substrate. Quercetin galactoside apparently decreased the absorbance, whereas quercetin glucoside, xyloside, arabinoside, and rhamnoside showed a slight increase. However, a significant increase in absorbance was observed in mixtures of phloretin glycosides/epicatechin (Figure 3). Both phloretin glucoside and xyloglucoside enhanced the rate of browning drastically. Results showed that both the mixture of epicatechin/phloretin glucoside and epicatechin/ phloretin xyloglucoside after 3 hrs oxidation gave absorbance values of 1.25 and 1.21, repectively (*29*). These values were several-fold higher than those measured for other phenolics.

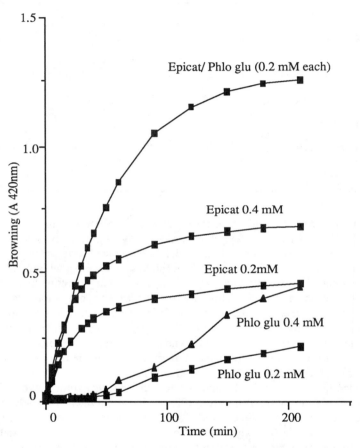

Figure 3. Rate of browning of single and mixture of epicatechin and phloretin glucoside.

A similar result was obtained when a chlorogenic acid and procyanidins mixture was used. The browning rate of a chlorogenic acid/phloretin glucoside mixture was several times higher than that of chlorogenic acid alone and was very close to the browning rate of the epicatechin/phloretin glucoside mixture. The addition of phloretin glucoside to procyanidins also resulted in a large increase in their browning rates. This synergistic effect of phloretin glycosides on the browning reactions cannot be explained. However, we believe that this synergistic effect is a very important factor in determining browning of apple products, since those glycosides are present in significant amounts in the skin and flesh (*30*).

Enzymatic browning in relation to the polyphenol compounds and the reaction products.

There have been recent reports that the enzymatically oxidized phenolic acid *o*-quinones oxidize other polyphenols, such as flavans, by a coupled oxidation reaction (*31*). The *o*-quinone formed by enzymatic or coupled oxidation can also react with a hydroquinone to yield a condensation product (*32*). Dimers or oligomers are reported to be generated by condensation of hydroquinone with quinone or by condensation of quinones of phenolic acids and catechin (*31*).

In order to study the sequence of enzymatic browning reactions among polyphenol compounds found in apples, chlorogenic acid and catechin were subjected to the enzyme in model systems and the reaction products of individual phenolic compound and the mixture of the two were monitored by HPLC.

We noticed more than 7 reaction products from chlorogenic acid after a 2 hr reaction, all of which showed a maximum absorbance at or near 320 nm, a value that is significantly different from 420 nm, which is the absorbance used for brown color measurement. Enzymatic oxidation of catechin also produced several compounds that had longer elution times than that of catechin. One which had a shorter elution time appeared to be procyanidin B3. On the basis of the oxidation reaction sequence proposed by Cheynier et al. (*31*) and Singleton (*32*), enzymatic oxidation of catechin may be able to produce the dimer, procyanidin B3. All reaction products showed a maximum absorbance near 420 nm and contributed more to the brown color than those produced by chlorogenic acid.

Enzymatic oxidation of the catechin-chlorogenic acid mixture produced very different reaction products (9 major peaks), five of which had shorter elution times than those of chlorogenic acid and catechin. All reaction products showed maximum absorptions near 280 and 320 nm. These new compounds may correspond to the copolymers of chlorogenic acid-catechin as proposed by Cheynier et al (*31*) for catechin-caffeoyl tartaric acid mixtures. This confirms the recent report of Cheynier et al (*33*), who found that caftaric acid-catechin copolymers were formed more readily than single caftaric acid oligomers in the oxidation of the caftaric acid-catechin mixture. The participation of catechin *o*-quinones in copolymerization with chlorogenic acid in our experiment appeared to slow down the formation of catechin polymers.

The kinetic study showed that the rate of catechin oxidation in the mixture was faster than that of catechin alone, while the rate of chlorogenic acid oxidation in the mixture was slower than that of chlorogenic acid alone. This might be due to the fact that chlorogenic acid quinones oxidize catechin by a coupled reaction mechanism (*31*), so that chlorogenic acid quinones are converted back to chlorogenic acid. Consequently, the degree of browning after a 2 hr reaction of the catechin-chlorogenic acid mixture was less than that of catechin alone.

In a similar study of phloretin glucoside oxidation in the presence of catechin and chlorogenic acid, we found that phloretin glucoside alone produced only two major reaction products which had shorter elution times than that of phloretin glucoside. However, the mixture of phloretin glucoside and chlorogenic acid produced several reaction products that had maximum absorption at 280 nm, which were probably the copolymerization products similar to those products of the catechin-chlorogenic acid mixture described above. When the same chromatogram of phloretin glucoside was monitored at 420 nm to detect the colored products, two additional peaks were observed. The size of these two peaks from phloretin glucoside alone was small compared to that of the mixture of phloretin glucoside-chlorogenic acid mixture. A similar observation was made in the mixture of phloretin glucoside-catechin. This indicated that chlorogenic acid and catechin have a synergistic effect by accelerating oxidation of phloretin glucoside and thus produce highly brown-colored products. We also noticed that increasing the amount of chlorogenic acid added accelerated the reaction; the concentrations of colored products were proportional to the concentration of chlorogenic acid added to phloretin glucoside. The reaction products that had maximum absorption at 280 nm are probably the hydroxylation products of phloretin glucoside and the other reaction products that had maximum absorption at 420 nm are the oxidation products of hydroxylated phloretin glucoside.

The hydroxylation of monophenols to *o*-diphenols and the oxidation of *o*-diphenols to *o*-quinones are commonly found with many polyphenol oxidase preparations. This hydroxylation can be explained by the lag period of the initial browning reaction of the phloretin glucoside oxidation measured at 420 nm. Table III shows the effects of catechin and chlorogenic acid on oxidation of phloretin glucoside expressed in the lag period. The lag period observed in oxidation of phloretin glucoside alone was 54 min. Addition of chlorogenic acid or catechin to phloretin glucoside decreased the lag period from 54 min to 6 and 2 min, respectively. After 2 hrs of oxidation, phloretin glucoside showed an absorbance (A420 nm) of 0.727, catechin, 0.833 and chlorogenic acid, 0.400. However, the mixture of phloretin glucoside and catechin (1:1 mM) showed an absorbance of 1.574, and the mixture of phloretin glucoside and chlorogenic acid (1:1 mM) was 1.568. Whitaker (*32*) explained that when a monophenol is added alone as a substrate of polyphenol oxidase, there is a delay in the reaction and this lag period is due to the need for an *o*-diphenolic compound. However, this lag period can be eliminated by addition of a small amount of an *o*-diphenol at the beginning of the reaction; in the absence of added *o*-diphenol, there is a lag period until the enzyme can build up a sufficient amount of the *o*-diphenol to permit the reaction to proceed.

This reaction mechanism explains the rapid change in the lag period of phloretin glucoside in the presence of *o*-diphenols. The synergistic effect between phloretin glucoside and chlorogenic acid or catechin confirms the earlier presentation

Table III. Effect of Catechin or Chlorogenic Acid Added to Phloretin
Glucoside on the Lag Period and Brown Color During Enzymatic Oxidation

Phloretin glucoside (mM)	Added Cat or Chlg (mM)	Catechin		Chlorogenic acid	
		Lag (min)	A420nm (after 2 h)	Lag (min)	A420nm (after 2 h)
1 mM	0	54	0.727	54	0.727
1 mM	0.01	23	1.389	27	1.020
1 mM	0.05	21	1.538	22	1.410
1 mM	0.10	19	1.555	20	1.514
1 mM	1.00	2	1.574	6	1.568

(Some data are from ref. 34.)

that showed a strong browning reaction in phloretin glucosides in the presence of chlorogenic acid or flavans. The oxidation rate constant of phloretin glucoside (207.6 x 10^{-4} mM/min) was higher than catechin (36.5 x 10^{-4} mM/min) and chlorogenic acid (188.5 x 10^{-4} mM/min). However, the mixture of phloretin glucoside and catechin or chlorogenic acid increased the oxidation rate proportionally to the increased concentration of the two polyphenols. This rapid reaction of the phenolic mixture containing monomeric and dimeric flavan-3-ol and gallic acid in the presence of grape o-DPO was also reported (*35*)

These findings clearly show that the rate of browning in apple products cannot be regarded as a linear function of the phenolic content, since the rate of browning is not an additive of particular o-DPO substrates. This may explain the difficulties in finding a correlation between total phenolic contents and the rate of browning of apple products. The synergistic effect of polyphenol compounds must be considered carefully in the study of fruit browning.

Conclusions

As we have seen here, the enzymatic browning in fruits is complex. The rate of enzymatic oxidation of polyphenols depends on the polyphenol oxidase activity as well as on the individual polyphenol compounds that show different degrees of browning: in general, the monomeric catechins and dimeric procyanidins brown more intensely than do other polyphenol compounds. Therefore, the contribution of a given polyphenol compound to browning not only depends on its concentration but also on the nature of it and other polyphenol compounds co-present in the tissue. It is also clear that some polyphenol compounds present in fruits and their quinones formed after oxidation may have a synergic or antagonistic effect.

Literature Cited

1. Lindet, M. *Compt. rend.* **1895**, *120*, 370.
2. Romeyer, F. M.; Sapis, J. C.; Macheix, J. J. *J. Sci. Food Agric.* **1985**, *36*, 728.
3. Sapis, J. C.; Macheix, J. J.; Cordonnier, R. E. *Am. J. Enol. Vitic.* **1983**, *34*, 157.
4. Traverso-Rueda, S.; Singleton, V. L. *Am. J. Enol. Vitic.* **1973**, *24*, 103.
5. Wisseman, K. W.; Lee, C. Y. *J. Food Sci.* **1981**, *46*, 506.
6. Coseteng, M. Y.; Lee, C. Y. *J. Food Sci.* **1987**, *52*, 985.
7. Bureau, D.; Macheix, J. J.; Rouet-Mayer, M. A. *Lebensm. Wiss. Technol.* **1977**, *10*, 211.
8. Golan, A., Kahn, V. ; Sadovski, A. Y. *J. Agric. Food Chem.* **1977**, *25*, 1253.
9. Goupy, P.; Fleuriet, A. *Bull. Liaison Groupe Polyphenols* **1986**, *13,* 455.
10. Lee, C. Y.; Jaworski, A. W. *Am. J. Enol. Vitic.* **1988**, *39*, 337.
11. Sapis, J. C.; Macheix, J. J.; Cordonnier, R. E. *J. Agric. Food Chem.* **1983**, *31*, 342.
12. Vamos-Vigyazo, L.; Nadudvari-Markus, V. *Acta Alimentaria* **1982**, *11*, 157.
13. Jaworski, A.; Lee, C. Y. *J. Agric. Food Chem.* **1987**, *35*, 257.
14. Wisseman, K. W.; Lee, C. Y. *Am. J. Enol. Vitic.* **1980**, *31*, 206.
15. Lee, C. Y.;Kagan, V.; Jaworski, A. W.; Brown, S. K. *J. Agric. Food Chem.* **1990**, *38*, 99.
16. Lee, C. Y.; Jaworski, A. *Am. J. Enol. Vitic.* **1987**, *38*, 277.
17. Romeyer, F. M.; Macheix, J. J.; Goiffon, J. P.; Reminiac, C. C.; Sapis, J. C. *J. Agric. Food Chem.* **1983**, *31*, 346.
18. Rossi, J. A. ; Singleton, V. L. *Am. J. Enol. Vitic.* **1966**, *17*, 231.
19. Simpson, R. F. *Vitis* **1982**, *21*, 233.
20. Singleton, V. L.; Kramling, T. E. *Am. J. Enol. Vitic.* **1976**, *27*, 157.
21. Lee, C. Y.; Jaworski, A. W. *Am. J. Enol. Vitic.* **1988**, *39*, 337.
22. Cheynier, V.; Osse, C.; Rigaud, J. *J. Food Sci.* **1988**, *53*, 1729.
23. Cheynier, V.; Rigaud, J.; Souquet, J. M.; Barillere, J. M.; Moutounet, M. *Am. J. Enol. Vitic.* **1989**, *40*, 36.
24. Durkee, A. B.; Poapst, P. A. *J. Agric. Food Chem.* **1965**, *13*, 137.
25. Macheix, J. J.; Fleuriet, A.; Billot, J. *Fruit Phenolics.* CRC Press, Inc. Boca Raton, FL. **1990**, pp 302-303.
26. Stelzig, D. A.; Akhtar, S.; Ribeiro, S. *Phytochem.* **1972**. *11*, 535.
27. Abdel-Rahman, A. H. Y. *Riv. Ital. Sost. Grasse.* **1985**, *62*, 147.

28. Araujo, J. M. A.; Pratt, D. E. *Ciencia Tech. Alim.* **1985**, *5*, 22.
29. Oleszek, W.; Lee, C. Y.; Price, K. R. *Acta Soc. Botan. Polon.* **1989**, *58*, 273.
30. Oleszek, A.; Lee, C. Y.; Jaworski, A. W.; Price, K. R. *J. Agric. Food Chem.* **1988**, *36*, 430.
31. Cheynier, V.; Osse, C.; Rigaud, J. *J. Food Sci.* **1988**. *53*, 1729.
32. Singleton, V. L. *Am. J. Enol. Vitic.* **1987**, *38*, 69.
33. Cheynier, V.; Basire, N.; Rigaud, L. *J. Agric. Food Chem.* **1989**, *37*, 1069.
34. Oszmianski, J; Lee, C. Y. *J. Agric. Food Chem.* **1991**, *39*, 1050.
35   Whitaker, J. R. *Principles of Enzymology for the Food Sciences.* Marcel DekkerInc., New York, **1972**, pp 572-574.
36. Oszmianski, J.; Sapis, J. C.; Macheix, J. J. *J. Food Sci.* **1985**, *50*, 1505.

RECEIVED November 7, 1991

Chapter 25

Inhibition of Polyphenol Oxidase by Phenolic Compounds

A. J. McEvily, R. Iyengar, and A. T. Gross

Opta Food Ingredients, Inc., 64 Sidney Street, Cambridge, MA 02139

Several inhibitors of polyphenol oxidase (PPO) were isolated from a plant extract. The inhibitors were purified and their relative potencies determined using commercially-available mushroom PPO in an *in vitro* assay system. Structural elucidation of the inhibitors showed them to be novel, plant secondary metabolites closely related to other phenolic compounds known to play important roles in flowering plants. Structure/activity relationships have been studied using synthetic analogs, several of which are potent PPO inhibitors. In addition to the *in vitro* studies, certain of these compounds inhibit enzymatic browning (melanosis) in foods such as shrimp, potatoes and apples as well as beverages such as grape juice. The inhibitors are water-soluble, stable, effective at low concentration, and have potential as functional alternatives to sulfites for the inhibition of melanosis. The chemical nature of the PPO inhibitors, kinetic studies, safety data, and results obtained on foods are discussed.

Enzymatic Browning

Browning or melanosis of foods and beverages is a cosmetic discoloration which has a negative impact on the appearance, consumer acceptability, commercial value, and in certain cases such as wines, the organoleptic properties of the food system. In most foods, the browning process has two components: non-enzymatic and enzymatic. Since the compounds described herein inhibit enzymatic browning, only this component will be addressed.

The causative agent which initiates the formation of brown pigments is the enzyme, polyphenol oxidase (PPO). This enzyme is endogenous to most foods and browning is not necessarily an indication of microbial contamination or spoilage. PPO is a mixed function oxidase which catalyzes the hydroxylation of monophenols to diphenols and in a second step, the oxidation of colorless diphenols to highly colored *o*-quinones. The *o*-quinones react spontaneously with other *o*-quinones and with many constituents of foods such as proteins, amino acids, reducing sugars, etc. to form high molecular weight polymers which precipitate yielding the dark pigments characteristic of browned foods.

The most widespread methodology used in the food and beverage industries for control of browning is the addition of sulfiting agents to foods susceptible to

0097–6156/92/0506–0318$06.00/0
© 1992 American Chemical Society

browning. Sulfites are currently used to inhibit melanosis (blackspot) in shrimp, browning of potatoes, mushrooms, apples, and other fruits and vegetables, as well as to reduce formation of polyphenolic polymers which can lead to off-flavors in juices and wines. The major effect of sulfites on enzymatic browning is to reduce the *o*-quinones produced by PPO catalysis to the less reactive, colorless diphenols thereby preventing the non-enzymatic condensations to high molecular weight pigments. Sulfites may also function as antimicrobial agents when used in sufficient concentration.

Although sulfites are very effective in the inhibition of both enzymatic and nonenzymatic reactions, there are several negative attributes associated with their use in foods and beverages. Sulfites are known to cause adverse health effects especially in sensitive individuals such as asthmatics. Several deaths have resulted due to consumption of sulfited foods among this highly sensitive group. According to the Food Safety and Applied Nutrition Health Hazard Evaluation Board of FDA "a four-ounce serving of shrimp containing 90 ppm sulfites presents an acute life-threatening hazard to health in sulfite sensitive individuals" (*1*). Sulfites can liberate sulfur dioxide gas and in enclosed areas such as the holds of fishing vessels, sulfur dioxide vapors have led to several deaths among fishermen (*2*). Sulfites are highly reactive, nonspecific reducing agents which are consumed during browning inhibition necessitating the use of relatively high concentrations. Also, in certain foods sulfite residuals are high enough to have a negative effect on the taste of the treated product. For more information on the use of sulfiting agents and associated health risks, the reader is referred to an excellent review by Taylor et al. (*3*).

In recent years, the Food and Drug Administration has banned sulfites for use on salad bars, moved to ban their use on fresh, peeled potatoes (*4*), increased surveillance and seizure of imported products with undeclared or excessive sulfite residuals (*5*), and has set specific limits on allowable sulfite residuals in certain foods. The negative connotations associated with sulfited foods has led to decreased consumer acceptance. According to the National Coalition of Fresh Potato Processors: "There are no substitutes" for sulfites available (*6*); however, the adverse health effects, increased regulatory scrutiny, and lack of consumer acceptance of sulfited foods have created the need for a practical, functional sulfite alternative.

Isolation and Characterization of Browning Inhibitors from Fig

Certain protease preparations especially ficin, the protease from fig (Ficus sp.) latex, appear to function as anti-browning agents in a host of food systems (*7,8*). The protease preparations employed in these studies were only partially purified, hence, the possibility existed that a non-protease component of the preparation was responsible for the observed browning inhibition. Indeed, preparations of heat-inactivated ficin and fig latex ultrafiltered to remove ficin were equally effective in PPO inhibition as the preparation containing active protease (*9,10*).

Partially-purified ficin preparations (Enzyme Development Corporation, New York, NY) were extracted with deionized water, centrifuged, and the supernatant was ultrafiltered using a YM5 (5,000 molecular weight cut-off) membrane (Amicon, Beverly, MA). The ultrafiltrate was concentrated by lyophilization, resolubilized in 2 mM sodium phosphate, pH 6.5, then subjected to conventional cation exchange liquid chromatography on SP-Sephadex^R (Pharmacia, Piscataway, NJ). The column was developed with a 0 to 0.2 M sodium chloride gradient and fractions were analyzed for absorbance at 214 and 280 nm, conductivity, and inhibition of commercially-available mushroom PPO (Sigma Chemical Co., St. Louis, MO) in a spectrophotometric assay (*10*). Three absorbance peaks associated with peaks of PPO inhibition were observed indicating the existence of at least three potential PPO inhibitors. A linear increase in inhibition was seen in the chromatogram due to increasing sodium chloride concentration which was used to elute bound materials. Sodium chloride is known to be a weak inhibitor of PPO activity (11); however, discrete and measurable peaks of PPO inhibition were observed above this background. The fractions containing significant absorbance and inhibition peaks were combined into three separate pools for further fractionation.

The three inhibitor pools from SP-Sephadex[R] chromatography were concentrated by lyophilization and further fractionated by reverse phase HPLC. One of these pools had not adsorbed to the SP-Sephadex[R] and thus was further fractionated on DEAE-Sephadex[R] (Pharmacia, Piscataway, NJ) prior to the HPLC procedure. The major absorbance peak in each HPLC run was concentrated and tested as an inhibitor of PPO. Subsequent analyses showed the recovered compounds to be nearly homogeneous preparations of three distinct inhibitors. The HPLC procedure was reproduced on a preparative scale and the recovered inhibitors were analyzed by methods such as mass spectrometry, ^{1}H NMR, ^{13}C NMR, and FTIR. Based on the resultant analytical data, the three purified inhibitors were found to be 4-substituted resorcinols and are novel, plant secondary metabolites (12). The compounds, shown in Figure 1, were identified as 2,4-dihydroxydihydrocinnamic acid (I), 2,4-dihydroxydihydrocinnamoyl putrescine (II), and *bis*-(2,4-dihydroxy-dihydrocinnamoyl)-spermidine (III). The structures of I and II were confirmed by total synthesis. Compounds II and III are novel whereas the synthesis of I *in vitro* had been described previously (13-15). However, I had not been previously isolated from a natural source. In subsequent research, compound I has also been isolated from the edible fig fruit, in addition to the fig latex from which the ficin preparation had been derived (unpublished results).

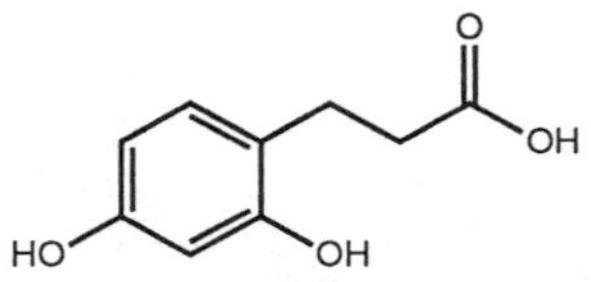

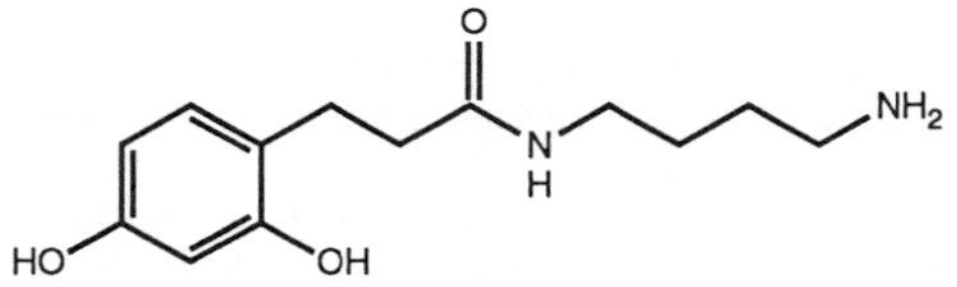

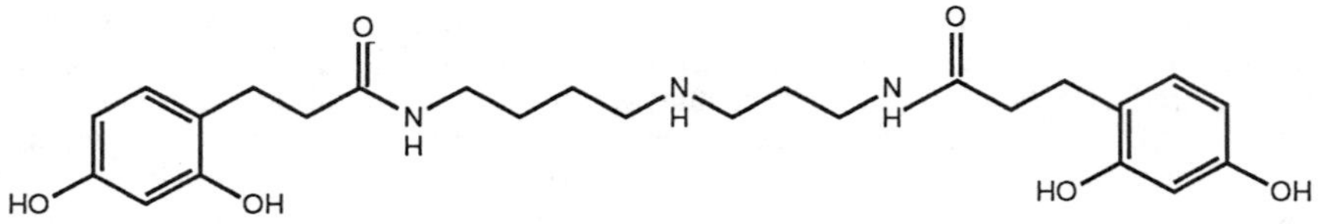

Figure 1. Structures of the naturally occurring browning inhibitors from fig latex: (I) 2,4-dihydroxydihydrocinnamic acid; (II) 2,4-dihydroxydihydro-cinnamoyl putrescine; and (III) *bis*-(2,4-dihydroxydihydrocinnamoyl)-spermidine.

A *bis*-substituted putrescine derivative, *bis*-(2,4-dihydroxydihydro-cinnamoyl)-putrescine was produced in a secondary reaction during the *in vitro* synthesis of **II** (Figure 2).

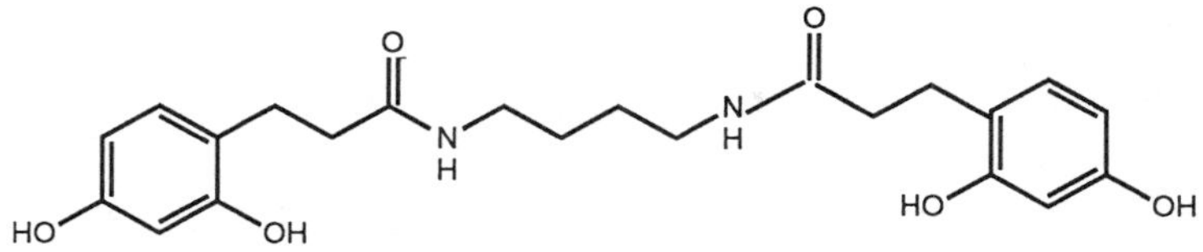

IV

Figure 2. Structure of synthetic browning inhibitor *bis*-(2,4dihydroxy-dihydrocinnamoyl)-putrescine (**IV**).

It is interesting to note the structural relationship between the amine derivative, **II**, and hydroxycinnamic acid amides (HCAs) such as caffeoyl putrescine (Figure 3). HCAs are known to have important physiological roles in the flowering of certain plants.

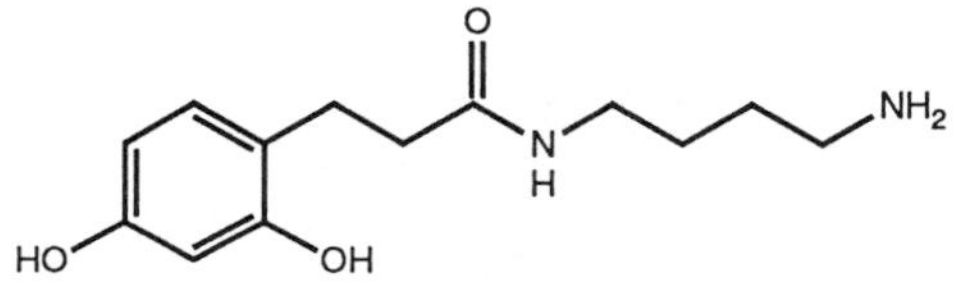

II

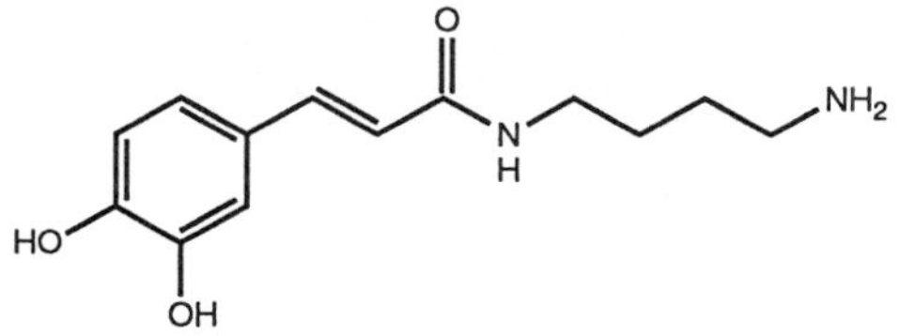

Caffeoyl putrescine

Figure 3. Structures of naturally occuring browning inhibitor from fig latex 2,4-dihydroxydihydrocinnamoyl putrescine (**II**) and the HCA, caffeoyl putrescine.

Screening of 4-Substituted Resorcinols as PPO Inhibitors

The I_{50} values for the naturally occurring inhibitors and compound **IV** were determined using mushroom PPO in the *in vitro* assay system (*10*). The I_{50} is defined as the inhibitor concentration at which 50% inhibition of PPO activity is obtained. The results are presented in Table I.

Since the resorcinol moiety was common to the natural inhibitors, commercially-available, synthetic 4-substituted resorcinols were screened for efficacy

Table I. I_{50} values for 4-substituted resorcinols as PPO inhibitors

Compound	I_{50}, μM
I	25
II	5
III	5
IV	5

as PPO inhibitors. Initial studies were performed to determine the I_{50} of the synthetic resorcinol derivatives to obtain information on their relative potencies. The results are summarized in Table II. Resorcinol is a poor inhibitor with an I_{50} in the millimolar range; however, substitutions in the 4-position yield decreased I_{50} values. The lowest values are obtained with hydrophobic substituents in the 4-position such as 4-hexyl-, 4-dodecyl- and 4-cyclohexylresorcinol with I_{50} values of 0.5, 0.3, and 0.2, respectively.

Table II. I_{50} values for synthetic 4-subsituted resorcinols as PPO inhibitors

R	I_{50}, μM
Hexanoyl	750
Carboxyl	150
Ethyl	0.8
Hexyl	0.5
Dodecyl	0.3
Cyclohexyl	0.2

Resorcinol derivatives with substitutions in the 5-, 2-, and 1,3-positions were also evaluated as PPO inhibitors. 5-Substituted resorcinols exhibited an inhibitory trend analogous to that seen with 4-substituted resorcinols: hydrophobic substituents of increasing chain length yielded inhibitors with decreasing I_{50} values (Table III). Although the 5-substituted resorcinols appeared to be effective PPO inhibitors *in vitro*, their use in food applications was not pursued due to the toxic and irritant properties associated with these compounds.

Substitutions in the 2- or 1,3-positions led to greatly increased I_{50} values relative to resorcinol (Table IV). These compounds exhibited only low levels of PPO inhibition even at the limit of their respective solubilities.

The use of 4-hexylresorcinol as a browning inhibitor was focused on due to its low I_{50} in the spectrophotometric assay system, positive preliminary results from tests on foods, and the fact that this compound has a long, safe history of human use in over-the-counter drugs. 4-Hexylresorcinol is the active ingredient in Sucrets[R] lozenges at 2.5 mg/2.5 g lozenge. Also, the numerous toxicological studies on 4-hexylresorcinol which are the subject of a recent review (*16*) show this compound to be safe and, therefore, a potential candidate for food use.

Table III. I_{50} values for 5-substituted resorcinols as PPO inhibitors

R	I_{50}, μM
Methyl	1000
Pentyl	250
Heptyl	350
Pentadecyl	7

Table IV. I_{50} values for 2- and 1,3-substituted resorcinols as PPO inhibitors

R	I_{50}, mM

| Methyl | >70 |

| Methyl | >5 |

Applications in Foods

Several 4-substituted resorcinols described herein inhibit PPO *in vitro* and have potential for widespead application in the food and beverage industries on products that exhibit problematic browning. Preliminary research shows that 4-substituted resorcinols have several advantages over sulfites for use on foods. Among others, these include: 1) these compounds are specific, potent polyphenol oxidase inhibitors allowing use at much lower concentrations than sulfites; 2) 4-substituted resorcinols do not "bleach" pigments as excess sulfites can, therefore, use of excessive concentrations is not encouraged; and 3) the 4-substituted resorcinols are more chemically stable relative to sulfites.

The initial food application targetted for intensive investigation was the prevention of shrimp melanosis (blackspot). The efficacy of 4-hexylresorcinol in maintaining the high quality of landed shrimp has been shown in both laboratory and field trials under a variety of process conditions (*17,18*). This highly potent inhibitor is substantially more effective than bisulfite on a weight-to-weight basis, should prove to be competitive with bisulfite on a cost basis, and will require no changes in the on-board or ex-vessel handling of the shrimp product.

4-Hexylresorcinol is a water soluble, stable compound which is non-toxic, non-mutagenic, and non-carcinogenic and is Generally Recognized As Safe (GRAS) for use in the prevention of shrimp melanosis (*16*). The use of 4-hexylresorcinol as a processing aid for the inhibition of shrimp melanosis has no negative impact on taste, texture, or color of the treated shrimp product due to very low residuals (<1 ppm) on shrimp meat (*19,20*).

Preliminary results from laboratory studies indicate that 4-hexylresorcinol inhibits browning of both fresh and hot-air dried apple and potato slices. Inhibition of avocado browning is seen upon incorporation of 4-hexylresorcinol into guacamole formulations. Positive results have also been obtained in liquid systems including apple and white grape juices (unpublished results).

Acknowledgments

The authors thank Craig W. Bohmont, Kathryn Frohlich, Barry Gosselin, and Joan M. King for outstanding technical assistance.

Literature Cited

1.   Anonymous. *Food Chemical News*, **1991**, *33*, 42.
2.   Anonymous. *Product Safety and Liability Reporter*, **1991**, *July*, 785.
3.   Taylor, S. L., Higley, N. A., and Bush, R. K. *Adv. Food Res.* **1986**, *30*, 1-76.
4.   Federal Register. **1990**, *55*, 9826.
5.   Anonymous. *Food Chemical News*, **1991**, *33*, 34.
6.   Anonymous. *Food Institute Report.* **1990**, *63*, 9.
7.   Labuza, T. P. *Cereal Foods World.* **1989**, *34*, 353.
8.   Taoukis, P. S., Labuza, T. P., Lillemo, J. H., and Lin, S. W. *Lebensm. Wiss. Technol.* **1990**, *23*, 52-54.
9.   Anonymous. *Prepared Foods.* **1990**, *August*, 114.
10.   McEvily, A. J. *United States Patent No. 4,981,708*, **1991**.
11.   Lerner, A. B. *Arch. Biochem. Biophys.* **1952**, *36*, 473-481.
12.   McEvily, A. J., Iyengar, R., and Gross, A. T. *United States Patent No. 5,059,438*, **1991**.
13.   Amakasu, T. and Sato, K. *J. Org. Chem.* **1966**, *31*, 1433-1436.
14.   Forchiassin, M. and Russo, C. *Heterocyclic Chem.* **1983**, *20*, 493-494.

15.   Langley, W. D. and Adams, R. *J. Am. Chem. Soc.* **1922**, *44*, 2320-2331.
16.   Frankos, V. H., Schmitt, D. F., Haws, L. C., McEvily, A. J., Iyengar, R., Miller, S. A., Munro, I. C., Clydesdale, F. M., Forbes, A. L., and Sauer, R. M. *Reg. Toxicol. and Pharmacol.* **1991**, *14*, 202-212.
17.   McEvily, A. J., Iyengar, R., and Otwell, W. S. *Food Technol.* **1991**, *45*, 80&82-86.
18.   Otwell, W. S., Iyengar, R., and McEvily, A. J. *J. Aquatic Food Product Technol.* **1991**, *1*, 53-65.
19.   Iyengar, R., Bohmont, C. W., and McEvily, A. J. *J. Food Comp. Anal.* **1991**, *4*, 148-157.
20.   King, J. M., McEvily, A. J., and Iyengar, R. *J. Assoc. Off. Anal. Chem.* **1991**, In press.

RECEIVED December 2, 1991

Author Index

Affiliation Index

Subject Index

Production: Anne Wilson
Indexing: Deborah H. Steiner
Acquisition: Barbara C. Tansill
Cover design: Pat Cunningham

Printed and bound by Maple Press, York, PA